Basic Mathematics with Hand-Held Calculator

A Work-Text

Basic Mathematics with Hand-Held Calculator
A Work-Text

Roy Dubisch

University of Washington

Vernon R. Hood

Portland Community College

The Benjamin/Cummings Publishing Company, Inc.
Menlo Park, California • Reading, Massachusetts
London • Amsterdam • Don Mills, Ontario • Sydney

Sponsoring editor: Adrian Perenon
Production editor: Margaret Moore
Book designer: Don Fujimoto
Cover designer: John Edeen
Artist: Georg Klatt

Library of Congress Cataloging in Publication Data

Dubisch, Roy, 1917-
Basic mathematics with hand-held calculator.

Includes index.
1. Arithmetic--1961- 2. Calculating-machines.
3. Problem solving. I. Hood, Vernon R., joint author.
II. Title.
QA107.D8 513'.028'54 78-57267
ISBN 0-8053-2341-4

abcdefghij–AL–782109

The Benjamin/Cummings Publishing Company, Inc.
2727 Sand Hill Road
Menlo Park, California 94025

Contents

To the Instructor

This book, together with a hand-held calculator (HHC), is intended for use in a semester or quarter course in basic arithmetic at either the secondary school or the college level. It emphasizes the *intelligent* use of an HHC in problem solving rather than a detailed review of arithmetic calculations with pencil and paper. Furthermore, since computational difficulties are essentially nonexistent in using an HHC, the kinds of problems presented can have realistic data rather than artificial data designed to keep the arithmetic simple. Relieved of the necessity of doing tedious arithmetic calculations, students can now concentrate on *methods* of solving problems. We also emphasize *estimation* since it is all too easy to press the wrong keys on an HHC and obtain completely unrealistic answers.

Although we assume that most of the calculations required will be performed with an HHC, we are careful to point out that there are many calculations where even a modest knowledge of the basic facts of arithmetic can result in a shorter computation time than can be achieved using an HHC. (As a result, we have found that students who have used this book in a preliminary version compare favorably in pencil and paper computation at the end of the course with students taking a non-HHC course in arithmetic.) Furthermore, although we have not emphasized the theoretic aspects of arithmetic, we have found it quite feasible to incorporate a good deal of this in a sort of incidental way. For example, students find that various properties of the four fundamental operations such as the distributive property, $a - (b + c) = a - b - c$, etc. turn out to be very useful in simplifying computations even when an HHC is used. (Even the prosaic facts that $a + 0 = a$, $a - 0 = a$, $a \times 1 = a$, and $a \div 1 = a$ can be useful when the wrong operation key is pressed!)

Finally, although our emphasis is on the practical side of the use of an HHC, we have found it possible to introduce a good deal of recreational type mathematics since, once delivered from the need for hand computation, students at this level become much more interested in such activities than experience with a non-HHC course at this level might indicate. It also provides for a change of pace for the class.

The format of the book is designed to encourage active participation by the student as he or she reads the material. Almost all of the text pages have an outer margin in which developmental exercises appear; students are referred to these at the appropriate times.

In addition to these developmental exercises, each section concludes with a set of exercises and each chapter concludes with a set of review exercises. All exercises can be removed from the book and, for quick grading, the answers in the Instructor's Guide are arranged to match these sheets. Over 3200 exercises are provided with emphasis on problems involving applications of arithmetic. Objectives of each section are stated at the beginning of the section.

Each chapter also contains a "Just for Fun" section involving work with patterns of numbers, puzzles, and games. This is followed by a section involving an in-depth analysis of such varied topics as computations used in sports and analysis of the cost of operating an automobile.

Only the simplest kind of four-function HHC is assumed. From time to time, comments are made concerning variations in the operation of various HHC's and there is an appendix on the use of a memory key. However, we do not attempt to discuss the operation of more sophisticated HHC's. Furthermore, although we do discuss various short-cuts in using HHC's (e.g., computing $a - (b + c)$ as $a - b - c$), we do not include all possible short-cuts. To do so, we feel, would result in confusion rather than in an increase in proficiency and understanding.

Answers are given at the back of the book for all of the exercises that appear in the narrow columns of the text since it is important that the student work all of these at the same time he or she reads the text. Answers are also given at the back of the book for the odd-numbered problems in the exercise sets and for the problems at the end of each chapter. When round-off is asked for in a problem (or is appropriate), we give the rounded-off answer followed by the answer in parentheses as it appears on an HHC. (Note, however, that the last digit in the HHC answer can sometimes differ from the answer obtained on a particular HHC because of differences in the internal logic of various HHC's.) Those problems that are marked with an asterisk are either somewhat more difficult than the others or involve material that has not been discussed in the text.

The material in this book has been used twice in trial classes at Portland Community College. The first trial was with a very rough draft that was then revised on the basis of classroom experience. The manuscript used for the second trial was then further revised on the basis of the second trial and on reviews of the material.

To the Student

Until fairly recently, tens of thousands of people in the United States earned their living at least in part by means of their ability to do arithmetic quickly and accurately. Grocery clerks added up customers' bills; bank tellers totaled their receipts and disbursements; and bookkeepers kept their books by performing calculations using paper and pencil. Today, of course, even a small "mom and pop" grocery store has at least a tape adding machine and the larger stores have registers that not only compute the bill, but indicate the change required; bank tellers are similarly equipped; and any kind of complicated bookkeeping is frequently done with the aid of computers that are capable of performing millions of calculations in a very few minutes.

Indeed, small hand-held calculators (HHC's) are so easily available today that an ever increasing number of people use them routinely for calculations around the home: for keeping track of checking accounts, figuring income tax, calculating interest payments, etc.

In the face of this increasing availability of HHC's, it would certainly be foolish to devote a large amount of time at any level of education in developing the ability to do long calculations with paper and pencil. On the other hand, it would also be foolish to have to depend entirely on an HHC to perform all calculations, particularly such simple ones as measuring the width of a room using a ten-foot tape measure and finding that it is the full length of the tape measure and two feet more!

Even if we have an HHC and rely on it to perform every arithmetic calculation, a basic knowledge of the *principles* of arithmetic is still necessary (or at least useful) for the following reasons:

1. (Most importantly) No HHC will tell you *what* arithmetic calculations to perform. If, for example, you want to find out what your gas mileage is and your car travels 164 miles on 8.3 gallons of gas, you must know that the answer is obtained by dividing 164 by 8.3 before you can use your HHC.

2. Many arithmetic calculations are more quickly done by "hand." For example, knowing the meaning of multiplication and of 1,000,000 enables us to conclude that

$$24 \times 1{,}000{,}000 = 24{,}000{,}000$$

faster than using an HHC to do the computation. Similarly, $\frac{2}{3} + \frac{3}{4}$ done by hand would be

$$\frac{2}{3} + \frac{3}{4} = \frac{8}{12} + \frac{9}{12} = \frac{17}{12}$$

whereas using an HHC for all operations, we would need to calculate 2×4, 3×3, $8 + 9$, and 3×4 (and write down the answer to 2×4 or keep it in our head unless our HHC had a "memory" key).

3. Knowing the basic principles of arithmetic enables us to increase the usefulness of an HHC. For example, suppose that we have 3545 items to be packed in boxes, each containing 12 items, and that we wish to know not only how many boxes will be needed, but also how many items will be left over when each box is completely filled. Using an HHC we divide 3545 by 12 to get 295.41666. This tells us that 295 boxes will be needed, but not how many items are left over. However, if we know some basic principles of arithmetic we can subtract 295 from 295.41666 and multiply the answer by 12 to find that 5 items are left over.

4. Knowing some of the basic principles of arithmetic (especially those involving fractions) is essential for success in an algebra course.

In using this text then, you will

1. learn how to use your HHC correctly and efficiently to perform arithmetic calculations,

2. learn how to apply such calculations to the solution of problems from business, industry, the trades, and the home.

3. review some of the basic principles of arithmetic that you studied previously.

In addition, the text includes problems involving various interesting properties of numbers and various games that can be played using an HHC which, we hope, will increase your interest in the study of mathematics for its own sake.

Answers are given at the back of the book for all of the exercises that are in the narrow columns of the text since it is important that you work all of these at the same time you read the text. Answers are also given at the back of the book for the odd-numbered problems in the exercise sets and for the problems at the end of each chapter. Here you will sometimes find two answers given. If so, the first one will be the preferred (rounded-off) answer and the second, in parentheses, the answer as it should appear on your HHC. (Occasionally, this second answer may differ from yours in the last digit because of differences in the internal logic of the various HHC's.) Those problems that are marked with an asterisk (*) are either somewhat more difficult than the others or involve material that has not been discussed in the text.

Introduction

Welcome to the world of the hand-held calculator! Through its use you will easily increase your ability to handle the mathematics required in your daily life.

Considerable variation exists among hand-held calculators, or HHC's. We will limit our discussion to those on which only the four basic operations of arithmetic (addition, subtraction, multiplication, and division) can be performed. If the one you use is a model with various additional features, you may occasionally need to consult the user's manual that came with it. The accompanying picture shows the kind of basic HHC that we will be discussing.

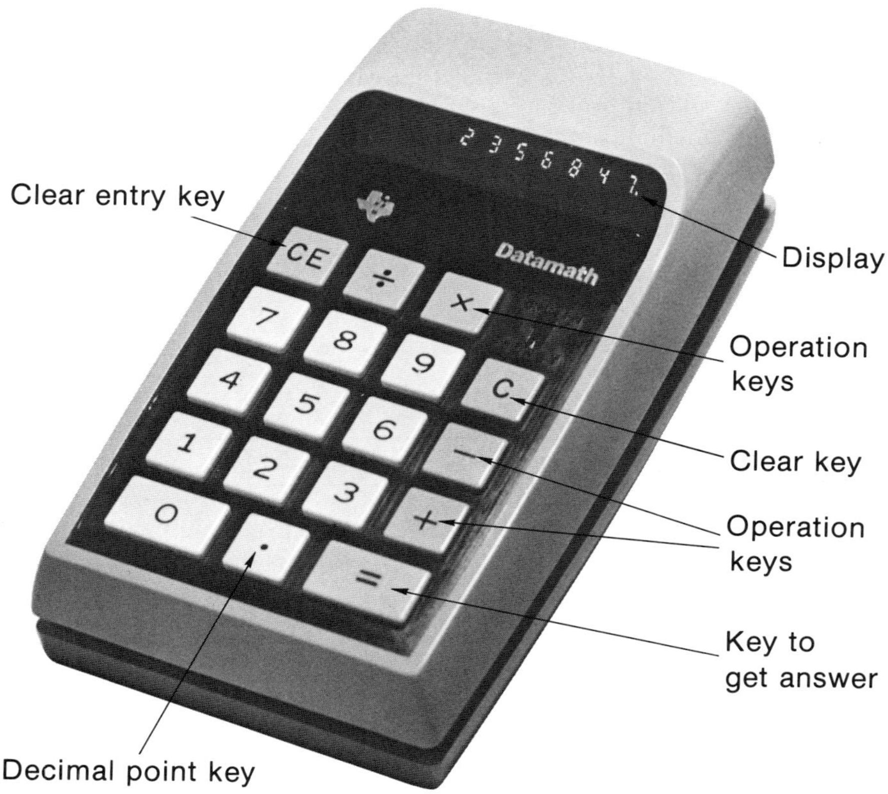

In Chapter 1 we will discuss details of the arithmetic of whole numbers. Later chapters will deal with decimals and fractions. Right now, however, let's just run through the basic operations with whole numbers using an HHC.

OBJECTIVES

You should be able to:

(a) Enter whole numbers on an HHC.

(b) Perform additions, subtractions, multiplications, and divisions involving whole numbers on an HHC.

(c) Understand the use of the clear and the clear entry key.

****

First of all, a whole number is entered by pressing the numbered keys in the same order, from left to right, as they appear in the numeral. For example, the entering of 525,089 on an HHC and what is shown on its display is the following:

Enter	5	2	5	0	8	9
Display	5.	52.	525.	5250.	52508.	525089.

(Notice that the comma used to separate the digits in 525,089 does not appear in the display.)

Do exercises 1 and 2 in the adjoining column.

Now for the operations.

ADDITION: 525 + 609 + 777 = ?

Enter	525	+	609	+	777	=
Display	525.	525.	609.	1134.	777.	1911.

Thus 525 + 609 + 777 = 1911. Notice that each number being added is displayed in turn so that you can see if you have entered the correct number, and that the intermediate sum of 525 and 609 (1134) is also displayed.

Now be sure to press the C (clear) key before beginning another computation.

Do exercises 3 and 4.

SUBTRACTION: 12,103 – 3148 = ?

Enter	12103	–	3148	=
Display	12103.	12103.	3148.	8955.

Thus 12,103 – 3148 = 8955.

Do exercises 5 and 6.

1. Enter 302,105 on your HHC.

2. Enter 16,003,214 on your HHC.

STOP

3. Check the answer to the following addition by doing it on your HHC.

$$\begin{array}{r} 5289 \\ 7156 \\ 2314 \\ +\,6723 \\ \hline 21482 \end{array}$$

4. Perform the following indicated additions:

(a) $\begin{array}{r} 31{,}925 \\ 40{,}116 \\ +\,61{,}106 \\ \hline \end{array}$

(b) 132 + 169 + 324

STOP

5. Check the answer to the following subtraction problem by doing it on your HHC.

$$\begin{array}{r} 52{,}314 \\ -\,13{,}289 \\ \hline 39{,}025 \end{array}$$

6. Perform the following subtractions:

(a) $\begin{array}{r} 310{,}217 \\ -\,214{,}625 \\ \hline \end{array}$

(b) 19,146 – 3289

STOP

MULTIPLICATION: 48 × 124 × 69 = ?

Enter	48	×	124	×	69	=
Display	48.	48.	124.	5952	69.	410688.

Thus 48 × 124 × 69 = 410,688. Note that the intermediate product of 48 and 124 (5952) is displayed.

Do exercises 7 and 8.

DIVISION: 29,440 ÷ 115 = ?

Enter	29440	÷	115	=
Display	29440.	29440.	115.	256.

Note that the answer is a whole number. Division involving decimals will be discussed later.

Do exercises 9 and 10.

If your HHC has a key labeled CE (clear entry) or CD (clear display), an incorrect entry of a number—or, on *some* HHC's, an operation—can be cleared without erasing previous entries. For example, suppose that in doing the addition

$$\begin{array}{r} 78{,}492 \\ 14{,}261 \\ +\ 79{,}186 \\ \hline \end{array}$$

you entered 78,492, +, 14,261, + correctly but then began entering 79,186 by pressing (incorrectly) the 9 key instead of the 7 key. If you noticed your error at this point, pressing the CE or CD key would allow you to continue by entering 79,186 without having to enter 78,492, +, 14,261, and + again [as you would have had to do if you had pressed the C (clear) key after noticing your mistake].

7. Check the answer to the following multiplication problem by doing it on your HHC.

$$\begin{array}{r} 537 \\ \times\ 126 \\ \hline 67{,}662 \end{array}$$

8. Perform the following multiplications:

(a) $\begin{array}{r} 93{,}116 \\ \times\ \ 223 \\ \hline \end{array}$

(b) 169 × 23

(c) 54 × 28 × 19

STOP

9. Check the answer to the following division by doing it on your HHC.

16,800 ÷ 96 = 175

10. Perform the following divisions:

(a) 1248 ÷ 16

(b) 1,522,224 ÷ 124

STOP

11. Check that both of the following procedures will give the correct answer to 59 + 84.

(a)

Enter	59	+	84	=
Display	59.	59.	84.	143

(b)

Enter	59	+	83	CE	84	=
Display	59.	59.	83.	0.	84.	143

12. Check to see whether both of the following procedures give the correct answer to 59 + 84 on *your* HHC.

(a)

Enter	59	+	84	=
Display	59.	59.	84.	143.

(b)

Enter	59	–	CE	+	84	=
Display	59.	59.	59.	59.	84.	?

If your answers to (a) and (b) are not the same, you can*not* use the CE key to correct a wrong operation entry.

Do exercise 11.

Similarly, suppose that in doing the subtraction

$$78{,}492 - 14{,}261$$

you entered 78,492 correctly but then pressed + instead of –. If you noticed your error at this point, pressing the CE or CD key (on some HHC's) would allow you to continue by then pressing the – key without having to enter 78,492 again (as you would have had to do if you had used the C key after noticing your mistake).

Do exercise 12.

Another way of handling the problem of a correction when the wrong operation key is pressed will be discussed at the end of section 1.6.

We will now, in Chapter 1, consider some of the details of computations with whole numbers and the applications of these computations to story problems.

1

Computations with Whole Numbers

1.1 Addition of Whole Numbers

CHECKING FOR ERRORS

Do exercise 1 in the adjoining column.

In exercise 1 you should have found that the answer to (a) is correct whereas the answer to (b) is incorrect.

It is very easy to do computations on an HHC; it is also very easy for you (not the HHC!) to make mistakes. In the second addition in (b) you probably got an answer of 14,843 and, indeed, this is the correct answer. The wrong answer of 14,833 is, however, all too easy to get—as the next three exercises illustrate.

Do exercises 2 through 4.

Exercises 2–4 show that the wrong answer of 14,833 could be obtained by entering 8113 instead of 8123, 3564 instead of 3574, or 3136 instead of 3146. (Of course there are other ways of getting the wrong answer!)

Clearly, when we do problems involving more numbers, the chance of making such errors increases. It will pay you to check the display after entering each number and to be careful about pressing the + key after each number is entered. Re-doing the problem is a good idea if the answer is important to you (as on a test, or in making out your income tax!).

OBJECTIVES

You should be able to:

(a) Use the HHC to add whole numbers.

(b) Be aware of some types of errors that are likely to occur.

(c) Recognize the associative and commutative properties of addition and know how to use these properties to shorten computation time.

(d) Handle overflow problems involving addition.

(e) Apply addition to the solution of problems.

****

1. Check the answers to the following addition problems by using your HHC.

   (a) 5623 + 2148 + 8954 = 16725 (b) 8123 + 3574 + 3146 = 14833

   $$\begin{array}{r} 5623 \\ 2148 \\ +\,8954 \\ \hline 16725 \end{array} \qquad \begin{array}{r} 8123 \\ 3574 \\ +\,3146 \\ \hline 14833 \end{array}$$

STOP

2. Add: $\begin{array}{r} 8113 \\ 3574 \\ +\,3146 \\ \hline \end{array}$

3. Add: $\begin{array}{r} 8123 \\ 3564 \\ +\,3146 \\ \hline \end{array}$

4. Add: 8123 + 3574 + 3136

STOP

5. Add:

```
    7184
     287
    1024
     239
    6144
     128
   10174
    2138
 +  3142
```

Do exercise 5.

MENTAL COMPUTATION

Many arithmetic calculations are more quickly done mentally without the use of the HHC as, for example,

$$20{,}000 + 50{,}000 = 70{,}000$$

where essentially the only arithmetic that needs to be done is to add 2 and 5. Also, you will find that knowing the basic principles of arithmetic will enable you to increase the efficiency and usefulness of your HHC.

Consider the calculation of the following sum:

```
   75
   49
 + 25
```

It takes about six seconds on a HHC to find this sum by proceeding in the usual way:

Enter	75	+	49	+	25	=
Display	75.	75.	49.	124.	25.	149.

If we observe, however, that

$$75 + 25 = 100 \quad \text{and} \quad 100 + 49 = 149$$

we can find the sum mentally in one or two seconds.

Do exercises 6 through 10.

In exercises 6–10 use observations as described in the text to perform the given addition mentally. Then check your answer by using your HHC and compare computation time with and without an HHC.

6.

```
   70
   83
 + 30
```

7.

```
  125
   86
 + 75
```

8. 150 + 40 + 60 + 50

9.

```
  4030
  2070
+ 1300
```

10.

```
  250
  129
+ 350
```

BASIC PROPERTIES OF ADDITION

In doing the addition 75 + 49 + 25 mentally by noting that

$$75 + 25 = 100 \quad \text{and} \quad 100 + 49 = 149$$

we were really re-writing

$$(75 + 49) + 25 \quad \text{as} \quad (75 + 25) + 49$$

where the parentheses indicate which addition is to be done first. Is this permissible? It is, because of the *commutative* and *associative properties of addition.* For all whole numbers a, b, and c:

$a + b = b + a$ Commutative property of addition (order property)

and

$(a + b) + c = a + (b + c)$ Associative property of addition (grouping property)

Do exercises 11 through 13.

OVERFLOW IN ADDITION

We now use these properties to show how an addition problem involving an *overflow* can be handled on an HHC. Overflow can occur because only numbers with eight or fewer digits can appear as a display on most HHC's. (More than eight digits are technically possible and are occasionally found on a HHC, but the standard today is eight.) When this limit is exceeded, all HHC's have some way of indicating that an overflow has occurred.

Do exercises 14 and 15.

One way of obtaining an answer to the problem of exercise 14 on an HHC is as follows: We first note that

$$58{,}249{,}342 \;+\; 89{,}714{,}328$$

is equal to

$$(58{,}240{,}000 + 9342) \;+\; (89{,}710{,}000 + 4328)$$

Now, using the commutative and associative properties of addition, we can write this sum as

$$(58{,}240{,}000 + 89{,}710{,}000) \;+\; (9342 + 4328)$$

Do exercises 16 through 18.

From the results of exercises 16–18, we see that

$$\begin{aligned}(58{,}240{,}000 + 89{,}710{,}000) + (9342 + 4328) &= 147{,}950{,}000 + 13{,}670 \\ &= 147{,}963{,}670\end{aligned}$$

where we have only to note that $5 + 1 = 6$ to get the answer.

(Note that although this process is an interesting example of the use of the commutative and associative properties of addition, it is clearly easier to do such additions with pencil and paper!)

In exercises 11–13 state which of the two properties of addition, commutative or associative, allows us to write the given equality.

11. $(75 + 49) + 25 = 75 + (49 + 25)$

12. $75 + (49 + 25) = 75 + (25 + 49)$

13. $75 + (25 + 49) = (75 + 25) + 49$

STOP

14. Attempt to do the following addition on your HHC.

$$\begin{array}{r} 58{,}249{,}342 \\ +\,89{,}714{,}328 \\ \hline \end{array}$$

15. What is the overflow indicator on your HHC?

STOP

16. Do the addition

$$\begin{array}{r} 5824 \\ +\,8971 \\ \hline \end{array}$$

17. Use the result of exercise 16 to do the addition

$$\begin{array}{r} 58{,}240{,}000 \\ +\,89{,}710{,}000 \\ \hline \end{array}$$

18. Do the addition

$$\begin{array}{r} 9342 \\ +\,4328 \\ \hline \end{array}$$

STOP

19. Compute

$$\begin{array}{r} 98{,}124{,}638 \\ +\ 41{,}789{,}247 \\ \hline \end{array}$$

by the procedure suggested in the text. Begin by writing

98,124,638 = (98,120,000 + 4638)

and

41,789,247 = (41,780,000 + 9247)

STOP

Do exercise 19.

SOLVING STORY PROBLEMS

In applications of mathematics outside the classroom, much of the information concerning a problem will arise naturally without a lot of words needing to be interpreted mathematically. In a textbook, however, most problems must be described in words. And no matter how clearly the problem is worded it may be necessary to read it carefully several times in order to be able to solve it.

In doing this reading, be on the lookout for certain words, phrases, or sentences that can provide the key to the solution. In particular, there are certain key words, phrases, or sentences that indicate the kind of arithmetic operations (addition, subtraction, multiplication, or division) that need to be used in solving the problem. For the operation of addition, some of these are: "Find the *total*"; "What is the value of *all* . . . ?"; Find the *sum*"; and "How many *altogether* . . . ?"

Example: You wish to buy a new sports car and find that the base price for model XK-750 is \$6750. Additional items that you want are the following: radio and tape deck at \$165, air conditioning at \$425, deluxe interior at \$187, and a special paint job at \$225. What would be the total price of the car before sales tax and license fees are added?

The key phrase is *total price* and it indicates that you should *add* the separate prices. Thus

$$6750 + 165 + 425 + 187 + 225 = 7752$$

tells us that the total price of the XK-750 sports car is \$7752 before sales tax and license fees are added.

20. The monthly operating costs of Ms. White's office are as follows: rent, \$215; payroll, \$550; electricity, \$25; janitorial services, \$30; telephone, \$185; and insurance, \$18. What is the total monthly cost of operating Ms. White's office?

21. Joe has four stamp albums. One is valued at \$224, one at \$162, one at \$76, and one at \$82. What is the value of all of his stamp albums together?

STOP

Do exercises 20 and 21.

Many times information is presented in a table such as the one that follows:

PULPWOOD–PRODUCTION AND CONSUMPTION: 1950 TO 1974

(In thousands of cords of 128 cubic feet. Excludes Alaska prior to 1955 and Hawaii prior to 1962. See *Historical Statistics, Colonial Times to 1970,* series L 166 and 168 for similar but not exactly comparable data.)

Year	Production[1]	Consumption			Year	Production[1]	Consumption		
		Total	Softwood	Hardwood			Total	Softwood	Hardwood
1950	20,712	23,627	20,406	3,221	1966	55,782	56,260	42,857	13,403
1955	30,948	33,356	27,728	5,628	1967	55,820	55,932	42,703	13,229
1960	40,011	40,485	32,061	8,424	1968	59,976	62,092	47,195	14,897
					1969	64,640	65,892	49,696	16,196
1961	40,272	42,191	33,095	9,097	1970	67,723	67,562	51,296	16,266
1962	42,772	44,070	34,193	9,877	1971	65,789	67,157	50,585	16,572
1963	44,708	46,435	35,661	10,774	1972	70,273	71,538	53,613	17,925
1964	49,219	50,659	39,329	11,330	1973	73,180	73,282	54,768	18,514
1965	52,022	51,970	39,676	12,294	1974	(NA)	75,030	55,270	19,760

NA Not available.
[1] Includes slabs, chips, and millwaste. Production is equal to receipts from domestic sources by woodpulp mills located in the United States.
Source: U.S. Bureau of the Census, *Current Industrial Reports,* series M26A.

In doing problems involving such tables, care must be used to take the information from the correct column and row of the table. It is also important to note the units of measurement used in the table. Thus, in this table, the numbers represent "thousands of cords of wood." So to find the number of cords indicated by an entry, the number in the table must be multiplied by 1000.

Example: Find the total consumption of cords of hardwood during the years 1970 through 1973.

Solution: From the table we obtain

$$\begin{array}{ccccccccc} (1970) & & (1971) & & (1972) & & (1973) & & \\ 16{,}266 & + & 16{,}572 & + & 17{,}925 & + & 18{,}514 & = & 69{,}277 \end{array}$$

Thus the addition tells us that 69,277 thousands of cords were consumed during the years 1970–73 or

$$69{,}277 \times 1000 = 69{,}277{,}000$$

cords.

Do exercise 22.

There is no easy road to success in problem solving, but practice and persistance are the keys to success.

22. Use the table to find the total number of cords of hardwood consumed during the years 1960 through 1963.

STOP

Exercises 1.1

Name ______________

Class ______________

In exercises 1–5 check the sum for each addition problem. If the sum is incorrect, give the correct sum.

1. 104 + 203 + 613 + 214 = 1134

$$\begin{array}{r} 104 \\ 203 \\ 613 \\ +\ 214 \\ \hline 1134 \end{array}$$

2.
$$\begin{array}{r} 1875 \\ 3143 \\ 5826 \\ +\ 10832 \\ \hline 21712 \end{array}$$

3.
$$\begin{array}{r} 14203 \\ 8173 \\ 1065 \\ +\ 10042 \\ \hline 20704 \end{array}$$

4. 214 + 1231 + 141 + 5000 = 6703

5. 6354 + 2145 + 2043 + 1111 + 3333 = 14,985

In exercises 6–16 use the ideas of order (commutative property), grouping (associative property), and sums of 10, 100, and 1000 to *compute the sums mentally.*

6.
$$\begin{array}{r} 30 \\ 113 \\ +\ 70 \\ \hline \end{array}$$

7.
$$\begin{array}{r} 60 \\ 17 \\ +\ 40 \\ \hline \end{array}$$

8.
$$\begin{array}{r} 92 \\ 25 \\ 108 \\ +\ 75 \\ \hline \end{array}$$

9. 75 + 86 + 25

10. 130 + 213 + 70

11. 113 + 87 + 70

12. 8 + 13 + 2 + 7 + 12

13.
$$\begin{array}{r} 25 \\ 175 \\ 300 \\ +\ 700 \\ \hline \end{array}$$

14.
$$\begin{array}{r} 6025 \\ 1500 \\ +\ 4075 \\ \hline \end{array}$$

15.
$$\begin{array}{r} 13 \\ 7 \\ 85 \\ 120 \\ 15 \\ +\ 80 \\ \hline \end{array}$$

16.
$$\begin{array}{r} 14 \\ 6 \\ 65 \\ 140 \\ 35 \\ +\ 60 \\ \hline \end{array}$$

In exercises 17–27 perform the indicated addition.

17.
$$\begin{array}{r} 4,699,698 \\ +\ 6,745,679 \\ \hline \end{array}$$

18.
$$\begin{array}{r} 8,469,876 \\ +\ 7,597,647 \\ \hline \end{array}$$

19.
$$\begin{array}{r} 5,339,563 \\ +\ 3,567,534 \\ \hline \end{array}$$

20. 2,498,376 + 1,357,362

21. 8,314,331 + 4647 + 48,416,732

1.
2.
3.
4.
5.
6.
7.
8.
9.
10.
11.
12.
13.
14.
15.
16.
17.
18.
19.
20.
21.

22.

22. The area in square miles of the six New England states is as follows: Maine, 33,215; New Hampshire, 9304; Vermont, 9609; Massachusetts, 8257; Rhode Island, 1214; Connecticut, 5009. Find their total area.

23.

23. The area in square miles of the states bordering the Pacific Ocean is as follows: Washington, 68,192; Oregon, 96,981; California, 158,693; Alaska, 586,412; Hawaii, 6450. Find their total area.

24.

24. The ordinary receipts of the treasury of the United States for the fiscal year ending June 30, 1886, were: from customs, $192,905,023; from internal revenue, $116,805,937; from direct taxes, $108,240; from sale of public land, $5,630,999; miscellaneous, $20,989,528. What was the total revenue?

25.

25. The ordinary receipts of the treasury of the United States for the fiscal year ending June 30, 1974, were: from individual income taxes, $40,700,000,000; from corporation income taxes, $21,500,000,000; from social insurance and employment taxes and contributions, $14,700,000,000; from excise taxes, $11,600,000,000; from estate and gift taxes, $1,600,000,000; from custom duties, $1,100,000,000; miscellaneous, $1,200,000,000. What was the total revenue? (Hint: Ignore the last six zeros in entering the numbers on your HHC.)

26.

26. In Maryland it is estimated that a person who keeps a standard size 1974 car for ten years and drives it 100,000 miles will have the following costs:

Depreciation	$4258
Maintenance	3521
Gasoline	2026
Insurance	1618
Garage and parking tolls	1960
Gas and oil taxes and registration fees	1509

What does it cost to operate such a car for the period of ten years and 100,000 miles?

27.

27. The Forest Service incorporated the following amounts in a sale-area betterment plan:

Precommercial thinning	$4125
Thinning slash	3000
Tree planting	8700
Indirect costs	6881

Determine the total cost of this sale-area betterment plan.

In exercises 28–32 refer to the following table.

GRAIN RECEIPTS FOR ONE WEEK (in bushels)

	Mon.	Tue.	Wed.	Thurs.	Fri.	Sat.
Corn	28,325	15,326	35,715	29,128	75,183	46,217
Wheat	35,719	41,719	50,108	32,546	59,275	81,126
Oats	12,136	9,237	18,265	7,268	6,950	17,230
Barley	18,230	15,738	21,375	15,828	19,263	13,637
Rye	5,275	6,829	7,201	11,325	7,825	13,261

28. What was the total number of bushels of wheat received during the week? 28.

29. What was the total number of bushels of barley received during the week? 29.

30. How many bushels of grain were received on Monday? 30.

31. How many bushels of grain were received on Thursday? 31.

*32. Find the total grain received for the week and prove the results are correct by showing that the sum of each day's receipts is the same as the total receipts for each type of grain. *32.

In exercises 33–36 use the following table (from the 1976 *Statistical Abstract*).

HOME LOANS

No. 1241. Residential Loans Underwritten by FHA and VA: 1950 to 1974

(In millions of dollars. FHA-insured loans represent gross amount of insurance written; VA-guaranteed loans, gross amount of loans closed. Figures do not take account of principal repayments on previously insured or guaranteed loans. See also *Historical Statistics, Colonial Times to 1970,* series N 291-297.)

Loan	1950	1955	1960	1965	1970	1971	1972	1973	1974
FHA-insured loans	4,343	3,807	6,293	8,689	11,982	14,689	12,319	7,592	5,865
Home mortgages:									
New properties	1,637	1,269	2,197	1,705	2,667	3,900	3,459	1,675	752
Existing properties	856	1,816	2,403	5,760	5,447	6,475	4,608	2,798	3,182
Project type mortgages	1,157	76	711	591	3,251	3,641	3,448	2,286	1,225
Property improvement loans[1]	694	646	982	634	617	674	805	832	706
VA-guaranteed loans[2]	3,072	7,156	1,985	2,652	3,440	5,961	8,293	7,416	8,205
Home mortgages:									
New properties	1,865	4,582	1,554	876	1,311	1,694	2,539	2,313	2,083
Existing properties	1,202	2,564	428	1,776	2,129	4,267	5,754	5,103	6,122

[1] Not ordinarily secured by mortgages.

[2] Includes refinancing loans, loans on mobile homes (beginning in 1971), and a small amount of alteration and repair loans. In the latter category, only loans of more than $1000 need be secured.

Source: Board of Governors of the Federal Reserve System. Current data in *Federal Reserve Bulletin.*

33. 33. What was the value of all home mortgages underwritten by the FHA and VA in 1974?

34. 34. What was the value of all home mortgages underwritten for new property by the FHA in 1970–74?

35. 35. What was the value of all home loans underwritten on new property by the VA in 1970–74?

36. 36. What was the value of all home mortgages underwritten by the FHA and the VA on existing properties from 1970 to 1974?

*37. *37. Suppose that as a prize an airline offered you a free trip of not more than 10,000 miles between the cities listed in the following table. Plan a series of flights that would come as close to 10,000 miles as possible (but less than 10,000) leaving from and returning to the city nearest your home.

AIR DISTANCES BETWEEN U.S. CITIES IN STATUTE MILES

Cities	Memphis	Miami	Minne-apolis	New Orleans	New York	Omaha	Phila-delphia
Birmingham, AL	217	665	862	312	864	732	783
Boston, MA	1137	1255	1123	1359	188	1282	271
Buffalo, NY	803	1181	731	1086	292	883	279
Chicago, IL	482	1188	355	833	713	432	666
Cleveland, OH	630	1087	630	924	405	739	360
Dallas, TX	420	1111	862	443	1374	586	1299
Denver, CO	879	1726	700	1082	1631	488	1579
Detroit, MI	623	1152	543	939	482	669	443
El Paso, TX	976	1643	1157	983	1905	878	1836
Houston, TX	484	968	1056	318	1420	794	1341
Indianapolis, IN	384	1024	511	712	646	525	585
Kansas City, MO	369	1241	413	680	1097	166	1038
Los Angeles, CA	1603	2339	1524	1673	2451	1315	2394
Louisville, KY	320	919	605	623	652	580	582
Memphis, TN	....	872	699	358	957	529	881
Miami, FL	872	....	1511	669	1092	1397	1019
Minneapolis, MN	699	1511	....	1051	1018	290	985
New Orleans, LA	358	669	1051	....	1171	847	1089
New York, NY	957	1092	1018	1171	....	1144	83
Omaha, NE	529	1397	290	847	1144	....	1094
Philadelphia, PA	881	1019	985	1089	83	1094	....
Phoenix, AR	1263	1982	1280	1316	2145	1036	2083
Pittsburgh, PA	660	1010	743	919	317	836	259
St. Louis, MO	240	1061	466	598	875	354	811
Salt Lake City, UT	1250	2089	987	1434	1972	833	1925
San Francisco, CA	1802	2594	1584	1926	2571	1429	2523
Seattle, WA	1867	2734	1395	2101	2408	1369	2380
Washington, DC	765	923	934	966	205	1014	123

Source: National Geodetic Survey.

*38. What natural numbers (1, 2, 3, 4, . . .) from 1 through 30 can be written as a sum of three consecutive natural numbers? For example 6 = 1 + 2 + 3 and 12 = 3 + 4 + 5 but 3 and 10 cannot be so written. Show these sums. Do you see a pattern? *38.

*39. Which is faster—the calculator or mental arithmetic? Try this experiment! *39.

Materials needed:

- a partner
- a calculator
- a timing device with second hand
- a set of cards on which are written 1-digit numbers (the set of numbered cards from a set of playing cards or a Rook deck, or you may make your own)

(a) Shuffle the card deck and deal your partner ten cards face down. Time how long it takes your partner to find the sum of the ten numbers using the calculator. Do this several times. (a)

(b) Repeat (a) except this time find the sum by mental arithmetic. (b)

(c) Compare the speed and accuracy of calculator addition and mental addition. What conclusions can you draw? (c)

(d) Repeat (a) through (c) for sets of twenty numbers. (d)

(e) Repeat (a) through (c) for sets of five numbers. (e)

(f) Do the observations you made in (c) depend upon the number of digits in the numbers being summed? (f)

(g) What final conclusions can you now draw about the speed and accuracy of calculator addition compared with mental addition? (g)

Use your calculator to answer the following question.

40. What does a farmer till from daybreak to sunset? To find out add the following numbers: 40.

$$2351 + 1107 + 2222 + 1425$$

If the calculator does not give the answer directly, then turn it upside down.

1.2 Our System of Numeration

In the previous section you had occasions to read the display on your HHC. As you work with numbers you will find it advantageous to be able to read and write number names correctly and to have an understanding of our system of numeration.

Our system of numeration is a *decimal* (*base ten*) system. It is called a decimal system because it is based upon grouping by tens. By using ten as a base and the idea of *place value*, we can write the name of any number using the ten symbols (*digits*) 1, 2, 3, 4, 5, 6, 7, 8, 9, and 0.

Suppose that the display on your HHC were as shown in the illustration. How would you read this display?

The following chart shows how this question and others like it may be answered.

OBJECTIVES

You should be able to:

(a) Read and write numerals.

(b) Demonstrate an understanding of the decimal place value system of numeration.

****

	MILLIONS			THOUSANDS			UNITS		
Place name	Hundred millions	Ten millions	Millions	Hundred thousands	Ten thousands	Thousands	Hundreds	Tens	Ones
Digits →			2	3	5	6	8	4	7

(Group name: top row)

7 × 1	=	7
4 × 10	=	40
8 × 100	=	800
6 × 1000	=	6,000
5 × 10,000	=	50,000
3 × 100,000	=	300,000
2 × 1,000,000	=	2,000,000
		2,356,847

To write a word name for a number we write word names for each group of digits. Thus the number displayed in the HHC on the previous page is read

Two million,	three hundred fifty-six thousand,	eight hundred forty-seven
millions group	thousands group	units group

Other examples are:

(a) 68,724 is read
Sixty-eight thousand, seven hundred twenty-four

(b) 635,432 is read
Six hundred thirty-five thousand, four hundred thirty-two

(c) 8,947,627 is read
Eight million, nine hundred forty-seven thousand, six hundred twenty-seven

Note that numbers such as 24, 32, and 27 are always written with a hyphen: twenty-four, thirty-two, twenty-seven, etc.

Shown below are all group and place names through a trillion.

TRILLION			BILLION			MILLION			THOUSAND			UNIT		
Hundred trillion	Ten trillion	Trillion	Hundred billion	Ten billion	Billion	Hundred million	Ten million	Million	Hundred thousand	Ten thousand	Thousand	Hundred	Ten	One

Notice that in each group we have the pattern of one-ten-hundred that is repeated over and over.

Do exercises 1 through 6 in the adjoining column.

In exercises 1–3 read the numeral orally and then write the name in words.

1. 284,960
2. 172,456,367
3. 106,070

In exercises 4–6 write a numeral for the given number.

4. Nine million, three hundred four thousand, seven hundred fifty
5. Thirty-five thousand, six hundred seventy-nine
6. One hundred seventy-four thousand twenty-six

STOP

The numeral 2,356,847 displayed on the HHC at the beginning of this section can also be thought of in the following ways by using the ideas of place value:

(a) $2{,}000{,}000 + 300{,}000 + 50{,}000 + 6{,}000 + 800 + 40 + 7$

(b) $2 \times 1{,}000{,}000 + 3 \times 100{,}000 + 5 \times 10{,}000 + 6 \times 1000 + 8 \times 100 + 4 \times 10 + 7 \times 1$

(c) $2 \times 10^6 + 3 \times 10^5 + 5 \times 10^4 + 6 \times 10^3 + 8 \times 10^2 + 4 \times 10^1 + 7 \times 10^0$

These are all examples of *expanded notation* and (c) is also an example of the use of *exponential notation.* Unless otherwise specified, the words *expanded notation* in this book will refer to form (b).

An *exponent* tells us to use a number such as 10 a given number of times as a factor. Thus

$$10^6 = 10 \times 10 \times 10 \times 10 \times 10 \times 10$$
$$10^5 = 10 \times 10 \times 10 \times 10 \times 10$$
$$10^4 = 10 \times 10 \times 10 \times 10$$
$$10^3 = 10 \times 10 \times 10$$
$$10^2 = 10 \times 10$$
$$10^1 = 10$$

and, as a special case, 10^0 is defined to be equal to 1.

In exercises 7 and 8 write the given numeral in expanded notation in the forms (b) and (c) shown in the text.

7. 1875

8. 368,492

STOP

Do exercises 7 and 8.

To help you see how each place in our system of numeration stands for a number that is 10 times the number represented by the place to the right, study the pattern found in doing exercise 9 on your HHC. Notice what happens to the "7" that you begin with, at each stage of the computation. (Use your HHC to do the computations even though you may be able to obtain the answers mentally.)

9. Using your HHC do (a) through (g), multiplying each time by 10 without clearing the calculator until the end and notice what happens to the original 7 in the display.

(a) $7 \times 10 =$ ______

(b) ______ $\times 10 =$ ______

(c) ______ $\times 10 =$ ______

(d) ______ $\times 10 =$ ______

(e) ______ $\times 10 =$ ______

(f) ______ $\times 10 =$ ______

(g) ______ $\times 10 =$ ______

STOP

Do exercise 9.

In our system of numeration, each digit has a *place value* and also a *value.* For example, for the numeral 53,732 we have

Ten thousand	Thousand	Hundred	Ten	One
5	3	7	3	2

In exercises 10–12 what is the place value of the digit 7?

10. 375

11. 6768

12. 758,692

In exercises 13–15 what is the value of the digit 7?

13. 3784

14. 8473

15. 27,898

STOP

Thus

5 has a *place value* of ten thousand (10,000)

and

5 has a *value* of 5 × 10,000 (50,000)

Furthermore,

7 has a *place value* of a hundred (100)

and

7 has a value of 7 × 100 (700)

"3" has two values depending on its position. The "3" on the left of the comma is in the thousands place and therefore

3 (left) has a *place value* of a thousand (1000)

and

3 (left) has a *value* of 3 × 1000 (3000)

Finally, the "3" on the right of the comma is in the tens place and therefore

3 (right) has a *place value* of ten (10)

and

3 (right) has a *value* of 3 × 10 (30)

Do exercises 10 through 15.

Exercises 1.2

Name ____________________

Class ____________________

In exercises 1–10 give the place value of the "6" in the given numeral.

1. 376
2. 367
3. 673
4. 45,623
5. 4106
6. 6401
7. 106,001
8. 263,241
9. 76,543,121
10. 235,678,001

1.
2.
3.
4.
5.
6.
7.
8.
9.
10.

In exercises 11–20 give the value of the "3" in the given numeral.

11. 32
12. 301
13. 123
14. 357
15. 63,074
16. 123,167
17. 105,131
18. 3,672,841
19. 167,376,505
20. 3,678,478,001

11.
12.
13.
14.
15.
16.
17.
18.
19.
20.

In exercises 21–28 write the given numeral in expanded notation in the forms (b) and (c) given in the text.

Example: (b) $273 = 2 \times 100 + 7 \times 10 + 3 \times 1$

(c) $273 = 2 \times 10^2 + 7 \times 10^1 + 3 \times 10^0$

21. 3567

21.

22.

23.

24.

25.

26.

27.

28.

29.

30.

31.

32.

22. 72,456

23. 284,960

24. 1030

25. 108,040

26. 805,706

27. 60,090

28. 206,070

In exercises 29-35 read orally, and then write, the name of each number.

29. 7849

30. 30,027

31. 71,071

32. 126,452

33. 70,603,049 | 33.

34. 9,304,750,024 | 34.

35. 60,004,084,600 | 35.

In exercises 36–45 write the numeral for the given number.

36. Two thousand, fifty | 36.

37. Sixteen thousand, three hundred four | 37.

38. Twenty thousand, sixty | 38.

39. Three hundred thousand, sixteen | 39.

40. Two hundred fifty-six thousand, three hundred ninety-four | 40.

41. Seven million, forty-two thousand, thirty-three | 41.

42. Fifteen million, nine hundred four thousand, sixty | 42.

43. Eighty billion, ninety million, forty thousand, twenty | 43.

44. One hundred ten billion, five million | 44.

45. Sixteen trillion, thirty-two billion, two hundred twenty-nine million, eight hundred thousand, ten | 45.

46. 46. Refer to the accompanying table. Read orally, and then write, the number name in English for the distance from the earth for the two nearest planets.

Planet	Minimum distance from earth
Mercury	50,974,000
Venus	24,564,000
Mars	34,389,000
Jupiter	367,320,000
Saturn	745,900,000
Uranus	1,606,600,000
Neptune	2,678,000,000
Pluto	2,669,900,000

47. 47. Give the largest number that uses all ten digits only once.

48. 48. Give the smallest number that uses all ten digits only once. (Zero is not to be used as the first digit on the left.)

49. 49. Write, in order of increasing size, all three digit numbers that can be formed using the digits 4, 5, and 6, each exactly once in each number.

50. 50. Give the largest three-digit number that uses both the digits 5 and 3 and only the digits 5 and 3.

51. 51. Give the smallest three-digit number that has only one zero digit.

52. 52. Give the largest six-digit number that has no two digits the same.

53. 53. What number comes just before one million?

54. 54. What is the largest number that can appear in the display of your calculator?

55. 55. How many thousands does it take to make a million?

56. 56. Which is larger? A million billion or a billion million?

57. The number 10^{100} is called a *googol.* How many places would you need to write the decimal numeral for one googol? 57.

*58. Determine the sum of all the three-digit numbers (no two digits the same) that can be formed from the digits 3, 5, 6, and 7. *58.

*59. The number of working parts in the human body is estimated as ten thousand million. Express this number as a power of ten. *59.

*60. If all the digits used to number the pages in a certain book were cut out and tossed into a box, there would be a total of 198. How many pages are there in the book? *60.

*61. Research Problem: Find out how groups above a trillion are named. Give the names of the next five groups. *61.

*62. What small animal gathers honey and has a stinger? For the answer, find the following sum. *62.

$$126 + 13 + 72 + 95 + 32$$

1.3 Subtraction of Whole Numbers

Do exercises 1 and 2 in the adjoining column.

CHECKING SUBTRACTIONS

The answer to a subtraction problem can be checked by doing an addition. Remember that, for example,

$$\begin{array}{r} 7 \\ -3 \\ \hline 4 \end{array} \quad \text{means that} \quad \begin{array}{r} 4 \\ +3 \\ \hline 7 \end{array}$$

That is,

$$7 - 3 = 4 \quad \text{means that} \quad 4 + 3 = 7$$

Thus to check the subtractions in exercises 1 and 2 we can perform the following additions:

$$\begin{array}{r} 9{,}181 \\ +\,39{,}842 \\ \hline 49{,}023 \end{array} \quad \text{and} \quad \begin{array}{r} 25{,}361 \\ +\,59{,}213 \\ \hline 84{,}574 \end{array}$$

Thus the answer given in exercise 1 is incorrect because $49{,}023 \neq 48{,}923$ ("$\neq$" is read *not equal to*) whereas the answer given in exercise 2 is correct because $84{,}574 = 84{,}574$.

Do exercises 3 through 5.

THE USE OF PARENTHESES

Consider the problem of calculating

$$(84{,}213 - 57{,}321) - 12{,}348$$

The *parentheses,* ( ), tell us that we are to first subtract 57,321 from 84,213 and then to subtract 12,348 from the resulting number. On an HHC, our work would be as follows:

Enter	84213	–	57321	–	12348	=
Display	84213.	84213.	57321.	26892.	12348.	14544.

Thus

$$(84{,}213 - 57{,}321) - 12{,}348 =$$

$$26{,}892 - 12{,}348 = 14{,}544$$

OBJECTIVES

You should be able to:

(a) Use the HHC to subtract whole numbers.

(b) Check your answer to a subtraction by doing an addition.

(c) Recognize basic properties of subtraction and addition.

(d) Simplify calculations on the HHC by using these properties of addition and subtraction.

(e) Apply subtraction to the solution of problems.

****

In exercises 1 and 2 perform the given subtractions and compare your answers with the ones shown.

1. $\begin{array}{r} 48{,}923 \\ -\,39{,}842 \\ \hline 9{,}181 \end{array}$

2. $84{,}574 - 59{,}213 = 25{,}361$

STOP

In exercises 3–5 do the subtraction and then check your work by doing an addition.

3. $\begin{array}{r} 45{,}327 \\ -\,27{,}148 \\ \hline \end{array}$

4. $\begin{array}{r} 5132 \\ -\,1462 \\ \hline \end{array}$

5. $10{,}001 - 9992$

STOP

6. (5214 – 1920) – 629

Do exercise 6.

Now consider the following subtraction problem:

84,213 – (57,321 – 12,348)

It involves the same numbers, 84,213, 57,321, and 12,348 as before and it again involves two subtractions. The difference in the two problems lies in the arrangement of parentheses. In the second problem we are asked first to subtract 12,348 from 57,321 and then to subtract the resulting number from 84,213.

7. Subtract 12,348 from 57,321.

Do exercise 7.

Unless your HHC has a memory key (see Appendix A), to complete this problem on an HHC you must either write down the answer to the subtraction 57,321 – 12,348 (44,973) or keep this number in your head and then subtract 44,937 from 84,213 after using the clear key.

8. Subtract 44,973 from 84,213.

Do exercise 8.

The results of exercises 7 and 8 tell us that

84,213 – (57,321 – 12,348) = 39,240

whereas, from our previous work, we have

(84,213 – 57,321) – 12,348 = 14,544

Thus we see that the grouping of numbers in a subtraction problem is important. This is unlike addition, where, because of the associative property, we can change the grouping of the numbers and still get the same sum.

9. Compute

(84,213 – 57,321) + 12,348

(This means to first subtract 57,321 from 84,213 and then add 12,348 to the resulting number. You can do this without using a memory key or writing down an intermediate result.)

Do exercise 9.

If you did exercise 9 correctly, you have found that

(84,213 – 57,321) + 12,348 = 39,240

which is exactly the answer we got for

84,213 – (57,321 – 12,343)

You will also have noted that the computation of

(84,213 – 57,321) + 12,348

is the easier of the two (even if a memory key is available).

Do exercises 10 and 11.

In exercises 10 and 11 compare your answers.

10. 7921 – (4922 – 3144)

11. (7921 – 4922) + 3144

SHORTCUTS IN COMPUTATIONS INVOLVING SUBTRACTION AND ADDITION

Exercises 9, 10, and 11 are instances which show us that, for all whole numbers a, b, and c.

(1) $$a - (b - c) = a - b + c$$

where we do not need to use parentheses on the right. This is because $a - b + c$ clearly indicates, as does $(a - b) + c$, that we are expected to proceed by first subtracting b from a and then adding c to this difference.

STOP

Do exercises 12 and 13.

In exercises 12 and 13 use (1) to simplify the computation.

12. 5196 – (3124 – 132)

13. 80,144 – (75,386 – 40,129)

There are other useful equalities similar to (1) that enable us to simplify calculations. In each of these which we now list, the given equality holds for all whole numbers a, b, c, and d.

(2) $$a - (b + c) = a - b - c$$

Example:

$$5212 - (1194 + 2169) = 5212 - 3363 = 1849$$

and also

$$5212 - 1194 - 2169 = 1849$$

STOP

Again, it is true that using the procedure suggested by the righthand side of (2) is easier than that suggested by the lefthand side.

Do exercises 14 and 15.

In exercises 14 and 15 use (2) to simplify the computation.

14. 6142 – (4324 + 1609)

15. 17,969 – (6143 + 8965)

(3) $$(a + b) - (c + d) = a + b - c - d$$

Example:

$$(4124 + 9322) - (5196 + 3224) = 13{,}446 - 8420 = 5026$$

and also

$$4124 + 9322 - 5196 - 3224 = 5026$$

more directly.

STOP

In exercises 16 and 17 use (3) to simplify the computation.

16. (4329 + 7812) – (1321 + 8692)

17. (14,129 + 29,207) – (17,283 + 6128)

STOP

In exercises 18 and 19 use (4) to simplify the computation.

18. (3296 – 1324) – (1396 – 969)

19. (60,368 – 5364) – (50,680 – 10,823)

STOP

20. On your HHC enter, in turn, 4392, –, 6197. What is the display after pressing the = key?

STOP

21. What would be a reasonable interpretation of negative and positive numbers when considering yards gained and yards lost in football?

STOP

Do exercises 16 and 17.

(4) $$(a - b) - (c - d) = a - b - c + d$$

Example:

$$(5969 - 3124) - (6322 - 5196) = 2845 - 1126 = 1719$$

and also

$$5969 - 3124 - 6322 + 5196 = 1719$$

more directly.

Do exercises 18 and 19.

NEGATIVE NUMBERS

Do exercise 20.

The display on your HHC after doing exercise 20 should be –1805, or it should indicate in some way that the answer is a *negative number* by showing –. We read –1805 as "negative 1805." Negative numbers are useful in, for example, contrasting temperatures below 0° with temperatures above 0°, debts versus assets, etc. Thus 3° would indicate a temperature of 3° above 0°, whereas –3° would indicate a temperature of 3° below 0°.

Naturally, a number such as 3079 is called a *positive number.* Thus 3079, 1, 50,000, $\frac{1}{2}$, 5.67, etc., are all positive numbers. The number 0 is neither positive nor negative.

Do exercise 21.

The fact that an HHC automatically allows for negative numbers permits us to carry out a sequence of additions and subtractions without worrying about the possibility of subtracting a larger number from a smaller one at some stage.

Example: Compute 5132 – 7389 + 7241.

Enter	5132	–	7389	+	7241	=
Display	5132.	5132.	7389.	–2257.	7241.	4984.

Thus 5132 – 7389 + 7241 = 4984 with the negative number –2257 displayed as an intermediate answer.

Do exercises 22 and 23.

Of course, some problems lead to negative answers. For example, suppose that you had \$273 in your checking account and later wrote two checks, one for \$178 and the other for \$183. If you had made no deposits and computed your balance with your HHC you would get

$$273 - 178 - 183 = -88$$

to indicate that you were overdrawn by \$88.

Do exercises 24 and 25.

SOLVING STORY PROBLEMS

Some of the key phrases that indicate a subtraction is involved in a problem are: "How many *more* . . . ," "How much *greater* . . . ," "How much *higher* . . . ," "How much *farther* . . . ," "What was the *increase* . . . ," and "What was the *decrease* . . ." That is, problems that involve finding out how much farther, how much higher, how much lower, how much greater, or how much less one amount is than another usually indicate that a subtraction is involved.

Do exercises 26 and 27.

In doing exercises 22 and 23 note that, although the final answers are positive numbers, negative numbers appear as intermediate answers.

22. Compute
5431 – 6723 + 1924

23. Compute
1671 – 82,946 + 92,104

STOP

24. Compute
1739 + 3964 – 7984

25. Suppose you had \$78 in your checking account and then made a deposit of \$17. If you then wrote checks for \$56 and \$42 and made no more deposits, by how much would you be overdrawn?

STOP

26. Mount Rainier in Washington is 14,410 feet high; Mount Shasta in California is 14,162 feet high. How much higher is Mount Rainier than Mount Shasta?

27. Mrs. Garcia's salary is \$15,300 a year and her husband's yearly salary is \$14,250. How much more per year does Mrs. Garcia make than her husband?

STOP

Exercises 1.3

Name ____________

Class ____________

In exercises 1–7 do the indicated subtraction and check by performing an addition.

1. $\begin{array}{r} 8456 \\ -\ 3243 \\ \hline \end{array}$

2. $\begin{array}{r} 54{,}809 \\ -\ 29{,}646 \\ \hline \end{array}$

3. $\begin{array}{r} 89{,}003 \\ -\ 20{,}908 \\ \hline \end{array}$

4. 2435 – 1649 Check:

5. 5913 – 2936 Check:

6. 40,046 – 36,759 Check:

7. 386,720 – 199,563 Check:

In exercises 8–18 perform the indicated operations.

8. 53,714 – 25,842 – 10,367

9. 310,465 – 102,141 – 36,785

10. 220 – 24 – 1156

11. 187 – 59 – 7465

12. 200 – 10 – 15

13. (81,043 + 1041) – 14,831

14. (504 – 754) + 826

15. (2345 – 678) + (8070 – 6040)

16. (3752 – 951) + (4321 – 4296)

17. 7090 – 508 + 9897 – 960 + 819

1.
2.
3.
4.
5.
6.
7.
8.
9.
10.
11.
12.
13.
14.
15.
16.
17.

18.

18. 1893 + 149 − 1325 + 490 − 1085

In exercises 19–25 use equalities (1) through (4) to simplify the computation.

19.

19. 3576 − (8672 − 1875)

20.

20. 20,482 − (18,216 − 21,248)

21.

21. (989 + 978) − (1001 + 101)

22.

22. (1982 + 7843) − (6722 + 89)

23.

23. (8469 + 9698) − (3674 + 7923)

24.

24. (9764 + 7692) − (664 − 6984)

25.

25. (758,692 − 60,092) − (68,728 − 84,969)

26.

26. Shanghai, China is the world's largest city with a population of 10,820,000, and New York is the world's fifth largest city with a population of 7,567,000. How many more people does Shanghai have than New York?

27.

27. The Great Salt Lake Desert in Utah is 80 miles long, 50 miles wide, and is at an elevation of 4500 feet. The Painted Desert in Arizona is 75 miles wide and is on a 5000 foot high plateau. How much wider and how much higher is the Painted Desert than the Great Salt Lake Desert?

28.

28. Yosemite Valley in California has the highest waterfall in the United States with a drop of 2425 feet, but it is only the fourth highest in the world. Angel Falls on a tributary of the Caroni River in Venezuela is the highest in the world with a total drop of 3212 feet. How much farther does the water drop in Angel Falls than at Yosemite?

29. The greatest weight at which an airplane has taken off is 820,700 pounds, achieved by the prototype Boeing 747-200 (747B) commercial transport in November 1970. The basic weight of the plane was 320,000 pounds, the remaining weight represented fuel, flight test equipment, and an artificial payload of sand and water. How much weight was used in fuel, flight test equipment, and artificial payload? 29.

30. The capacity of a coliseum is 12,225. If a crowd of 10,687 attended a basketball game there, how many empty seats should there have been? 30.

31. A coat on sale for $239 originally sold for $318. How much was saved by buying the coat on sale? 31.

32. How many feet must a pilot flying on instruments in the clouds at 7400 feet climb so that he is 500 feet higher than a mountain 11,235 feet high? 32.

33. A 100-foot roll and 75-foot roll of fencing were purchased. If lengths of 85 feet and 35 feet of the property were fenced, how much fencing was left over? 33.

34. A man drove to a town 365 miles away and returned by a route that was 47 miles shorter but two hours longer. If his odometer (mileage indicator) read 28,186 when he finished, what was it when he started? 34.

35. On January 1, Ms. Rogers had a balance of $565 in her checking account. During the month of January she deposited checks in amounts of $57 and $225, and wrote checks for $22, $37, $123, $247, and $7. How much did she have left in her checking account at the end of January? 35.

In exercises 36–40 refer to the following table.

PER CAPITA PERSONAL INCOME IN THE U.S., 1950 to 1974

Year	Amount	Year	Amount	Year	Amount	Year	Amount	Year	Amount
1950	$1496	1955	$1876	1960	$2222	1965	$2785	1970	$3966
1951	1652	1956	1975	1961	2274	1966	3001	1971	4195
1952	1733	1957	2045	1962	2381	1967	3188	1972	4537
1953	1804	1958	2067	1963	2469	1968	3457	1973	5023
1954	1785	1959	2167	1964	2603	1969	3733	1974	4448

Source: Bureau of Labor Statistics.

36.

36. How much did the per capita income in the United States increase between 1950 and 1974?

37.

37. For what ten-year period (1950–60, 1960–70) was the increase in the per capita income the greater? How much was this increase?

38.

38. For the years from 1970 through 1974, in what year was the increase in per capita income the largest? How much was this increase?

39.

39. For the years from 1955 through 1959, in what year was the increase in per capita income the smallest? How much was this increase?

*40.

***40.** For what period of ten years was the increase in per capita income the largest? How much was the increase?

In exercises 41–45 refer to the following table.

ENERGY CONSUMPTION—TOTAL AND PER CAPITA: 1920 TO 1973
(Total in trillions, per capita in millions, of British thermal units.)

	All Energy		Natural Gas[1]		Coal[2]		Crude Petroleum[3]	
Year	Total	Per capita	Total	Per capita	Total	Per capita	Total	Per capita
1920	19,782	186	827	8	15,504	146	2,634	25
1925	20,899	180	1,212	10	14,706	127	4,156	36
1930	22,288	181	1,969	16	13,639	111	5,652	46
1935	19,107	150	1,974	16	10,634	84	5,499	43
1940	23,908	181	2,726	21	12,535	95	7,487	57
1945	31,541	238	3,973	30	15,972	121	9,619	73
1950	34,153	226	6,140	41	12,913	85	12,706	84
1955	39,956	243	9,232	56	11,703	71	16,328	99
1960	44,816	249	12,736	71	10,414	58	18,608	103
1965	53,969	278	16,097	83	12,358	64	21,566	111
1970	67,143	330	22,029	108	12,693	62	29,537	145
1971	68,698	338	22,819	111	12,043	58	30,570	148
1972	72,108	346	23,125	111	12,495	60	32,965	158
1973	75,561	360	23,558	112	13,520	64	34,689	165

[1]Dry gas only. Marketed production minus shrinkage caused by liquids extraction (3½ cubic feet per gallon produced). [2]Includes bituminous coal and lignite and anthracite coal. [3]Includes petroleum products. Source: The Library of Congress, *Energy Facts,* November 1973.

41. What was the increase in total energy consumption between 1920 and 1970? 41.

42. What was the increase in per capita energy consumption between 1920 and 1970? 42.

43. For what ten-year period (1920–30, 1930–40, 1940–50, 1950–60, or 1960–70) was the increase in total consumption of crude petroleum the greatest? How much was the increase? 43.

44. What five-year period from 1950 through 1970 saw the greatest increase in per capita use of natural gas? How much was this increase? 44.

45. In 1973 how much energy was derived from sources other than natural gas, coal, or crude petroleum? 45.

In exercises 46–49 use the following information: In 1790 the population of the United States was 3,929,214; in 1800 it was 5,308,483 and in 1810 it was 7,239,881.

46. What was the increase in population from 1790 to 1800? 46.

47. What was the increase in population from 1800 to 1810? 47.

48. What was the increase in population from 1790 to 1810? 48.

49. How much more did the population increase in the period 1800 to 1810 than in the period 1790 to 1800? 49.

***50.** Find the missing digits in each addition exercise. *50.

(a)

$$\begin{array}{rcccc} & 1 & \_ & 0 & 6 \\ & & 3 & 8 & \_ \\ & & 3 & \_ & 7 \\ (+) & & & 4 & 2 \\ \hline & \_ & 4 & 4 & 2 \end{array}$$

(b)

$$\begin{array}{rcccc} & & 4 & 5 & 6 \\ & & \_ & 7 & \_ \\ & 1 & 4 & \_ & 3 \\ (+) & \_ & 5 & 8 & 4 \\ \hline & 6 & 4 & 4 & 1 \end{array}$$

*51.

*51. Find the missing digits in each subtraction problem.

(a)

```
     6 _ 1 _
(−)  3 4 _ 9
     -------
     2 8 1 8
```

(b)

```
     _ _ 0 _ 6 _
(−)      1 _ 0 _ 8
     -------------
       9 9 9 6 6
```

In exercises 52–54 complete the magic squares. (A magic square is a square in which the sum of each row, column, and diagonal is the same number.)

*52.

*52.

12	9	6	
7	19	13	
18		11	14
	15		5

*53.

*53.

18		12	26
	25	19	14
24			
16	21	23	11

*54.

*54.

4		12		20	45	28
35	11		19	44	27	
10		18	43	26		34
41	17		25		33	9
16		24	7	32	8	40
		6	31	14	39	15
22	5		13	38	21	46

***55.** Use the numbers 1 through 9 and place them in the square below to form a magic square. Is there more than one way to do this?

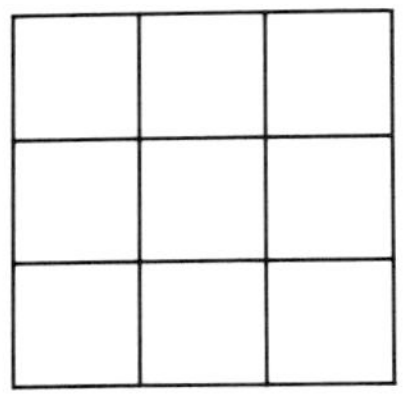

*55.

***56.** Use your HHC to perform the indicated computation and complete the following statement.

$$(2376 - 462) - (785 + 521)$$

Cranberries are grown in a ________ .

*56.

1.4 Multiplication of Whole Numbers

OBJECTIVES

You should be able to:

(a) Use the HHC to multiply whole numbers.

(b) Check your answer to a multiplication by doing a division.

(c) Multiply mentally by 10, 100, 1000, 10,000, etc.

(d) Recognize the commutative and associative properties of multiplication and know how to use these properties to shorten computation time.

(e) Use the HHC when an overflow occurs in some multiplication problems.

(f) Use exponential notation.

****

BASIC PRINCIPLES

Remember that 5 × 7 relates to 5 sets of 7 things joined together. Thus

$$5 \times 7 = 7 + 7 + 7 + 7 + 7 = 35$$

Similarly,

$$62 \times 35 = 35 + 35 + 35 + \ldots (62 \text{ times}) \ldots = 2170$$

In the past, to multiply 62 by 35 we used one of the following *algorithms* (procedures):

```
   35
 × 62
 ----
   10  ←   2 × 5
   60  ←   2 × 30
  300  ←  60 × 5
 1800  ←  60 × 30
 ----
 2170
```

or the short cut

```
   35
 × 62
 ----
   70  ←   2 × 35
  210  ←  60 × 35
 ----
 2170
```

The calculator will do all the messy work of the algorithm and give the product directly.

CHECKING MULTIPLICATIONS

Do exercises 1 and 2 in the adjoining column.

In exercises 1 and 2 perform the given multiplication and compare your answers with those shown.

1.
```
    4321
  ×   59
 -------
 254,949
```

2. 7245 × 62 = 449,190

STOP

In exercises 1 and 2 you should have found that the answer to exercise 2 is correct, whereas the answer to exercise 1 is incorrect.

Can we check a multiplication other than by re-doing the problem? Our check for subtraction involved an addition—and this procedure is reasonable even for hand calculations. Likewise we can check a multiplication by a division. Remember that

$$\text{if } 6 \times 3 = 18, \quad \text{then } 18 \div 3 = 6, \quad \text{and } 18 \div 6 = 3$$

In exercises 3–5 perform the multiplication and check your answer by doing a division.

3. 489 × 732

Check:

____________ ÷ 732

4. 12,432 × 84

Check:

5. 13,296 × 124

Check:

STOP

In exercises 6–9 do the multiplication mentally. Check your answer by using your HHC and compare the computation times with and without using the HHC.

6. 69 × 10,000

7. 4 × (967 × 25)

8. 40 × (322 × 25)

9. 50 × (192 × 20)

STOP

Using hand calculations such a checking procedure would be completely unreasonable for a multiplication such as 4321 × 59. On the other hand, with an HHC such a checking procedure is quite feasible. For example, using this idea to check the answer to exercise 1 we find that

$$\text{if} \quad 4321 \times 59 = 254{,}949, \quad \text{then} \quad 254{,}949 \div 59 = 4321$$

But performing the division of 254,949 by 59 on an HHC we have, as you should check,

Enter	254949	÷	59	=
Display	254949.	254949.	59.	4321.1695

Because 4321.1695 ≠ 4321, we know that the answer of 254,949 is incorrect.

Similarly, to check the answer to exercise 2, we use the fact that

$$\text{if} \quad 7245 \times 62 = 449{,}190, \quad \text{then} \quad 449{,}190 \div 62 = 7245$$

Since, as you should check, it is true that 449,190 ÷ 62 = 7245, we conclude that the answer of 449,190 is correct.

Do exercises 3 through 5.

Consider now a multiplication such as 43 × 10. We hope you agree that it would be foolish to perform this multiplication on an HHC. We have

$$43 \times 10 = 430$$
$$43 \times 100 = 4300$$
$$43 \times 1000 = 43{,}000$$

and so on. All we need to do is to add zeros to the numeral 43.

Even problems such as

$$(25 \times 487) \times 4$$

are most easily handled without an HHC by using a little thought. We have

$$4 \times 25 = 100 \quad \text{and} \quad 487 \times 100 = 48{,}700$$

Do exercises 6 through 9.

In doing the multiplication

$$(25 \times 487) \times 4$$

by noting that

$$25 \times 4 = 100 \quad \text{and} \quad 487 \times 100 = 48{,}700$$

we were really re-writing

$$(25 \times 487) \times 4$$

as

$$487 \times (25 \times 4) = 487 \times 100$$

Is this permissible? It is, because of the *commutative* and *associative properties of multiplication:*

$a \times b = b \times a$ — Commutative property of multiplication (order property)

and

$a \times (b \times c) = (a \times b) \times c$ — Associative property of multiplication (grouping property)

for all numbers a, b, and c.

Thus

$$(25 \times 487) \times 4 = (487 \times 25) \times 4$$

by the commutative property of multiplication and

$$(487 \times 25) \times 4 = 487 \times (25 \times 4)$$

by the associative property.

Do exercises 10 through 14.

OVERFLOW IN MULTIPLICATION

It is very easy to get an overflow indication in multiplication on an HHC unless it has scientific notation.

Do exercise 15.

Handling overflow in a multiplication such as 32,894 × 5367 is rather difficult. Such problems will be discussed in sections 1.6 and 7.2. However, a multiplication such as 32,894 × 10,000 that would cause overflow should certainly be done mentally: 32,894 × 10,000 = 328,940,000. Furthermore, a multiplication such as 32,894 × 125,000 can easily be done on an HHC by writing

$$32{,}894 \times 125{,}000 = 32{,}894 \times (125 \times 1000) \quad \text{(mentally)}$$

$$= (32{,}894 \times 125) \times 100 \quad \text{(associative property of addition)}$$

$$= 4{,}111{,}750 \times 100 \quad \text{(using an HHC)}$$

$$= 411{,}175{,}000 \quad \text{(mentally)}$$

In exercises 10–14 state what property of multiplication is involved in going from the lefthand side of the given equality to the righthand side.

10. 92 × 116 = 116 × 92

11. 512 × (28 × 19) = (512 × 28) × 19

12. 251 × (41 × 487) = (251 × 41) × 487

13. (251 × 14) × 487 = (14 × 251) × 487

14. (41 × 251) × 487 = 487 × (41 × 251)

STOP

15. Attempt to multiply 32,894 by 5367 on your HHC. Again note what the overflow indicator is on your HHC.

STOP

16. Use the procedures just described to perform the following multiplications.

(a) $437 \times 125{,}000{,}000$

(b) $61{,}700{,}000 \times 924$

(c) $32{,}100 \times 57{,}300$

STOP

Do exercise 16.

EXPONENTIAL NOTATION

In section 1.2 we used exponents with the number 10 to show the meaning of our system of numeration. Exponents can be used with any number. Thus 2^5, for example, is a shorthand way of writing

$$2 \times 2 \times 2 \times 2 \times 2 \qquad \text{(5 factors of 2)}$$

Similarly,

$$5^4 = 5 \times 5 \times 5 \times 5 \quad \text{and} \quad 15^2 = 15 \times 15$$

We read

3^5 as "three to the fifth,"

5^4 as "five to the fourth"

etc. But when the exponent is 2 or 3 we usually use the words "squared" and "cubed," respectively. Thus

15^2 is read "fifteen squared"

and

5^3 is read "five cubed"

(This is because the area of a square of side a is $a \times a = a^2$ and the volume of a cube of side a is $a \times a \times a = a^3$.)

As special cases we make the definitions

$$a^1 = a \quad \text{for all numbers } a$$

and

$$a^0 = 1 \quad \text{for all numbers } a \neq 0$$

A number such as 225 that is the square of another number ($225 = 15^2$) is called a *perfect square.* Likewise, since $125 = 5^3$, 125 is an example of a *perfect cube.*

Do exercises 17 through 19.

17. Compute 5^4 and 4^5. Which is larger?

18. Compute 6^5 and 5^6. Which is larger?

19. Compute 7^6 and 6^7. Which is larger?

STOP

From the results of exercises 17–19, what do you think is true about n^{n-1} and $(n-1)^n$ for any whole number $n > 1$? ("$>$" is read "greater than.")

SOLVING STORY PROBLEMS

Recognizing what kind of action is involved in story problems is a very important part of solving them. Multiplication is usually associated with the joining together of sets of things that have the same number of things in each set.

Example: Mr. White, in taking inventory in his warehouse, found that he had 13 stacks of tires with 18 tires in each stack. How many tires did he have in his warehouse?

Solution: With 13 sets and 18 in each set, the total number of tires is found by a multiplication:

$$13 \times 18 = 234$$

There are 234 tires in the warehouse.

Do exercises 20 and 21.

Many times the cost or amount for each unit is expressed in such terms as "\$20 *per* month," "34 miles *per* gallon," or 18 cents *per* can." If you know the cost or amount for each unit, often expressed by the word *per*, you can find the number associated with more than one unit by a multiplication.

Example: Jan spends \$160 per month on rent for her apartment. How much rent does the apartment cost her for a nine-month school year?

Solution:

Cost per unit is \$160	(cost per month)
Number of units is 9	(number of months)

Total amount is found by a multiplication:

$$9 \times 160 = 1440$$

Total rental cost for the apartment for nine months is \$1440.

Do exercises 22 and 23.

20. A grocer sees that she has 7 shelves of soup with 48 cans on each shelf. How many cans of soup are on her shelves?

21. What is the cost of 123 tickets to a rock concert if each ticket costs \$6?

STOP

22. Bill spends \$4 per week to ride the metro bus system to work and back. If he rides the bus 48 weeks out of the year, how much does he spend each year on bus fare riding to and from work?

23. Jack gets 35 miles per gallon in his new car. How far can he travel on a tank of gas if his tank holds 12 gallons?

STOP

Exercises 1.4

Name ____________________

Class ____________________

In exercises 1–4 perform the indicated multiplication.

1. 7369 × 526

2. 2507 × 704

3. 8096 × 608

4. 5973 × 4863

In exercises 5–8 check each of the products by doing a related division problem. If the product is incorrect, find the correct product.

5. 375 × 76 = 25,120 Check:

6. 29 × 64 = 1856 Check:

7. 438 × 39 = 17,082 Check:

8. 738 × 146 = 114,318 Check:

In exercises 9–14 use the procedures described in the text to find the product.

9. 137 × 123,000,000

10. 2461 × 139,300

11. 506,000 × 231,000

12. 673,000,000 × 127

13. 148,995 × 940,000

14. 754,632,000 × 24,000

1.
2.
3.
4.
5.
6.
7.
8.
9.
10.
11.
12.
13.
14.

In exercises 15–26 find the products without using your calculator by using the commutative and associative properties of multiplication and multiplying by 10, 100, 1000, etc.

15.

15. $(62 \times 4) \times 25$ 16. $(506 \times 250) \times 4$

16.

17.

17. $(4 \times 75) \times 25$ 18. $(8 \times 25) \times 498$

18.

19.

19. $2 \times (637 \times 500)$ 20. $(50{,}706 \times 2) \times 50$

20.

21.

21. $(485 \times 250) \times 4$ 22. $(7982 \times 500) \times 2$

22.

23.

23. $846 \times (20 \times 50)$ 24. $250 \times (937 \times 4)$

24.

25.

25. $(5 \times 3625) \times 20$ 26. $(40 \times 800) \times 25$

26.

In exercises 27–40 perform the indicated computations.

27.

27. 25^2 28. 37^2

28.

29.

29. 15^3 30. 22^4

30.

31.

31. $13^2 + 17^3$ 32. $35^2 - 25^2$

32.

33.

33. $18^3 + 15^2 - 19^3$ 34. $(9^3)^2$

34.

35.

35. 137^0 36. $(7^4)^2$

36.

37.

37. $(7^2)^4$ 38. $17^3 - 17^2$

38.

39. $9^3 - 9^2 + 9$

40. $13^3 + 13^2 - 13$

41. A woman who collects owl figurines had a collection that took 15 shelves and there were 25 owls on each shelf. How many owls were in her collection?

42. A theater has 65 rows of seats with 35 seats in each row. How many people can be seated in the theater?

43. A store ordered 27 cases of canned goods and each case contained 24 cans. How many cans were ordered?

44. If the rent of an apartment is \$167 per month, what is the cost for one year?

45. If the per capita consumption of eggs in the United States is 285 eggs per year, how many eggs would a city of 300,000 need in a year?

46. A car gets 33 miles per gallon of gas. If the gas tank of the car holds 21 gallons, how far might you expect the car to go on a tank of gas?

In exercises 47-49 refer to the following information.

Automobile pollution emissions for a 1967 car averaged 87 grams of carbon monoxide per mile while a 1974 car averaged 28 grams per mile.

47. If you drove a 1967 car 15,300 miles, what amount of carbon monoxide pollutant did your car produce?

48. If you drove a 1974 car 47,500 miles, what amount of carbon monoxide did your car produce?

49. What would be the difference in carbon monoxide emission produced by a 1967 car and a 1974 car both driven for 25,000 miles?

39.

40.

41.

42.

43.

44.

45.

46.

47.

48.

49.

In exercises 50–52 refer to the following information.

It is estimated that the total water used in the United States in 1900 amounted to 525 gallons per person and that in 1980 this will have increased to 1675 gallons per person. The population of the United States in 1900 was 76,000,000 and the estimated population in 1980 is 220,000,000.

50.

50. What was the total amount of water used in the United States in 1900?

51.

51. What will be the total amount of water used in the United States in 1980 according to the estimates given above?

52.

52. How much more water will be needed in 1980 than in 1900?

In exercises 53–58 refer to the following information.

A man has a gross monthly salary of $1247 and a take-home pay of $845. Each month on payday he puts $55 in the bank, pays $125 in child support, makes a car payment of $85, and pays $155 for rent.

53.

53. How much money would he bank if he does this for ten years?

54.

54. How much money does he pay in one year for child support?

55.

55. If he makes car payments for three years, how much money did he put out for his car in monthly payments?

56.

56. How much money does he have left each month after he makes all the indicated payments?

57.

57. A local school has, in a certain period, 8 classes with 42 students, 15 classes with 33 students, and 19 classes with 27 students. How many students are registered in school for that period?

58.

58. If the earth has a radius of 3963 miles at the equator and the diameter of the sun is 109 times that of the earth, what is the diameter of the sun in miles? (Remember that the diameter is equal to twice the radius.)

In exercises 59–61 find the missing digits in each multiplication problem.

*59.

```
        4 _ 7
  (X)     _ 9
  -----------
    3 _ 4 3
  1 _ _ 1
  -----------
  1 _ 6 _ 3
```

*60.

```
        _ 9
  (X)   6 _
  ---------
      1 9 6
    2 _ _
  ---------
    3 _ 3 _
```

*59.

*60.

*61.

```
          _ 0 _ _
  (X)       8 _ 5
  ---------------
      2 0 3 3 5
    _ _ _ 4
  _ _ _ _ _
  ---------------
  _ _ _ _ _ 7 5
```

*61.

* 62. (a) Check that

$15^2 = 225$ and $2 = 1 \times (1 + 1)$

$25^2 = 625$ and $6 = 2 \times (2 + 1)$

$35^2 = 1225$ and $12 = 3 \times (3 + 1)$

(b) Using the above information, formulate a simple rule for squaring a number ending in 5.

(c) Test your rule on the following:

45^2 = ____________ 55^2 = ____________

85^2 = ____________ 95^2 = ____________

115^2 = ____________ 475^2 = ____________

*62. (a)

(b)

(c)

*63. Do the computation below and use your HHC to find out what the statement refers to.

$$(59 \times 203) - (27 \times 157)$$

"Many churches have this item in their steeple."

*63.

1.5 Division of Whole Numbers

THE MEANING OF DIVISION

Just as multiplication can be thought of in terms of the joining of sets with the same number of elements, division can be thought of in terms of separating a set into subsets each containing the same number of elements. Consider the following example:

> A manufacturer of electronic equipment is ready to ship 525 items to a customer. If each box holds 15 items, how many boxes will he need to ship the order?

To solve this problem in pre-HHC days we would have divided 525 by 15 by one of the following methods:

```
      35
15 ) 525                       15 ) 525
     45                             300    20
     ---                            ---
      75          or                225
      75                            150    10
      --                            ---
       0                             75
                                     60     4
                                     --
                                     15
                                     15     1
                                     --    --
                                           35
```

Therefore, 35 boxes are needed to ship 525 items.

The most useful key on your HHC is the division key. The preceding problem (525 ÷ 15) and others like it can be very tedious to perform with pencil and paper, but with an HHC, doing division is just as easy as doing addition, subtraction, or multiplication.

There is, however, one basic difference between division and the other three operations as applied to whole numbers: If the numbers involved in an addition, a subtraction, or a multiplication are whole numbers, the answer will always be a whole number (possibly a negative number in a subtraction). In division, on the other hand, we do not always get a whole number for an answer: 1 ÷ 2 = 0.5, for example. In this chapter we will consider only divisions for which the answer is a whole number; divisions involving decimals will be considered in Chapter 2.

Do exercises 1 and 2 in the adjoining column.

In exercise 1 you should have found that the answer is correct, whereas the answer to exercise 2 is incorrect.

OBJECTIVES

You should be able to:

(a) Use the HHC to perform divisions involving whole numbers when the answer is also a whole number.

(b) Check your answer to a division problem by doing a multiplication.

(c) Recognize the basic properties of division along with multiplication and know how to use these properties to shorten computation time.

(d) Apply division to the solution of problems.

****

In exercises 1 and 2 check the answers to the division problems by doing them on your HHC.

1. 2,439,748 ÷ 529 = 4612

2. 4,612,272 ÷ 748 = 5893

STOP

In exercises 3–6 perform the indicated division and check your answer by a multiplication.

3. $22{,}932 \div 98$

   Check:

4. $238{,}784 \div 112$

   Check:

5. $506{,}142 \div 927$

   Check:

6. $56{,}242{,}452 \div 6923$

   Check:

STOP

7. $(20{,}160 \div 840) \div 24$

   $= \underline{\hspace{6em}} \div 24$

   $= \underline{\hspace{6em}}$

8. $20{,}160 \div (840 \div 24)$

   $= 20{,}160 \div \underline{\hspace{6em}}$

   $= \underline{\hspace{6em}}$

STOP

9. Compute
   $(234 \times 926) \div 926$

10. Compute
    $(234 \times 926) \div 234$

STOP

CHECKING DIVISIONS

To check a division, we can perform a multiplication (as is customary even for pencil and paper calculations). Thus

$$18 \div 3 = 6$$

is checked by observing that

$$3 \times 6 = 18$$

So we can check the division of exercise 1,

$$2{,}439{,}748 \div 529 = 4612$$

by calculating that

$$4612 \times 529 = 2{,}439{,}748 \quad \text{(answer is correct)}$$

and the division of exercise 2,

$$4{,}612{,}272 \div 748 = 5893$$

by calculating that

$$5893 \times 748 = 4{,}407{,}964 \neq 4{,}612{,}272 \quad \text{(answer is incorrect)}$$

Do exercises 3 through 6.

PROPERTIES OF DIVISION AND MULTIPLICATION

Do exercises 7 and 8.

What conclusions do you think you can draw from your answers to exercises 7 and 8 as to the associativity of the operation of division? That is, is $(a \div b) \div c = a \div (b \div c)$ for all whole numbers a, b, and c with $b \neq 0$ and $c \neq 0$?

Do exercises 9 and 10.

From your answer to exercise 9, what do you think you can conclude about $(a \times b) \div b$ for a and b any numbers with $b \neq 0$?

From your answer to exercise 10, what do you think you can conclude about $(a \times b) \div a$ for any numbers a and b with $a \neq 0$?

Do exercises 11 and 12.

From your answers to exercises 11 and 12, what do you think you can conclude about $(a \div b) \times c$ and $(a \times c) \div b$ for all numbers a, b, and c with $b \neq 0$?

DIVISION BY ZERO

Do exercise 13.

Going back to the meaning of division, we see that

$$\text{if } a \div b = c, \quad \text{then } b \times c = a$$

Thus

$$24 \div 4 = 6 \quad \text{means that} \quad 4 \times 6 = 24$$

and

$$196 \div 0 = ? \quad \text{means that} \quad 0 \times ? = 196$$

Clearly, there is no number which when multiplied by zero will give us 196. Thus we can say that there is no answer to the problem $196 \div 0$, or, for that matter, any number $n \neq 0$ divided by 0. That is,

$$n \div 0 \text{ has no answer for any number } n \neq 0$$

What about $0 \div 0$? Try this on your HHC. Again, if

$$0 \div 0 = ?, \text{ then } 0 \times ? = 0$$

Because any number times zero is zero, we see that there is no one answer to $0 \div 0$. For example, $1 \times 0 = 0$ so we would conclude from this that $0 \div 0 = 1$. But also $2 \times 0 = 0$ so that we would conclude from this that $0 \div 0 = 2$ and so on. That is, any number could be the answer to $0 \div 0$. Thus

$$0 \div 0 \text{ has no one answer}$$

Because of these two results, $n \div 0$ has no answer and $0 \div 0$ has no one answer. We say that division by 0 is *undefined*; we cannot divide by 0.

Do exercise 14.

FINDING AVERAGES

One common application of division occurs in the finding of averages. Suppose a car took 4 hours to go 180 miles. If we wanted to find out how fast the car went, we would have to know a great deal more than the fact that it took 4 hours to go 180 miles. Sometimes it might have gone very fast and at other times very slow as would happen if the traffic backed up along a freeway during the rush hour.

11. Compute
$(50{,}052 \div 194) \times 74$

12. Compute
$(50{,}052 \times 74) \div 194$

STOP

13. Try to divide 196 by 0 on your HHC. What happens? Can you explain why?

STOP

14. Perform the following divisions on your HHC:

$0 \div 1 =$

$0 \div 20 =$

$0 \div 37 =$

$0 \div 234 =$

From your answers to these computations, what do you think you can conclude about $0 \div a$ for all numbers $a \neq 0$?

STOP

15. The sum of 3 numbers is 273. What is the average of the 3 numbers?

16. An airplane flew 1125 miles in 5 hours. What was the average speed of the plane?

17. John received scores of 70, 75, 60, 90, and 100 on 5 tests. What was his average score?

18. The wages paid to 6 employees in a small factory were, per month, \$512, \$540, \$548, \$558, \$596, and \$606. What was the average wage paid?

STOP

If we divided 180 by 4 to get 45, we would simply have the number of miles per hour that the car would have had to travel at that same speed in order to cover the 180 miles in 4 hours. We call this number, 45, the *average* speed. We also say that the car went 45 miles per hour *on the average* and that the car *averaged* 45 miles per hour.

We can think of the 180 miles as the sum of 4 numbers: the first would be the number of miles the car traveled the first hour, the second the number of miles traveled in the second hour, the third the number of miles traveled in the third hour, and the fourth the number of miles traveled in the fourth hour. Whenever we have a sum and know how many numbers were added together to get this sum, we can find an average:

$$\text{Average} = (\text{sum of the numbers}) \div (\text{number of numbers})$$

For example, the average of the 4 numbers 25, 32, 15, and 20 is

$$(25 + 32 + 15 + 20) \div 4 = 92 \div 4 = 23$$

Do exercises 15 through 18.

SOLVING STORY PROBLEMS

The need for using a division in a story problem is indicated by a situation in which an amount must be separated into a number of parts of the same size.

Example: The Hotspark Electronic Company needs to ship 345 distributor caps to an auto parts store chain. If each shipping package contains 15 caps, how many packages must be shipped?

Solution: 345 caps have to be separated into sets of 15. The number of sets of 15 is found by doing the division

$$345 \div 15 = 23$$

Thus 23 packages must be shipped.

Do exercises 19 and 20.

Many division problems arise when we are considering unit costs or unit amounts. When we are given a number associated with more than one unit and are asked to find the number related to one unit, we can solve such a problem using division. We have either

$$(\text{total number}) \div (\text{number of units}) = \text{amount per unit}$$

or

$$(\text{total number}) \div (\text{amount per unit}) = \text{number of units}$$

Example 1: If Jack's car goes 444 miles on 12 gallons of gas, what is his gas mileage (miles per gallon)?

Solution:

$$(\text{total number}) \div (\text{number of units}) = \text{amount per unit}$$

$$444 \text{ miles} \div 12 \text{ gallons} = 37 \text{ miles per gallon}$$

His mileage is 37 miles per gallon.

Do exercise 21.

Example 2: If a store owner paid \$17 each for several barbecue sets and his invoice indicates that he is to pay \$221 for them, how many barbecue sets should he have received?

Solution:

$$(\text{total amount}) \div (\text{amount per unit}) = \text{number of units}$$

$$221 \div 17 = 13$$

He should have received 13 barbecue sets.

Do exercises 22 and 23.

19. Sally has a collection of 171 bottles. If her collection takes up 9 shelves and there are an equal number of bottles on each shelf, how many bottles are on each shelf?

20. Mr. Buck stated in his will that the money received by the sale of his 2 cars was to be divided equally among his 7 children. Upon his death, the 2 cars were sold for \$8645. How much money did each of his children then receive?

STOP

21. If Marcia's yearly salary is \$19,500, what salary does she receive per month?

STOP

22. If a gross (12 dozen) of Gizmos costs \$1584, what is the cost of one Gizmo?

23. A certain typist can type 1125 words in 15 minutes. How many words per minute can he type?

STOP

Exercises 1.5

Name ______________

Class ______________

In exercises 1–5 perform the indicated division and check your answer by performing the proper multiplication.

1. $7280 \div 130$
2. $63{,}104{,}373 \div 10{,}629$
3. $24{,}884{,}574 \div 49{,}082$
4. $53{,}803{,}998 \div 10{,}629$
5. $58{,}161{,}020 \div 1357$

In exercises 6–9 perform each of the indicated computations. Notice the groupings and the resulting answers to your computations.

6. (a) $144 \div (12 \div 3)$

   (b) $(144 \div 12) \div 3$

7. (a) $(52{,}488 \div 972) \div 54$

   (b) $52{,}488 \div (972 \div 54)$

8. (a) $134{,}784 \div (936 \div 72)$

   (b) $(134{,}784 \div 936) \div 72$

9. (a) $3{,}007{,}584 \div (25{,}488 \div 59)$

   (b) $(3{,}007{,}584 \div 25{,}488) \div 59$

10. From the results of exercises 6–9, what conclusion can you draw regarding the grouping of numbers and the operation of division on these numbers?

Answers:

1.

2.

3.

4.

5.

6. (a)

   (b)

7. (a)

   (b)

8. (a)

   (b)

9. (a)

   (b)

10.

In exercises 11–14 perform the given computations. Notice the order of the numbers, the groupings, and the resulting answers to your computations.

11. (a) $(3842 \times 26) \div 26$

(b) $(3842 \times 26) \div 3842$

12. (a) $(18{,}217 \times 124) \div 124$

(b) $(18{,}217 \times 124) \div 18{,}217$

13. (a) $(126{,}498 \div 58) \times 176$

(b) $(126{,}498 \times 176) \div 58$

14. (a) $(1{,}525{,}915 \div 5003) \times 53$

(b) $(1{,}525{,}915 \times 53) \div 5003$

15. What conclusion can you draw after working exercises 11–14?

16. How many servings of 4 ounces each can be obtained from 172 ounces of fruit juice?

17. What weekly salary will give a yearly salary of $10,660?

11. (a)

(b)

12. (a)

(b)

13. (a)

(b)

14. (a)

(b)

15.

16.

17.

18. A grocer bought 18 sacks of potatoes weighing 98 pounds each. How many small bags of 12 pounds each could he fill from the 18 sacks? 18.

19. A balance of $1595, including interest, is due on a car after a down payment of $225 has been made. How long will it take to pay off this balance at the rate of $55 a month? 19.

20. How much, on the average, will you have to save each week to pay cash for a car costing $1924 if you have just a year to save the money? 20.

21. For each 330 feet that you go above sea level, the average temperature generally becomes one degree Fahrenheit lower. If the temperature at sea level is 66 degrees, what would you expect the temperature to be at 5280 feet (one mile high)? 21.

22. Three people rented a cabin at the beach for 6 weeks at the cost of $125 per week. What was the share of the total rent for each person? 22.

23. A ticket sale was organized by distributing 2625 tickets to a concert equally among 7 chairpersons, each of whom divided these among 15 ticket sellers. How many tickets did each seller receive? 23.

24. Your budget allows $336 for a 7-day vacation. On the average, how much would this allow you for each day? 24.

25. You filled the tank of your car with gas with the odometer reading 39,853 miles. Later you stopped for gas with the odometer reading 40,308 and it took 13 gallons to fill your tank. How many miles was your car averaging on a gallon of gas? 25.

26. At a certain time the moon is 239,040 miles from the earth. If a train locomotive could travel to the moon at the rate of a mile a minute for 24 hours each day, how many days would it take to make the trip? 26.

27. A plane left airport A at 1:00 PM and reached airport B at 8:00 PM. Find the average speed of the plane if the distance between the two airports is 3325 miles. 27.

28.

28. If you typed 4655 words in 35 minutes, what would be your average rate per minute?

29.

29. If your pace averages 24 inches, how many steps will it take you to walk a mile (1 mile = 5280 feet)?

30.

30. If Jacque's pace averages 50 centimeters (cm), how many steps will he take in walking one kilometer (km)? (100 cm = 1 meter (m) and 1 km = 1000 m)

31.

31. Three boys decided to share the cost of the refreshments during a basketball game. If they bought 9 drinks at 50 cents each and 7 hot dogs at 75 cents each, how much was each boy's share?

32.

32. A basketball team had scores of 121, 97, 113, 99, and 105 in 5 games. What was their average score for those 5 games?

33.

33. A football team scored 253 points in 11 games. What was their average score per game?

34.

34. Sam made the following scores on 4 tests: 85, 83, 97, and 95. What must he make on his fifth test for him to have a 90-point average for the 5 tests?

35.

35. A refreshment committee made 171 sandwiches for a picnic. If 57 people attended and all the sandwiches were eaten, what was the average number of sandwiches eaten by each person?

36.

36. The Oregon State Department of Forestry furnishes fire protection for 16,299,997 acres of forest. In one year 1057 fires were controlled with a loss of 15,855 acres. What was the average number of acres of forest burned per fire in that year?

37.

37. Running at 5.7 miles per hour uses up 12 calories each minute. A $2\frac{3}{4}$ oz. candy bar contains about 420 calories. How long would you have to run to use up the calories after eating a candy bar?

In exercises 38–41 use the information in the following table to answer each question.

U.S. MONEY IN CIRCULATION BY DENOMINATION[1]
(In millions of dollars.)

Denomination	1975[4]	1974	1973	1972	1970	1968	1961	1950	1945	1939
Coin	8,496	8,331	7,759	7,287	6,281	5,691	2,582	1,554	1,274	590
$1[2]	2,616	2,720	2,639	2,523	2,310	2,049	1,588	1,113	1,039	559
$2	135	135	135	135	136	136	92	64	73	36
$5	3,571	3,718	3,614	3,449	3,161	2,993	2,313	2,049	2,313	1,019
$10	10,239	10,503	10,226	9,827	9,170	8,786	6,878	5,998	6,782	1,772
$20	26,798	26,197	23,915	21,883	18,581	16,508	10,935	8,529	9,201	1,576
$50	7,671	7,444	6,514	5,868	4,896	4,186	2,869	2,422	2,327	460
$100	21,280	20,298	17,288	15,118	12,084	10,068	6,106	5,043	4,220	919
$500	177	179	185	193	215	244	242	368	454	191
$1,000	206	209	216	225	252	292	300	588	301	425
$5,000	2	2	2	2	3	3	3	4	7	20
$10,000	4	4	4	4	4	4	10	12	24	32
Total[3]	81,196	79,723	72,497	66,516	57,093	50,961	33,918	27,741	28,515	7,598

[1] End of year.
[2] Paper currency only. $1 coins reported under coin.
[3] Includes unassorted currency.
[4] June 30.
Source: Department of the Treasury.

38. How many 100-dollar bills were in circulation in 1975? 38.

39. How many more two-dollar bills were in circulation in 1945 than in 1950? 39.

40. How many more 1000-dollar bills were in circulation in 1950 than in 1975? 41.

41. How many more fifty-dollar bills were in circulation in 1975 than in 1950?

42. Give an example in which division is associative. That is, find a replacement for each blank below which will make a true statement. (Blanks need not be replaced by the same number.) 42.

(____ ÷ ____) ÷ ____ = ____ ÷ (____ ÷ ____)

43.

43. Is it possible to find an example in which division is commutative? If your answer is "yes," give the example.

*44.

***44.** Take any three-digit number. Repeat the digits to get a six-digit number called a *tautonym*. Divide the tautonym by 7, divide that quotient by the original number, and then divide that quotient by 11. What result do you get? Try several examples; do you always get the same result? What conclusion do you think you can draw?

Example: Number, 123; tautonym, 123,123

123,123 ÷ 7 = ______

______ ÷ 123 = ______

______ ÷ 11 = ______

*45.

***45.** Form a six-digit tautonym of the form ABABAB (repeating two digits three times). Divide this number and each quotient respectively by 3, 7, 13, and 37. Try several examples. Do you always get the same results? What conclusion do you think you can draw?

Example: Number, 32; tautonym, 323,232

323,232 ÷ 3 = ______

______ ÷ 7 = ______

______ ÷ 13 = ______

______ ÷ 37 = ______

*46.

***46.** A jeweler bought 100 gems for $10,000, paying $1000 each for diamonds, $300 each for emeralds, and $50 each for rubies. How many of each kind did she purchase?

*47.

***47.** Do the following computation on your HHC and then use your result to answer the question:

$$(5{,}918{,}904 + 78{,}824{,}706) \div 2345$$

What color was the new coat that Mr. Herbert's wife brought home?

1.6 Combining Operations with Whole Numbers

THE ORDER OF OPERATIONS

No great amount of thought is required to perform any *one* of the four fundamental operations of addition, subtraction, multiplication, or division on an HHC. All that is required is to enter the numbers correctly and press the proper operation key. However, things get more complicated when more than one operation is involved.

Consider, for example, the problem of calculating

$$234 \times 529 + 982$$

First of all, is it obvious what is asked for? Are we to multiply 234 by 529 and then add 982, or are we to add 529 to 982 and them multiply this result by 234?

Do exercises 1 and 2 in the adjoining column.

The results of exercises 1 and 2—124,768 and 353,574, respectively—indicate that we need to agree on which operation to do first if we are to know which of the answers is correct for 234 × 529 + 982.

Throughout mathematics, the agreement is that, unless parentheses (or other signs of grouping such as [ ]) tell us otherwise, we are to perform the operations in the order of multiplications and divisions before additions and subtractions. (MDAS—sometimes recalled as "My Dear Aunt Sally!") Thus

$$234 \times 529 + 982 = (234 \times 529) + 982 = 124{,}768$$

Parentheses and other signs of grouping are, however, often used for clarity even in cases when the MDAS rule applies. Consider, for example,

$$128 \times 56 - 6076 \div 124 + 416 \times 13$$

The MDAS rule tells us to

1. Compute 128 × 56 which is equal to 7168.
2. Compute 416 × 13 which is equal to 5408.
3. Compute 6076 ÷ 124 which is equal to 49.
4. Compute 7168 – 49 + 5408 = 12,527.

Regardless of the MDAS rule, however, this problem is easier to understand if it is written with parentheses:

$$(128 \times 56) - (6076 \div 124) + (416 \times 13)$$

and would ordinarily be so written. Then

$$(128 \times 56) - (6076 \div 24) + (416 \times 13) = 7168 - 49 + 5408 = 12{,}527$$

OBJECTIVES

You should be able to:

(a) Understand the rules in regard to order of operations and the use of parentheses.

(b) Perform computations with whole numbers involving one or more of the basic operations of addition, subtraction, multiplication, and division.

(c) Understand the distributive properties and how to use them, along with other properties of the operations of arithmetic, to simplify computations.

(d) Know how to estimate answers to problems.

(e) Handle overflow problems in multiplication.

(f) Use the properties of 0 and 1.

(g) Apply combinations of the four basic operations of arithmetic to the solution of story problems.

****

In exercises 1 and 2 perform the computation by doing those within the parentheses first. Then compare the answers.

1. (234 × 529) + 982

2. 234 × (529 + 982)

STOP

In exercises 3–5 use parentheses and the MDAS rule to rewrite the problems in a form that is easier to understand, and then perform the computation.

3. 8288 − 13 × 572

4. 128 × 14 + 69 − 18,208 ÷ 32

5. 437 × 56 + 538 × 172 + 38 × 59

STOP

In exercises 6–9 do the computation mentally and then check your answer by using your HHC. Compare the time it took you to do it each way.

6. (25 × 10) + 6

7. (75 + 25) × 69

8. 364 × (115 − 15)

9. (624 × 173) ÷ 173

STOP

10. Add to 1284 the product of 691 and 3214.

11. Add 1284 to the product of 691 and 3214.

STOP

Problems such as these require writing down answers to intermediate steps in the computation unless your HHC has a memory key. Thus, without such a key, a procedure like the following must be followed to compute (128 × 56) − (6076 ÷ 124) + (416 × 13):

1. Enter 128
2. Press ×
3. Enter 56
4. Press =
5. Write down 7168
6. Enter 6076
7. Press ÷
8. Enter 124
9. Press =
10. Write down 49
11. Enter 416
12. Press ×
13. Enter 13
14. Press +
15. Enter 7168
16. Press −
17. Enter 49
18. Press =

(See Appendix A for the use of a memory key.)

Do exercises 3 through 5.

MENTAL COMPUTATIONS

We have noted several times that the use of an HHC is inefficient in doing such computations as 756 × 10, 8000 + 7000, etc. This also may be the case in computations involving more than one operation.

Do exercises 6 through 9.

PROPERTIES OF THE OPERATIONS OF ARITHMETIC

Do exercises 10 and 11.

Did you have to read exercises 10 and 11 very carefully? Your answers to these exercises should be the same, since

$$1284 + (691 \times 3214) = (691 \times 3214) + 1284$$

What property of addition is involved in this equality? Which of the two procedures used in exercises 10 and 11 is the faster?

Using the fact that, for all numbers a, b, and c,

$$a + (b \times c) = (b \times c) + a$$

can shorten some computations.

Do exercises 12 through 14.

We have previously considered the two basic properties of addition and multiplication: the commutative and associative properties. Now we consider another basic property that relates these two operations.

Do exercises 15 and 16.

The fact that the answers to exercises 15 and 16 are the same illustrates the *distributive property of multiplication over addition* (commonly referred to simply as the distributive property):

$$a \times (b + c) = (a \times b) + (a \times c)$$

for all numbers a, b, and c. Also, by the commutative property of multiplication, we have

$$(b + c) \times a = (b \times a) + (c \times a)$$

Clearly it is easier to compute (with pencil and paper or with an HHC) $a \times (b + c)$ than it is to compute $(a \times b) + (a \times c)$. Thus the computation asked for in exercise 15 is easier to do than the computation asked for in exercise 16.

In exercises 12–14 use procedures like those suggested by exercises 10 and 11 to perform the computation in the easiest possible way.

12. $97{,}432 + (931 \times 123)$

13. $123 + (2773 \div 59)$

14. $2142 - 1765 + (89 \times 67)$

STOP

15. Compute

$262 \times (1431 + 963)$

16. Compute

$(262 \times 1431) + (262 \times 963)$

and compare your answer to that of exercise 15.

STOP

In exercises 17–20 use the distributive property to perform the indicated computation in the easiest possible way.

17. $(241 \times 116) + (241 \times 848)$

18. $1023 \times 59 + 1023 \times 68$

19. $7214 \times 136 + 7214 \times 269$

20. $14{,}142 \times 62 + 14{,}142 \times 38$

21. Compute
$(24{,}492 + 4888) \div 26$

22. Compute
$(24{,}492 \div 26) + (4888 \div 26)$

23. Which of the computations of exercises 21 and 22 is easier to perform?

STOP

24. Compute
$58 \times (4321 - 1962)$

25. Compute
$(58 \times 4321) - (58 \times 1962)$

26. Which of the computations of exercises 24 and 25 is easier to perform?

STOP

Do exercises 17 through 23.

From the results of exercises 21 and 22, what conclusion do you think you can draw as to the validity of the (*righthand*) distributive property of division over addition,

$$(a + b) \div c = (a \div c) + (b \div c)$$

for all numbers a, b, and c with $c \neq 0$?

Do exercises 24 through 26.

From the results of exercises 24 and 25, what conclusions do you think you can draw as to the validity of the property

$$a \times (b - c) = (a \times b) - (a \times c)$$

for all numbers a, b, and c? What would be a good name for this property? What about

$$(b - c) \times a = (b \times a) - (c \times a)?$$

Do exercises 27 through 29.

27. Compute
$(7192 - 1984) \div 124$

28. Compute
$(7192 \div 124) - (1984 \div 124)$

29. Which of the computations of exercises 27 and 28 is easier to perform?

From the results of exercises 27 and 28, what conclusions do you think you can draw as to the validity of the property

$$(a - b) \div c = (a \div c) - (b \div c)$$

for all numbers a, b, and c with $c \neq 0$? What would be a good name for this property?

Let us summarize the properties of the operations of arithmetic that we have noted so far. We have, for all numbers a, b, c, and d:

$$\mathbf{a + b = b + a, \qquad a \times b = b \times a}$$

(Commutative properties of addition and multiplication)

$$\mathbf{a + (b + c) = (a + b) + c, \qquad a \times (b \times c) = (a \times b) \times c}$$

(Associative properties of addition and multiplication)

$$a - (b - c) = a - b + c$$

$$a - (b + c) = a - b - c$$

$$(a + b) - (c + d) = a + b - c - d$$

$$(a - b) - (c - d) = a - b - c + d$$

$$(a \times b) \div b = a \quad (b \neq 0)$$

$$(a \div b) \times b = a \quad (b \neq 0)$$

$$a + (b \times c) = (b \times c) + a$$

$$(a \times b) + (a \times c) = a \times (b + c)$$

(Distributive property of multiplication over addition)

$$(a \times b) - (a \times c) = a \times (b - c)$$

(Distributive property of multiplication over subtraction)

$$(a \div c) + (b \div c) = (a + b) \div c \quad (c \neq 0)$$

(Distributive property of division over addition)

$$(a \div c) - (b \div c) = (a - b) \div c \quad (c \neq 0)$$

(Distributive property of division over subtraction)

All of these properties can, as you have seen, be used to simplify calculations whether by mental calculation, paper and pencil work, or the use of an HHC.

In exercises 30–43 use the properties just summarized to simplify the computation.

30. $20 \times (287 \times 50)$ (mentally)

31. $1262 - (997 - 431)$

32. $5369 - (573 + 89)$

33. $(469 + 872) - (891 + 67)$

34. $(468 - 322) - (189 - 72)$

35. $(542 \times 896) \div 896$ (mentally)

36. $(669 \times 544) \div 669$ (mentally)

37. $392 + (68 \times 27)$

38. $(92 \times 68) + (92 \times 112)$

39. $(215 \times 89) + (1122 \times 89)$

40. $(97 \times 118) - (97 \times 105)$

41. $(41 \times 52) - (39 \times 52)$

42. $(2832 \div 24) + (360 \div 24)$

43. $(672 \div 42) - (462 \div 42)$

STOP

44. Attempt to carry out the multiplication

$85{,}239 \times 1492$

on your HHC and note how an overflow is indicated.

STOP

Do exercises 30 through 43.

OVERFLOW IN MULTIPLICATION

Do exercise 44.

Although the multiplication $85{,}239 \times 1492$ cannot be performed directly on an HHC, it can be done with only a small amount of hand calculation. To do this, first note that

$$\begin{aligned} 85{,}239 \times 1492 &= (85{,}000 + 239) \times 1492 \\ &= (85{,}000 \times 1492) + (239 \times 1492) \end{aligned}$$

by the distributive property. Furthermore, an HHC can be used to find that

$$239 \times 1492 = 356{,}588$$

Now we note that

$$\begin{aligned} 85{,}000 \times 1492 &= (85 \times 1000) \times 1492 \\ &= (85 \times 1492) \times 1000 \end{aligned}$$

by the commutative and associative properties of multiplication.

Using our HHC once again, we have

$$85 \times 1492 = 126{,}820$$

so that

$$\begin{aligned} 85{,}000 \times 1492 &= (85 \times 1492) \times 1000 \\ &= 126{,}820 \times 1000 = 126{,}820{,}000 \end{aligned}$$

Putting all of this information together, we have

$$\begin{aligned} 85{,}239 \times 1492 &= (85{,}000 + 239) \times 1492 \\ &= (85{,}000 \times 1492) + (239 \times 1492) \\ &= (85 \times 1000) \times 1492 + (239 \times 1492) \\ &= (85 \times 1492) \times 1000 + 356{,}588 \\ &= 126{,}820 \times 1000 + 356{,}588 \\ &= 126{,}820{,}000 + 356{,}588 \end{aligned}$$

Finally, although we can't carry out this addition on an HHC without overflow (indeed, we cannot even enter 126,820,000!), it is easy to carry out the addition by hand to get

$$85{,}239 \times 1492 = 126{,}820{,}000 + 356{,}588 = 127{,}176{,}588$$

Do exercise 45.

CHECKING COMPUTATIONS

The more steps that are involved in a problem, the more opportunities there are to make errors. We have previously indicated ways of checking any *one* of the four basic operations, but what about computations involving more than one operation, such as

$$(343 \times 179 + 2450 - 329) \div 698?$$

One way of checking, of course, is simply to repeat the entire computation. If the same answer is obtained, the chances are reasonably good that this answer is correct. Another way is to check each stage of the computation; e.g.,

$$343 \times 179 = 61{,}397$$

can be checked by the fact that

$$61{,}397 \div 179 = 343$$

Clearly, however, such a check is tedious since the various intermediate answers must be written down (or kept in one's head).

Do exercises 46 through 49.

A third kind of check involves *estimation*, and this process is important in its own right, since we are very often not so much concerned with the exact answer as we are with an approximate answer: the distance is *about* 50 miles, the tank will hold *about* 50 gallons, etc. That is, learning how to estimate is useful in general as well as an aid in detecting any *major* errors in computation.

Using estimation as a check requires *rounding off* the numbers involved in such a way that the resulting computation can easily be done with pencil and paper, or even mentally. For example, in the computation of

$$(343 \times 179 + 2450 - 329) \div 698$$

we might round as follows:

$$343 \xrightarrow{\text{down to}} 300$$

Then, since we have rounded 343 down,

$$179 \xrightarrow{\text{up to}} 200$$

45. Fill in the blanks to calculate

288,932 × 567

where a direct approach on an HHC causes an overflow.

288,932 × 567

= (288,000 + ______) × 567

= (288,000 × ______) + (932 × 567)

= (288 × 1000 × 567) + ____________

= (288 × 567) × _______ + 528,444

= ____________ × 1000 + 528,444

= 163,296,000 + 528,444

= (163,000,000 + ____________) + 528,444

= 163,000,000 + (296,000 + ____________)

= 163,000,000 + ____________)

= ________________
(final answer)

STOP

In exercises 46–49 check the correctness of the answer given for the calculation.

46. 5924 − (342 + 974) = 4608

47. 978 + 344 × 82 = 108,404

48. 874 × 23 + 982 × 23 = 42,688

49. 54,689 − (49 × 72 − 49 × 62) = 55,179

STOP

50. Compute
$(343 \times 179 + 2450 - 329) \div 698$

STOP

51. Estimate the answer to the computation of
$69 \times 524 \times 1022$
by rounding 69 to 70, 524 to 500, and 1022 to 1000. Then compute the exact answer and compare it with your estimate.

Estimate:

____

Computed answer:

____

In exercises 52 and 53 make your own approximations to the numbers involved in the given computation to obtain an estimate of the answer. Then compute the exact answer and compare it with your estimate.

52. $489 \times 512 \times 316$

Estimate:

____

Computed answer:

____

53. $(962 + 514) \times 863$

Estimate:

____

Computed answer:

____

STOP

Similarly,

$$2450 \xrightarrow{\text{up to}} 2500$$

$$329 \xrightarrow{\text{down to}} 300$$

and

$$698 \xrightarrow{\text{up to}} 700$$

Then the entire computation is simplified, for estimation purposes, to

$$(300 \times 200 + 2500 - 300) \div 700 = (60{,}000 + 2200) \div 700$$

$$62{,}200 \div 700 = 622 \div 7$$

Then, using the symbol $\approx$ to mean *approximately equal to,* we have

$$622 \div 7 \approx 630 \div 7$$

where we have rounded off 622 to 630 because it is easy to divide 630 by 7. Thus

$$(343 \times 179 + 2450 - 329) \div 698 \approx 90$$

Do exercise 50.

The result of exercise 50 shows that our estimate of 90 is quite close to the actual answer of 91. However, using the rounding off procedures as we did could easily produce an estimate that was considerably further off the exact answer. But even with a much cruder estimate, what would almost always be detected is any *gross* error in the computation. For example, if we had pressed the + key instead of the × key in computing 343 × 179 and then obtained, as you should check, 3.7865329 as an answer to exercise 50 (if no further errors were made in entering numbers and operations), we could readily see that 3.7865329 is nowhere close to 90 and thus conclude that a mistake must have been made someplace. That is, a check by estimation helps to avoid silly answers!

Do exercises 51 through 53.

USING THE PROPERTIES OF 0 AND 1

The facts that, for all numbers a,

$$a + 0 = a, \qquad a - 0 = a$$

$$a \times 1 = a, \qquad a \div 1 = a$$

can be used in place of a clear entry (CE) key when the wrong operation key is pressed (or when the CE key does not clear an operation entry).

Do exercises 54 through 56.

You should have found that the display in exercise 55 shows the sum, 543 + 286, of exercise 54, since the calculation performed in exercise 55 is

$$(543 - 0) + 286$$

which is equal to 543 + 286 because 543 − 0 = 543. This means that if, in calculating 543 + 286, you had mistakenly used the − key instead of the + key, you could have simply pressed the 0 key and then the + key to avoid having to press the clear key and starting the problem over again.

Do exercise 57.

Similarly, exercise 57 shows how the fact that $a + 0 = a$ can be used if the + key is pressed when the − key should have been:

$$543 - 196 = (543 + 0) - 196$$

Do exercises 58 and 59.

Finally, exercises 58 and 59 show how the fact that $a \times 1 = a$ and $a \div 1 = a$ can be used, respectively, when the × key is pressed instead of the ÷ key or the ÷ key instead of the × key:

$$1215 \div 27 = (1215 \times 1) \div 27 = 45$$

and

$$89 \times 113 = (89 \div 1) \times 113 = 10{,}057$$

SOLVING STORY PROBLEMS

When more than one of the operations of arithmetic are involved in a problem, things can get rather complicated. Here are some steps you should follow which we will illustrate by the following problem:

A hardware store whose monthly rent is $750 had a special on electric drills at $15 and sold 27 of them in the morning and 35 in the afternoon. How much did the store take in that day from the sale of these drills?

1. *Read the problem carefully.* In particular note, and perhaps underline, the key numbers and words in the problem. Also note, and perhaps cross out, any parts of the statement of the problem that are *not* relevant to the solution of the problem.

   In our example, the key items are $15, 27, 35, and "total"; the fact that the store is a hardware store whose monthly rent is $750 is not relevant.

---

54. Calculate 543 + 286 and record your answer.

55. On your HHC enter the following in order:

543	−	0	+	286	=

What is the final display on your HHC?

56. Compare your answer to exercise 54 with your answer to exercise 55.

STOP

57. Calculate 543 − 196 by using the following entries on your HHC:

543	+	0	−	196	=

STOP

58. Calculate 1215 ÷ 27 by using the following entries on your HHC:

1215	×	1	÷	27	=

59. Calculate 89 × 113 by using the following entries on your HHC:

89	÷	1	×	113	=

STOP

In exercise 60 determine the key items and also indicate what parts of the statement of the problem are not relevant.

60. Pete enjoys making necklaces out of stones he collects and polishes. He makes one necklace and sells it to a neighbor for \$16. Then he makes 18 more that he sells for \$17 each. How much did he receive altogether from the sale of these necklaces?

STOP

61. Write a mathematical sentence that describes the problem of exercise 60.

STOP

62. In exercise 61 you should have found that the mathematical sentence involved was

$16 + (17 \times 18) =$ total receipts

Rewrite this for easier computation on an HHC.

STOP

Do exercise 60.

2. *Determine a mathematical sentence (equation) that describes the problem.* (This is the most important step—as well as the most difficult.)

   In our hardware store example we have

Morning receipts	Total	Afternoon receipts	
↓	↓	↓	
(27×15)	$+$	(35×15)	= Total receipts

Do exercise 61.

3. *Simplify, if possible, the mathematical sentence for easier computation.* Sometimes the sentence can be re-written by using various properties of numbers in such a way that the computations can be performed in fewer steps or without writing down the intermediate steps.

   In our example, we can use the distributive property to write

$$(27 \times 15) + (35 \times 15) = (27 + 35) \times 15$$

Do exercise 62.

4. *Estimate the answer.* This step, while not essential, does help to detect errors in key punching when an HHC is used.

   In our example, we might round off both 27 and 35 to 30 and have

$$(30 + 30) \times 15 = 60 \times 15 = 900$$

Do exercise 63.

63. Estimate the answer to

$(17 \times 18) + 16$

STOP

5. *Perform the computation and compare it with your estimate.* If your answer is very far off from your estimate, check your computation.
   In the example, we obtain

$$(27 + 35) \times 15 = 930$$

which is reasonably close to our estimate of 900.

Do exercise 64.

64. Compute the answer to the computation of exercise 63 and compare it with your estimate.

STOP

6. *Answer the question asked in the problem.* Sometimes the answer to your computation in step 5 doesn't quite answer the question.
   In our example all we have to do to answer the question is to note that we want the answer in dollars and so say "The store took in $930 from the sale of the drills."

Do exercise 65.

65. Answer the question in exercise 60.

STOP

Summarizing:

1. (Underline the key items and strike out those that are not relevant.) A ~~hardware~~ store ~~whose monthly rent is $750~~ had ~~a special on electric~~ drills at $15 and sold 27 of them in the morning and 35 in the afternoon. How much did the store take in that day from the sale of these drills?

2. (Write an equation.)

$$(27 \times 15) + (35 \times 15) = \text{Total receipts}$$

3. (Simplify the equation.)

$$(27 + 35) \times 15 = \text{Total receipts}$$

4. (Estimate the answer.)

$$(27 + 35) \times 15 \approx (30 + 30) \times 15 = 60 \times 15 = 900$$

5. (Calculate and compare with estimate.)

$$(27 + 35) \times 15 = 930 \approx 900$$

6. (Answer the question.) The store took in $930 from the sale of the drills.

Do exercises 66 through 71.

In our example, step 6 (writing the answer to the problem) followed easily from step 5 (doing the computation). This is not always the case, as the following problem illustrates:

A car travels 30,800 feet in seven minutes. What is its speed in miles per hour?

At first this seems like a simple division problem such as we considered in section 1.5:

$$\text{(total number)} \div \text{(number of units)} = \text{amount per unit}$$

$$30{,}800 \quad \div \quad 7 \quad = \quad 4400$$

However, this gives us 4400 *feet* per *minute* and the question concerns *miles* per *hour*. So we have to note that, since there are 60 minutes in an hour,

$$4400 \text{ feet per minute} = (4400 \times 60) \text{ feet per hour}$$

and then, since there are 5280 feet in a mile,

$$(4400 \times 60) \text{ feet per hour} = (4400 \times 60) \div 5280 \text{ miles per hour}$$

Thus, since

$$(4400 \times 60) \div 5280 = 50$$

we conclude that the answer to the question of the problem is "50 miles per hour."

Do exercise 72.

Exercises 66–71 refer to the following problem: John has $265 in his savings account in the Pacific National Bank. He then works for part of a week (39 hours) at $4 an hour but his employer deducts $35 for taxes. If John deposits in his bank what he receives for his work, how much will he then have in his savings account?

66. In the statement above, underline the key words and strike out those that are not relevant.

67. Write a mathematical sentence describing the problem.

68. Rewrite

$265 + (39 \times 4) - 35$

in a form more suitable for calculation with an HHC.

69. Estimate your answer.

70. Compute your answer and compare it with your estimate.

71. Answer the question.

STOP

72. If 12 cubic feet of a certain kind of wood weighs 540 pounds, how much will one cubic yard weigh? (1 cubic yard = 27 cubic feet)

Exercises 1.6

Name ____________

Class ____________

In exercises 1–5 determine if the given statement is correct. If it is not, give the correct answer.

1. $30 + 60 \times 40 = 2430$

2. $(26 + 5) \times 13 = 403$

3. $34 \times (34 - 1) = 1122$

4. $47 + 52 \times 17 - 39 = 41{,}509$

5. $76 - 36 \times 126 + 13 = 5560$

In exercises 6–15 perform the given computation.

6. $49 + 487 \times 52$

7. $80 \times 22 + 866$

8. $760 - 255 \div 51$

9. $87 \times 822 - 147$

10. $4 + 18 \times 51 \div 17$

11. $9458 - 78 \times 89 - 45$

12. $69 - 1362 \div 454 + 29$

13. $480 + 1219 \div 53 + 946$

14. $33 + 58 \times 47 - 3 + 39 \div 3$

15. $987 - 77 \div 7 + 9 + 49 - 25$

In exercises 16–25 place parentheses (if needed) in the given expression so as to obtain a true statement.

Example: $4 \times 3 + 8 = 44 \Rightarrow 4 \times (3 + 8) = 44$

16. $12 \times 9 + 8 = 116$

17. $12 \times 9 + 8 = 204$

18. $12 \times 9 - 8 = 12$

19. $12 \times 9 - 8 = 100$

20. $12 \times 9 + 4 \times 13 = 160$

21. $12 \times 9 + 4 \times 13 = 2028$

1.
2.
3.
4.
5.
6.
7.
8.
9.
10.
11.
12.
13.
14.
15.
16.
17.
18.
19.
20.
21.

22.
23.
24.
25.
26.
27.
28.
29.
30.
31.
32.
33.
34.
35.
36.
37.
38.
39.
40.
41.
42.
43.
44.
45.

22. $12 \times 9 + 4 \times 13 = 1456$

23. $12 \times 9 - 4 \times 13 = 780$

24. $12 \times 9 - 4 \times 13 = 56$

25. $23 + 16 \times 15 + 12 = 455$

In exercises 26–35 use a distributive property to simplify your computation.

26. $(52 \times 97) + (73 \times 97)$

27. $(864 \times 55) + (864 \times 80)$

28. $(62 \times 57) - (32 \times 57)$

29. $(542 \times 268) - (542 \times 243)$

30. $(1798 \div 62) + (6882 \div 62)$

31. $(10{,}904 \div 188) - (6768 \div 188)$

32. $(707 \times 467) + (707 \times 893)$

33. $(232 \times 187) - (232 \times 45)$

34. $(19{,}285 \div 203) - (13{,}804 \div 203)$

35. $(56{,}325 \div 75) + (27{,}900 \div 75)$

In exercises 36–45 make an approximation to each number involved in the given computation and give an estimate of the answer. Then do the computation and compare it with your estimate.

36. $42 \times 67 \times 21$

37. $32 + 68 \times 39$

38. $711 \times (981 + 166)$

39. $69 \times 42 \div 414$

40. $(51 \times 528) + (47 \times 528)$

41. $114 \times 71 \times 95$

42. $453 \times 12 + 69{,}124$

43. $124 + 25 \times 103 - 97$

44. $(21{,}021 \times 685) - (992 \times 685)$

45. $630 \times 405 + 15{,}010 \div 158$

In exercises 46–60, (1) write down the method you would use to solve the problem, (2) simplify, if possible, that method for easier computation, (3) estimate the answer, and (4) perform the computation to answer the question.

46. A farmer has 175 orange trees and 210 lemon trees to spray. If he uses three gallons of spray on each tree, how many gallons of spray does he need? 46.

47. Joan works in a department store. One day she sold 3 dresses at \$45 each and 7 dresses at \$17 each. What were her total sales for that day? 47.

48. John obtained a summer job paying \$3 per hour. If he works 40 hours per week for 12 weeks, how much will the job pay him? 48.

49. Ms. Lathrop has a yearly salary of \$10,470. If she has \$234 deducted from her paycheck each month, what is her yearly take-home pay? 49.

50. Mr. and Mrs. Rogers bought a new house. They paid \$7500 in cash as a down payment and agreed to pay the rest in monthly payments of \$225 for 17 years. By the time they have made all these payments, how much will the house have cost them? 50.

51. What is the profit on 35 electric appliances if the cost of each is \$169 and the selling price is \$215? 51.

52. Jones Appliance Company purchased 25 sets of washers and dryers at a special price of \$125 per dryer and \$175 per washer and sold them for \$155 and \$205, respectively. How much profit did they make on the transaction? 52.

53. Sound travels 1110 feet per second at 10° Celsius and its speed increases 2 feet per second for each degree of Celsius rise in temperature. What is the speed of sound at 30° Celsius? 53.

54. A barometer is a device for measuring atmospheric pressure. In going from a low elevation to a higher elevation the pressure decreases at the rate of 1 millimeter (mm) on the barometer for every 11 meters (m) in height. What is the height of a mountain on which a barometer reads 766 millimeters at its base and 469 millimeters at the top? (10 mm = 1 cm, 100 cm = 1 m.) 54.

55.

55. A certain brand of tuna fish sells regularly for 55 cents per can. How much is saved by buying a dozen cans at the special price of 43 cents?

56.

56. A merchant was able to buy 25 dolls at \$3 each and two weeks later he was able to purchase 25 more of the same dolls at \$2 each. If he sold them for \$4 each, how much profit did he make?

57.

57. Blacktop for a driveway 129 feet long was quoted at \$3 per foot. If the labor costs were \$250, what would be the total cost?

58.

58. Which has the greater seating capacity and by how much greater: an auditorium which has 40 rows of 15 seats each and 35 rows of 24 seats each; or an auditorium which has 25 rows of 14 seats each and 72 rows of 18 seats each?

59.

59. Which job will be paying more at the end of the seventh year: one with a starting salary of \$9000 with a raise of \$150 each year, or one with a starting salary of \$8400 and a \$250 raise each year? How much more will it pay? At what year will the two salaries be the same amount?

60.

60. How many seconds are there in one year (365 days)?

*61.

*61. If light travels 186,262 miles per second, how far does it travel in one year (365 days)?

*62.

*62. The velocity needed for a rocket to escape the earth is 25,200 miles per hour. How many feet per second is this?

*63. Think of a number: multiply it by 6, add 12, divide by 3, subtract 2, divide by 2, subtract the original number, and add 9. Is the result 10? Will it always be 10, regardless of what number you start with? Do you know why?

*63.

*64. Select any two numbers—each less than 10—choose either number and multiply by 5, add 7, double the result, and add this to the other number. From the result on your HHC, Melvin the Mystic can tell you the two numbers you selected and which one you multiplied by 5. How does he do it? Will the procedure he used work for any such two numbers?

*64.

*65. Do the following computation and use your calculator to answer the question:

$$(6767 \times 3747) + (7111 \times 3747) - (4757 \times 120 + 50{,}968{,}651)$$

What is the vehicle called in which Santa Claus travels?

*65.

Review Exercises 1

In exercises 1–15 perform the indicated computation. If possible, do some computations mentally and/or rearrange the way the operations are done to simplify your computation.

Name ________________

Class ________________

1. $2175 + 1300 + 25 + 200$

2. $6{,}547{,}946 + 90{,}998{,}375$

3. $59{,}688{,}932 + 3{,}299{,}277$

4. $(6378 - 5495) - 3316$

5. $(71{,}546 + 42{,}249) - (13{,}812 + 27{,}846)$

6. $(4 \times 25) + 1768 - 500$

7. $8102 + (5 \times 200) + (60 \times 10)$

8. $7621 - 5815 \times 375$

9. $4224 \times (10{,}153 + 2089)$

10. $68{,}215 - 599 \times 27$

11. $24{,}543 \div 243 - 142$

1.

2.

3.

4.

5.

6.

7.

8.

9.

10.

11.

12.

12. (157 × 82) + (157 × 422)

13.

13. (599 × 27) − (421 × 87)

14.

14. 491 − (209 × 84) + 5929 + (760 × 85)

15.

15. (1462 ÷ 43) ÷ 17 + 2091 − 136 × 13

In exercises 16–20 write the given numeral in expanded exponential notation.

16.

16. 2149

17.

17. 42,832

18.

18. 27,098

19.

19. 6,332,709

20.

20. 622,982

In exercises 21–25 give the place value of the "7" in the given numeral.

21.

22.

23.

24.

25.

21. 171

22. 7533

23. 3735

24. 61,137

25. 76,360,481

In exercises 26–30 give the value of the "7" in each numeral.

26. 474

27. 6788

28. 89,753

29. 79,309

30. 53,017

26.
27.
28.
29.
30.

In exercises 31–35 write the name of the given numeral.

31. 27,155

31.

32. 769,119

32.

33. 25,013

33.

34. 502,252,007

34.

35. 70,091

35.

In exercises 36–40 write the numeral for the given number.

36. Thirty-seven thousand two hundred twenty-seven

36.

37. Eight thousand seventy-two

37.

38. Eight hundred twenty-six thousand one hundred ninety-seven

38.

39. Twenty-seven thousand nine

39.

40. Seventy million ninety-one thousand fifty-six.

40.

In exercises 41–60 (a) write a mathematical statement which indicates the method you would use to solve the problem, (b) simplify, if possible, that method for easier computation, (c) estimate your answer, and (d) do the computation necessary to answer the question.

41.

41. If Sally earns \$165 per week, how much does she earn in one year? (Assume that 1 year is 52 weeks.)

42.

42. If it costs an average of \$550 per student to provide a certain educational program, how much does it cost for a school with 2500 students?

43.

43. An industrial plant produces an average of 2324 machine parts of a certain kind in every 8-hour shift. If the plant operates on 2 shifts a day, 5 days a week, how many machine parts are produced every 4 weeks?

44.

44. In a large city high school there are 3528 pupils. If there are 84 "rege" rooms in the school, what is the average number of pupils assigned to each registration room?

45.

45. The monthly payroll of a company that employs 1386 persons is \$1,219,680. What is the average monthly wage per person?

46.

46. During 1976 the sales for a small business were \$353,692 compared with \$431,932 in sales for 1977. What was the average monthly increase in sales?

47.

47. A school band raised \$7834 in a campaign to buy new uniforms. If they wanted to purchase 84 uniforms at \$74 each, did they have enough money? What was the difference between what they raised and what they needed?

48.

48. How much will it cost to purchase 4 tires at \$47 each and to have them balanced at \$3 for each wheel?

49.

49. How far will you travel in 7 hours at a constant speed of 55 miles per hour if you took a 10-minute break at the end of each full hour of driving time?

50. Jane traveled 1368 miles by car. If she traveled 403 miles the first day, 386 miles the second day, and 305 miles the third day, how many miles did she travel the fourth day to complete the trip? What was her average mileage per day?

50.

51. A school district estimated the following costs for building a new school: construction: $1,225,465; land $65,750; equipment, $104,785. What was the cost to the district if the state provided $700,000?

51.

In exercises 52–56 refer to the following table.

PRINCIPAL AGRICULTURAL EXPORTS–VALUE: 1951 TO 1974
(In millions of dollars. Includes Puerto Rico. Excludes reexports of foreign products. Includes shipments under foreign aid programs.)

Item	1951–1955, avg.	1956–1960, avg.	1961–1965, avg.	1966–1970, avg.	1965	1970	1972	1973	1974
Agricultural exports	3314	4264	5644	6480	6229	7358	9414	17,682	21,996
Meats and meat products	59	104	162	179	170	188	263	462	399
Eggs and dairy products	117	130	146	130	192	137	143	56	67
Animal fats and oils	172	182	195	179	224	244	207	330	580
Hides and skins	37	64	89	140	109	144	292	375	337
Bread grains and preparations	700	821	1268	1197	1214	1144	1490	4278	4678
Coarse grains and prepn., exc. rice	324	478	841	1082	1162	1099	1559	3598	4727
Rice	119	117	178	311	244	314	389	539	853
Fodders and feeds	34	65	179	386	278	496	596	1266	1287
Vegetables and preparations	83	115	134	169	151	181	213	313	398
Fruits and preparations	151	245	285	311	312	330	422	424	585
Vegetable oils, fats and waxes, oilseeds, nuts and preparations	254	499	774	1182	1029	1642	1852	3417	4865
Tobacco, unmanufactured	315	355	393	506	383	488	639	681	832
Cotton, unmanufactured	761	778	639	408	495	377	508	940	1353
All other	191	312	362	299	265	571	841	1003	1035

Source: U.S. Bureau of the Census, *Foreign Commerce and Navigation of the United States; Quarterly Summary of Foreign Commerce of the United States; U.S. Exports, Schedule B Commodity and Country,* RT 410, monthly.

52. What was the increase in the total value of agricultural exports from 1965–74?

52.

53. What was the total value of exports of bread grains and preparations in the years 1972–74?

53.

54.

54. What was the value of rice exports for the years 1951–60?

55.

55. How much more was the value of cotton exports in the years 1956–60 than in 1966–70?

56.

56. Compare the values of exports of meats and meat products with the values of eggs and dairy products for the years 1951–70. Which was exported more and how much more?

57.

57. Suppose in a small factory the following wages are paid: 3 unskilled workers receive $140 per week, 10 skilled workers get $300 per week, and a foreman gets $388 per week. What is the average salary at the factory?

58.

58. One Saturday Jed was paid for 5 days' work at $28 per day and $42 for working on Saturday. His deductions for the week were $56. How much was his take-home pay for that week?

59.

59. The maximum temperatures in degrees Fahrenheit for a certain week were as follows: 68, 75, 83, 77, 79, 72, and 78. What was the average temperature for that particular week?

60.

60. Max bought a car for $3280. If he paid $272 cash and arranged to pay off the balance in 8 equal monthly payments, what was the amount of each payment?

61.

61. A man bought a horse for $160, sold it for $170, then bought it back again for $180, and finally sold it again for $190. How much money did he make or lose in all?

*62.

*62. The sun moves 26 miles per second in space. How many miles does the sun travel in one year of 365 days?

***63.** Find a four-digit numeral so that the number of hundreds plus the number of thousands is one, the number of tens is 7 times the number of thousands, and the number of ones is 5 times the number of hundreds. *63.

***64.** Write a ten-digit numeral so that the first digit (from the left) tells how many zeros there are in the entire numeral, the second digit tells how many ones are in the numeral, the third digit tells how many two's, and so on. *64.

***65.** Find six numbers having only 1 for their digits whose sum is 123,456. *65.

***66.** What one number will go in the boxes below to make a true statement? *66.

$$(16 \times \square) + 267 = \square + 462$$

***67.** There is a number which is the square of one number and the cube of another, that when added to each of the numbers 161 and 225 makes them perfect squares. What is that number? *67.

***68.** Do the following computation and use your HHC to answer the question. *68.

$$(6414 \times 459) - (821 \times 459) + (763 \times 459) - (6344 \times 459)$$

What is the title of the person who is in charge of a job?

Just for Fun

A. PATTERNS AND SEQUENCES

Write the next five numbers in each sequence.

1. 2, 6, 18, 54, ____, ____, ____, ____, ____.
2. 0, 1, 4, 9, 16, ____, ____, ____, ____, ____.
3. 1, 2, 3, 5, 8, ____, ____, ____, ____, ____.
4. 8, 7, 10, 9, 12, ____, ____, ____, ____, ____.

B. MORE PATTERNS

For those items indicated by an asterisk (*) you are to examine the pattern indicated by the previous computations and then write what you think is the result.

5.

$0 \times 9 + 1 =$ ____________

$1 \times 9 + 2 =$ ____________

$2 \times 9 + 3 =$ ____________

$3 \times 9 + 4 =$ ____________

$4 \times 9 + 5 =$ ____________

$*5 \times 9 + 6 =$ ____________

$*6 \times 9 + 7 =$ ____________

$*7 \times 9 + 8 =$ ____________

$*8 \times 9 + 9 =$ ____________

$*9 \times 9 + 10 =$ ____________

6.

$1 \times 8 + 1 =$ ____________

$12 \times 8 + 2 =$ ____________

$123 \times 8 + 3 =$ ____________

$1234 \times 8 + 4 =$ ____________

$*12345 \times 8 + 5 =$ ____________

$*123456 \times 8 + 6 =$ ____________

$*1234567 \times 8 + 7 =$ ____________

$*12345678 \times 8 + 8 =$ ____________

$*123456789 \times 8 + 9 =$ ____________

C. MISSING DIGITS

Find the missing digits in the following addition problems.

7.

```
        6 3 _
        _ 7 5
  (+)   2 5 3
      -------
      _ 3 _ 6
```

8.

```
        _ 7 5 2
        3 _ 5 8
        4 9 _ 5
  (+)   2 4 0 _
      ---------
      _ 5 7 8 8
```

D. TRY THESE

9. The Guinness Book of Records indicates that Tyrus (Ty) Cobb scored 2224 runs in his baseball career (1905–28). How many yards did he run in scoring these runs?

10. Jack and Jill have an 8-liter container of root beer. They wish to divide the beverage equally but only have an empty 5-liter jar and an empty 3-liter container. How can they do it?

11. Place +, –, ×, ÷, or ( ) between the digits, so as to arrive at the answer of 1. (Do not change the order of the digits). See how many ways you can do this.

$$1 \quad 2 \quad 3 \quad 4 \quad 5 \quad 6 \quad 7 \quad 8 \quad 9 \quad = \quad 1$$

One solution:

$$1 + 2 + 3 + 4 - 5 + 6 + 7 - 8 - 9 = 1$$

E. MAGIC SQUARE

12. Complete the magic square at the right.

154	77	
99		
110		

F. TRIANGLE PUZZLE

13. The digits 1, 2, and 3 have been placed in the circles at each vertex of the triangle. Using only the numbers 4 through 9 (each one just once), place these numbers in the remaining circles so that the sum of the numbers on each side of the triangles is 17.

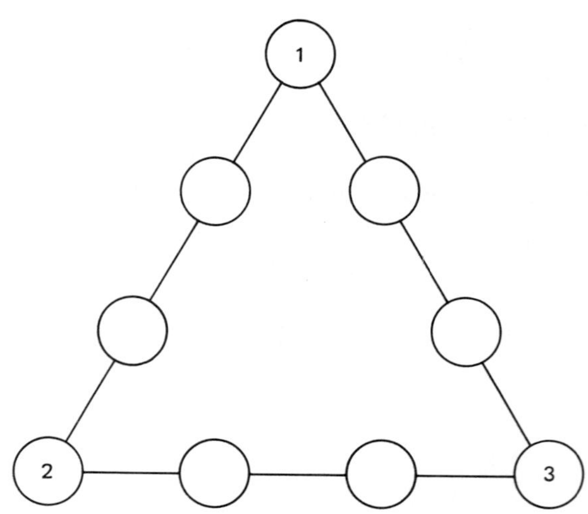

G. CALCULATOR GAME

14. THE GAME OF NIM. The ancient game of Nim—usually played with small sticks or pebbles—can be played on your HHC. To play the game of Nim on an HHC, one person enters 1, 2, or 3 on the HHC and then the two players take turns adding either 1, 2, or 3. The first person to reach 21 or more loses.

Sample game:			*Display shows:*
Player A:	Enters	1	1
B:	Adds	3	4
A:	Adds	2	6
B:	Adds	1	7
A:	Adds	3	10
B:	Adds	2	12
A:	Adds	3	15
B:	Adds	3	18
A:	Adds	2	20
B:	loses:	20 + 1 = 21, 20 + 2 = 22, 20 + 3 = 23	

(There is a strategy which insures that the second player will always win.)

(a) Play Nim with a friend. See if you can find the winning strategy.

(b) The game of Nim can also be played when the numbers to be used are 1, 2, 3, 4, 5, and 6 and the loser is the person who totals 50 or more. Play this game and try to find the winning strategy for the second player.

H. THE TALKING CALCULATOR

15. Perform the calculation below and use your calculator to find the name of a musical instrument played in a symphony orchestra.

$$([(35 \times 2) + 30] \times 8 - 50 + 250) \times 77 \div 25$$

Mathematics and Recreation

A. BOWLING

The game of bowling is becoming a very popular recreation in our country. Perhaps many of you have bowled. Keeping score is usually shared by all participants of the game. For those that have not been bowling, a brief description of the game will be given.

Each player bowls ten *frames.* This is equivalent to stating that each bowler has ten turns. The number of times a player rolls the ball during his turn is determined by the number of pins he knocks down on his first roll. If he knocks down all of them, a total of 10, on his first roll, he has rolled a *strike* and the turn or frame is done. The strike is shown by a ☒ on the score sheet. If all pins are not knocked down on the first roll, the bowler rolls the ball a second time and tries to knock down the remaining pins. If this is accomplished on the second roll, the player has rolled a *spare* in that particular frame. The frame is now completed. A spare is symbolized as ◩ on the score sheet. If the bowler fails to knock down the remaining pins on his second roll, the frame is completed and the score for the frame is the total number of pins knocked down in the two rolls of the ball. All except the tenth frame will require one or two rolls of the ball. The exception in the tenth frame will be explained later.

Suppose you bowled the following. (The first number is the number of pins knocked down on the first roll and the second number is the number of pins knocked down on the second roll, if there was one.)

Frame		Frame	
Frame 1	(10, __)	Frame 6	(9, 1)
2	(7, 2)	7	(10, __)
3	(9, 1)	8	(10, __)
4	(8, 2)	9	(9, 0)
5	(10, __)	10	(10, __)
		Extra 1	7
		Extra 2	2

Scoring is accomplished in this manner:

(a) A strike counts ten plus the number of pins knocked down on the next two rolls. A strike in the last frame gives you two extra rolls.

(b) A spare counts ten plus the number of pins knocked down on the next roll. A spare in the last frame gives you one extra roll.

(c) The total score is entered on the score sheet up to the frame you are scoring.

Your game would be scored as follows:

	1	2	3	4	5	6	7	8	9	10	Total
YOUR NAME	☒ 19	☐ 28	◩ 46	◩ 66	☒ 86	◩ 106	☒ 135	☒ 154	⊟ 163	☒ 182	182

Study the scoring until you feel that you can score the game correctly.

(1) Fred and Barney bowled the following two games:

Game 1:	Fred	Barney		Fred	Barney
Frame 1	(10, __)	(10, __)	Frame 6	(8, 2)	(10, __)
2	(8, 1)	(6, 3)	7	(9, 1)	(7, 2)
3	(9, 1)	(9, 1)	8	(10, __)	(10, __)
4	(10, __)	(7, 2)	9	(6, 4)	(8, 2)
5	(5, 4)	(10, __)	10	(8, 1)	(6, 3)

Game 2:	Fred	Barney		Fred	Barney
Frame 1	(8, 2)	(7, 3)	Frame 6	(10, __)	(10, __)
2	(10, __)	(3, 7)	7	(10, __)	(10, __)
3	(6, 3)	(8, 2)	8	(8, 2)	(3, 5)
4	(5, 5)	(10, __)	9	(7, 1)	(8, 2)
5	(10, __)	(10, __)	10	(10, __)	(9, 1)
			Extra 1	5	5
			Extra 2	4	

(2) Score Fred and Barney's games in the following spaces.

Game 1:	1	2	3	4	5	6	7	8	9	10	Total
Fred											
Barney											

Game 2:	1	2	3	4	5	6	7	8	9	10	Total
Fred											
Barney											

(3) What was Fred's bowling average for the day?

(4) What was Barney's bowling average for the day?

(5) A perfect game consists of all strikes. What would be the score for a perfect game?

*(6) Go bowling and bring your score sheet to your instructor.

B. GOLF

It is estimated that there are three million golfers in the United States and more starting every day. A standard golf course measures approximately 6000 yards in length and is broken up into holes from 100 to 600 yards. Each hole is given a certain *par* that depends on the length of the hole. Par is the number of strokes which a first-class player would ordinarily require to put the ball in the cup. Generally holes that measure less than 220 yards are par three, 220 to 440 yards are par four, and all others are par five. So a player can compete against par or against another individual. The score is the sum of the strokes for the game. A game may consist of 9 holes, 18 holes, or further multiples of 9.

Consider "Slice" Akin's score for 18 holes of golf.

Hole	Yards	Par	Strokes	Hole	Yards	Par	Strokes
1	230		4	10	280		5
2	350		5	11	530		4
3	520		4	12	180		3
4	120		3	13	410		6
5	410		6	14	380		4
6	320		5	15	580		5
7	540		5	16	210		4
8	480		4	17	460		4
9	200		4	18	360		4
Total				Total			

1. List par for each of the holes.

2. What is a par game for 18 holes of golf where Slice is playing?

3. Find the total number of yards for the first 9 holes. Find the total number of yards for the second 9 holes. What is the total number of yards for 18 holes?

4. What is Slice's score for 18 holes?

5. Was Slice's score equal to, less than, or greater than par? If the score is greater or less than par, give the amount over or under.

6. How many par holes did Slice have?

7. A *birdie* is a score on a hole one less than a par. How many birdies did Slice have?

8. A *bogey* is a score on a hole one greater than par. How many bogies did Slice have?

*9. Talk to a golfer and find out what an *eagle* is.

2

Computations with Decimals

2.1 Extending Our System of Numeration

OBJECTIVES

You should be able to:

(a) Understand the meaning of our system of numeration and its extension to decimals.

(b) Read and write decimal numerals.

(c) Understand the relation of common fractions to decimals.

(d) Round off whole numbers and decimals.

****

PLACE VALUE

In our place value system of numeration, as you have seen, the value of each place is 10 times the value of the place on its immediate right. This is illustrated in the following figure.

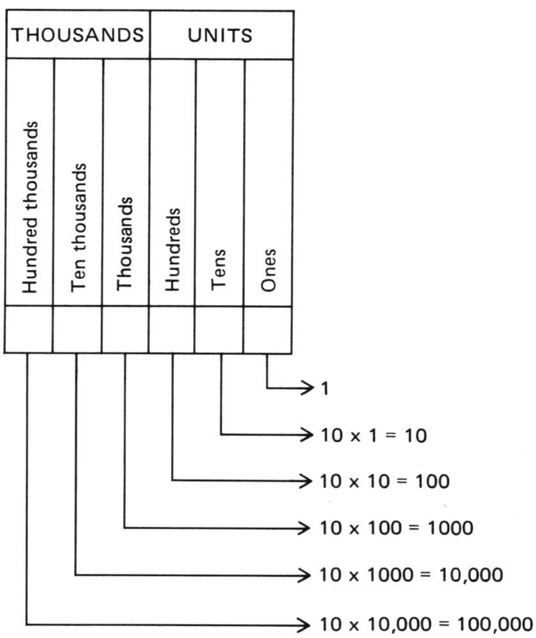

Do exercise 1 in the adjoining column.

1. Fill in the blanks below using your HHC. Notice the pattern that appears.

1 X 10 = ____

____ X 10 = ____

____ X 10 = ____

____ X 10 = ____

____ X 10 = ____

Because every multiplication problem leads to a division problem, we can observe that in our system of numeration the value of any place is equal to the value of the place on its immediate left divided by 10. This is illustrated in the following figure.

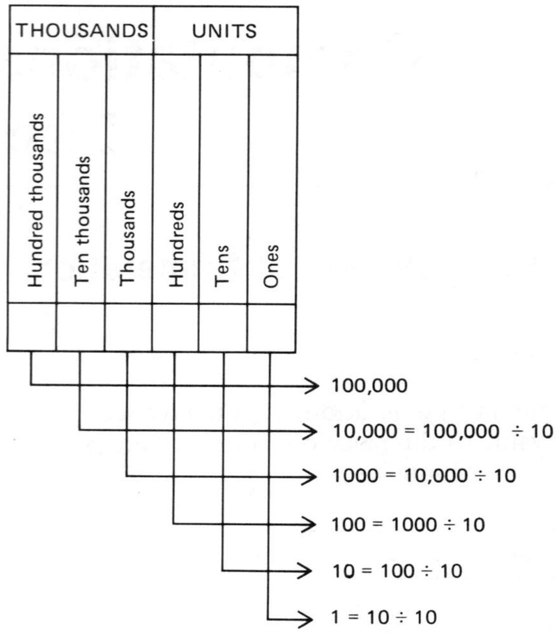

What happens if we continue to the right?

Do exercise 2.

2. Fill in the blanks below using your HHC. Notice the pattern that appears.

1 ÷ 10 = ______

______ ÷ 10 = ______

______ ÷ 10 = ______

______ ÷ 10 = ______

______ ÷ 10 = ______

STOP

Because of the results of these divisions we extend our system of numeration to the right in the manner shown in the following figure. Notice the symmetry obtained about the ones place.

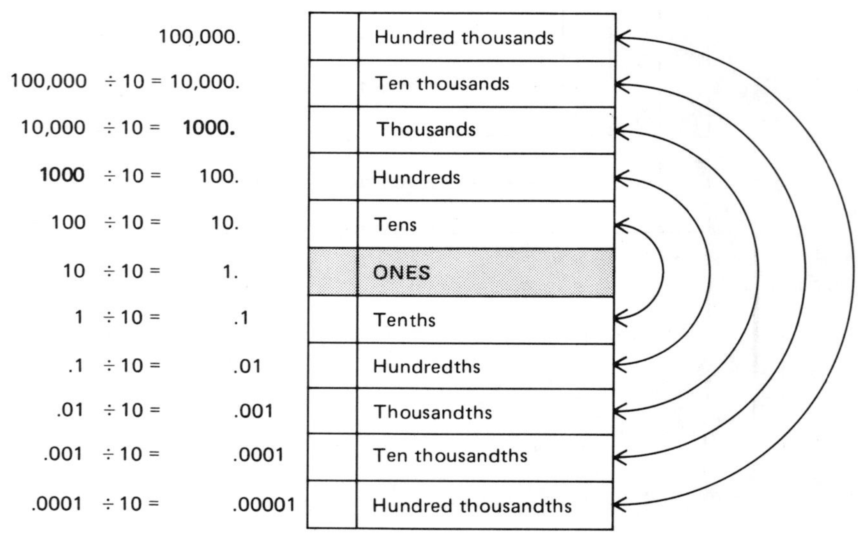

READING DECIMALS

The numeral

$$743{,}215.68349$$

written in expanded notation is

(a) $700{,}000 + 40{,}000 + 3000 + 200 + 10 + 5$
$+ 0.6 + 0.08 + 0.003 + 0.0004 + 0.00009$

or

(b) $7 \times 100{,}000 + 4 \times 10{,}000 + 3 \times 1000 + 2 \times 100 + 1 \times 10 + 5 \times 1$
$+ 6 \times 0.1 + 8 \times 0.01 + 3 \times 0.001 + 4 \times 0.0001 + 9 \times 0.00001$

The numeral is read as "seven hundred forty-three thousand two hundred fifteen and sixty-eight thousand three hundred forty-nine hundred thousandths." (However, if we were to read off this number to someone, it would be easier for this person to copy it down if it were read "seven, four, three, two, one, five, *point*, six, eight, three, four, nine.")

In our system of numeration for *whole* numbers we used the first place on the right as the ones place and that numeral as the ones digit. If this system is to be extended to the right, the first place on the right is no longer the ones place, and so some method of marking the ones place is needed. For this purpose we use the *decimal point* and read it as "and."

In the numeral

$$743{,}215.68349$$

the "4" on the *left* of the decimal point has a place value of ten thousand (10,000) and a value of forty thousand (40,000). On the other hand, the "4" on the *right* of the decimal point has a place value of ten thousand*ths* (0.0001) and a value of four ten thousandths (0.0004).

Similarly, the "3" on the left of the decimal point has a place value of one thousand (1000) and a value of three thousand (3000), whereas the "3" on the right of the decimal point has a place value of one thousandth (0.001) and a value of three thousandths (0.003).

Do exercises 3 through 8.

Exercises 3–8 refer to the numeral 2346.579.

3. Write the numeral in expanded notation in both the form (a) and the form (b).

4. What is the place value of the "3"?

5. What is the value of the "3"?

6. What is the place value of the "7"?

7. What is the value of the "7"?

8. Read orally and write the number name.

STOP

MULTIPLYING AND DIVIDING DECIMALS BY POWERS OF TEN

You have seen that

$$27 \times 10 = 270$$
$$27 \times 100 = 2700$$
$$27 \times 1000 = 27{,}000$$

and so on

9. Use your HHC to compute

$2.7 \times 10 =$ ______;

$2.7 \times 100 =$ ______;

$0.27 \times 10 =$ ______;

$0.27 \times 1000 =$ ______;

$0.027 \times 100 =$ ______;

$0.027 \times 1000 =$ ______;

$0.027 \times 10{,}000 =$ ______;

STOP

10. Without using your HHC, compute

$25.6 \times 100 =$ ______

$0.1923 \times 100 =$ ______

$1.14329 \times 1000 =$ ______

$8.26 \times 10{,}000 =$ ______

STOP

11. Use your HHC to compute

$27 \div 10 =$ ______

$27 \div 100 =$ ______

$27 \div 1000 =$ ______

$0.27 \div 10 =$ ______

$0.27 \div 1000 =$ ______

STOP

12. Without using your HHC, compute

$25.6 \div 100 =$ ______

$256 \div 10{,}000 =$ ______

$11.349 \div 1000 =$ ______

$0.0321 \div 100 =$ ______

STOP

13. Write each of the following decimals as a fraction with denominator a power of 10.

(a) 4.8

(b) 0.56

(c) 0.021

(d) 0.00257

STOP

That is, when multiplying a whole number by 10, 100, 1000, . . . , we can obtain the product by simply "adding on" zeros to the numeral.

What about multiplying a decimal such as 2.7, 0.27, 0.027, . . . by 10, 100, 1000, . . . ?

Do exercise 9.

The results of exercise 9 indicate that in multiplying a decimal by 10, 100, 1000, . . . , we can obtain the answer by moving the decimal point one place to the *right* for each zero in 10, 100, 1000, . . . until we have a whole number. Then, if necessary, we add zeros as we did before for whole numbers.

Do exercise 10.

What about division?

Do exercise 11.

The results of exercise 11 indicate that in dividing a decimal by 10, 100, 1000, . . . , we can obtain the answer by moving the decimal point to the *left,* one place for each zero in 10, 100, 1000, . . . until we have a number with the decimal point in front of the last non-zero digit of the given number. Then, if necessary, we insert additional zeros on the left and move the decimal point further to the left in front of these additional zeros.

Do exercise 12.

DECIMALS AS FRACTIONS

Decimals are really just a way of writing special kinds of fractions—those whose denominators are powers of 10, that is, 10^n for some whole number $n > 0$. Thus

$$2.7 = \frac{27}{10}, \quad 0.27 = \frac{27}{100}, \quad 0.027 = \frac{27}{1000},$$

$$\frac{2}{5} = \frac{4}{10} = 0.4, \quad \frac{4}{25} = \frac{16}{100} = 0.16$$

and so on.

Do exercise 13.

Fractions written in the form $\frac{a}{b}$ where a and b are whole numbers with $b \neq 0$ are sometimes called *common* fractions and decimals are sometimes called *decimal* fractions. Thus

$$\frac{2}{3}, \quad \frac{1}{2}, \quad \frac{7}{3}, \quad \frac{5}{2}, \quad \text{and} \quad \frac{17}{10}$$

are common fractions, whereas

2.7, 0.12, 0.001, and 314.2

are all decimal fractions. In this text, however, we will, for brevity, refer to common fractions simply as fractions and to decimal fractions simply as decimals.

ROUNDING OFF NUMBERS

Many times we do not want to use all the information we have about a number. Often we will want to approximate the number, or use a number that is close to a given number. For example, many of the large numbers appearing in newspapers, magazines, almanacs, etc., are expressed only approximately, being rounded off to the nearest thousand, or to the nearest million, or to the nearest billion. Thus from the statement

> The number of immigrants to the United States from 1820 to 1974 was 46,712,725.

we can say that the approximate number of immigrants to the United States from 1820 to 1974 was

50,000,000	Rounded off to the nearest ten million (46,712,725 is closer to 50 million than to 40 million)
47,000,000	Rounded off to the nearest million (6,712,725 is closer to 7 million than to 6 million)
47,700,000	Rounded off to the nearest hundred thousand (712,725 is closer to 700 thousand than to 800 thousand)
47,710,000	Rounded off to the nearest ten thousand (12,725 is closer to 10 thousand than to 20 thousand)
47,713,000	Rounded off to the nearest thousand (2725 is closer to 3 thousand than to 2 thousand)
47,713,700	Rounded off to the nearest hundred (725 is closer to 7 hundred than to 8 hundred)
47,712,720 or 47,712,730	Rounded off to the nearest ten (25 is exactly half way between 20 and 30)

There are two possible answers in rounding off this number to the nearest ten, depending on which rule you use for a number ending in 5. Here are the two commonly used rules:

Rule A. *If the digit is 5, round the digit to the left to the next higher number.*

So, using this rule, 47,612,725 would be rounded off to the nearest ten as 47,612,730, and 47,712,735 would be rounded off to the nearest ten as 47,712,740.

Note that a modification of rule A (and also rule B given on the next page) is needed if the digit to the left of the 5 is 9. If this is the case, the 9 becomes a 0 and the next digit (if it is not also a 9) is increased by 1. For example,

$$2395 \rightarrow 2400 \text{ (to the nearest ten)}$$

and

$$34{,}952 \rightarrow 35{,}000 \text{ (to the nearest hundred)}$$

In exercises 14-16 underline the number that is closest to the circled number.

14. 70 (74) 80

15. 400 (478) 500

16. 50,000 (50,600) 51,000

In exercises 17-20 circle the number that correctly completes the sentence.

17. 69,374 rounded to the nearest ten is (69,370) (69,380).

18. 750,378 rounded to the nearest hundred is (750,300) (750,400).

19. 18,500 rounded to the nearest thousand is (19,000) (18,000).

20. 76,083 rounded to the nearest hundred is (76,100) (76,000).

In exercises 21-23 round the numeral 27,569,553 as indicated.

21. To the nearest hundred:

22. To the nearest thousand:

23. To the nearest ten:

STOP

Similarly, if more than one 9 is involved, we continue to replace them by zeros until, if possible, we reach a digit not equal to 9 as in

39,952 → 40,000 (to the nearest hundred)

and

47,999,500 → 48,000,000 (to the nearest thousand)

Finally, we may need to add the digit 1 as in

9995 → 10,000 (to the nearest ten)

and

999,952 → 1,000,000 (to the nearest hundred)

Rule B. *If the digit is 5 and the digit to the left of it is odd, round it off to the next higher number; if it is even, leave it unchanged.*

So, using this rule, 47,712,725 would be rounded off to 46,712,720 since 2 is an even number. But 46,712,735 would be rounded off to 46,712,740 since 3 is an odd number.

Which rule should be used depends basically upon which rule is used by the people using the approximations. For our purposes in this text, we will use rule A and hence always round 5 or more to the next higher digit.

Do exercises 14 through 23.

The same kind of round-off procedures can be used with numerals extended to the right of the decimal point (ones place) as is illustrated by the following example:

> Up to now, the fastest lap ever made at the Indianapolis speedway was a practice lap of 199.071 miles per hour by Johnny Rutherford on May 12, 1973.

199.071 could be approximated (rounded off) in the following ways:

199.071 becomes

199.07	rounded to the nearest hundredth
199.1	rounded to the nearest tenth
199	rounded to the nearest one
200	rounded to the nearest ten

Do exercises 24 through 30.

In exercises 24–30 approximate the number 299.3856 as indicated.

24. To the nearest thousandth:

25. To the nearest hundredth:

26. To the nearest tenth:

27. To the nearest one:

28. To the nearest ten:

29. To the nearest hundred:

30. Round 99.995 to the nearest hundredth.

Exercises 2.1

Name ______________________

Class ______________________

In exercises 1–6 write the given numeral in both forms of expanded notation.

1. 138.23

2. 143.2176

3. 307.0205

4. 0.0039

5. 203.008

6. 1000.001

1.

2.

3.

4.

5.

6.

In exercises 7–12 express the given sum in decimal notation.

7. $40 + 9 + 0.7 + 0.01 + 0.006$

8. $8 + 0.3 + 0.07 + 0.008$

7.

8.

9. 9. $7 \times 10 + 3 \times 1 + 5 \times 0.1 + 4 \times 0.01 + 3 \times 0.001$

10. 10. $6 \times 1000 + 3 \times 10 + 8 \times 0.01 + 5 \times 0.00001$

11. 11. $3 \times 10{,}000 + 3 \times 1000 + 3 \times 10 + 3 \times 0.01$

12. 12. $2 \times 0.1 + 7 \times 0.01 + 3 \times 0.001 + 4 \times 0.0001$

In exercises 13–18 read orally and then write out the number name for the given numeral.

13. 13. 0.57

14. 14. 0.0806

15. 15. 21.136

16. 16. 37.0207

17. 17. 6715.8609

18. 18. 0.007958

In exercises 19–24 write the decimal numeral for the given number.

19. 19. Thirteen and eight hundredths

20. 20. Seventeen ten thousandths

21. Five hundred three and forty-seven thousandths

22. Thirty-two thousand and seven hundredths

23. Twenty-four millionths

24. Thirteen hundred and thirteen hundredths

In exercises 25–30 give the place value of the "5" in the given numeral.

25. 75.81

26. 0.31546

27. 70.54

28. 304.58

29. 32.852

30. 25,762.01

In exercises 31–36 give the value of the "8" in the given numeral.

31. 9.81

32. 28.34

33. 1.0875

34. 6.00825

35. 65.3281

36. 9.0208

In exercises 37–42 change the given fraction to a decimal.

37. $\frac{35}{100}$

38. $\frac{108}{10}$

39. $\frac{53}{1000}$

40. $\frac{1631}{1000}$

41. $\frac{73}{10,000}$

42. $\frac{645}{100}$

21.
22.
23.
24.
25.
26.
27.
28.
29.
30.
31.
32.
33.
34.
35.
36.
37.
38.
39.
40.
41.
42.

In exercises 43–48 change the given decimal to a fraction with a power of 10 (10, 100, 1000, etc.) as a denominator.

43. 43. 0.019 44. 0.1367

44.

45. 45. 0.0405 46. 0.0019

46.

47. 47. 2.31 48. 7.003

48.

In exercises 49–54 round off the given number as indicated.

49. 49. 6.10426 to the nearest thousandth

50. 50. 6.00050 to the nearest thousandth

51. 51. 8.00708 to the nearest hundredth

52. 52. 26.6666 to the nearest tenth

53. 53. 26.6666 to the nearest hundredth

54. 54. 13.9999 to the nearest tenth

*55. *55. Does your calculator round off or simply drop (*truncate*) the extra digits that cannot be shown on the display? Try 2 ÷ 3. What do you get, 0.66666666 (truncate) or 0.66666667 (round-off)?

*56. *56. Try this computation: (1 ÷ 3) × 9. Did you get 2.9999997, 2.9999999, or 3? What does your calculator do about rounding off?

*57. *57. Try 1 ÷ 57. Did you get 0.0175438, 0.0175435, or something else? Now multiply your answer by 100. Did you get some additional digits or keep the same ones?

*58. *58. Write the numeral for thirteen thousand thirteen hundred thirteen.

***59.** Investigate ways in which such businesses as supermarkets, banks, etc., round off prices or interest.

*59.

***60.** Do the computation and complete the following statement:

$(7816 + 4248) - (3754 + 2222) + (26 \times 764) + (26 \times 8391) + 134{,}691$

The only word in a turkey's vocabulary is ________________ .

*60.

2.2 Computations with Decimals

USING AN HHC WITH DECIMALS

In the past, when doing computations with decimals many problems arose in placing the decimal point. Now one of the very nice features of an HHC is that the decimal point in the answer to a problem involving decimals is located correctly automatically—we don't have to use any rule to locate it (as we do in hand calculations). Furthermore, there is really nothing new to learn in order to do computations with decimals on an HHC beyond what you have learned with whole numbers; just be sure to press the decimal point key at the proper time!

Thus to enter 1.2 we have

Enter	1	.	2
Display	1.	1.	1.2

Then the computation

$$1.2 \times 3.5 + 0.32$$

is done as follows:

Enter	1.2	×	3.5	+	.32	=
Display	1.2	1.2	3.5	4.2	0.32	4.52

Note that it is not necessary to press the 0 key to enter 0.32. Indeed, we could write 0.32 simply as .32. It is, however, customary to include the 0 for a decimal less than 1 to emphasize that we have not just forgotten to put a non-zero number there.

Do exercises 1 through 4 in the adjoining column.

ROUNDING OFF ANSWERS

A curious fact about computations with decimals on an HHC (especially division problems) is that the HHC sometimes supplies us with what is really a silly answer!

Do exercise 5.

Doing exercise 5 by correctly dividing 370 by 15 gives us as the answer, 24.666666.* But certainly you would agree that a report that a car averaged 24.666666 miles per gallon would sound utterly ridiculous and that it would be much more meaningful to say that it averages 24.7 miles per gallon; or

*Or 24.666667. The answer to 7 decimal places is 24.6666666. Some HHC's then round this to 24.666667. (See exercises 55 and 56 of Exercises 2.1.) As a result of this difference in HHC's, your answer to a problem involving a full display in your HHC may sometimes differ in the last digit from the text answer.

OBJECTIVES

You should be able to:

(a) Perform the four basic operations of arithmetic with decimals.

(b) Estimate and place the decimal point in a problem using decimals.

(c) Order a set of decimal numbers from smallest to largest or from largest to smallest.

(d) Apply computation with decimals to problems involving only one of the basic operations.

(e) Realize that answers in applied problems often should be rounded off.

****

1. Add 2.314, 0.018, and 13.573.

2. Subtract 13.018 from 49.314.

3. Multiply 4.329 by 0.0432.

4. Divide 0.15159 by 0.342.

STOP

5. A certain make of car travelled 370 miles on 15 gallons of gas. How many miles per gallon did the car get?

STOP

better, considering variations in driving conditions and drivers, that it gets about 25 miles to the gallon. In other words, the "666666" part is not really meaningful, and in this particular situation, the 24.666666 should be rounded off to 24.7 or even 25.

A more detailed discussion of rounding off for various kinds of measures is rather complicated and is given in Appendix B. All we ask of you here in doing problems dealing with applications is that you use common sense in rounding off answers that you get by using your HHC, as illustrated by exercise 5.

Do exercises 6 and 7.

In exercises 6 and 7 do the computations necessary to solve the problem on your HHC and then round off your answer, if necessary, in a reasonable fashion.

6. The area of a circle of radius r is given approximately by $3.14 \times r^2$ where r is the radius. If the radius of a circle is measured as 5.1 centimeters, find its area.

7. A dealer sold 37 cars for a total of $205,452. What was the average price of the cars he sold?

STOP

PLACING THE DECIMAL POINT IN MULTIPLICATION OF DECIMALS

In Chapter 1 we observed several times that it is easier to solve some problems mentally than to use an HHC. For example, problems like $2 + 3$, 4×5, and 8×1000, are done easier and quicker mentally. Similar situations arise in computations with decimals. In particular, even though an HHC will place the decimal point in the correct position in a problem involving multiplication of decimals, it is useful to be able to determine the correct placement without using an HHC.

Do exercise 8.

In exercise 8 do the multiplications by using your HHC and compare your answers.

8. 23×45

   23×4.5

   2.3×4.5

   0.23×45

   0.023×45

   0.23×0.45

   0.023×0.045

   0.0023×0.045

STOP

In exercise 8 you should have found that, except for the placement of the decimal point, the answers to each part of the exercise were the same; that is, 1035, 103.5, 10.35, and so on.

The results of this exercise also indicate the following rule for locating the decimal point in the product of two or more decimals (or decimals and whole numbers):

1. Count the number of digits to the right of the decimal point in each of the numbers. (This number is 0 for a whole number.)
2. Add the numbers found from the digit counts in (1).
3. Then the number of digits (using 0's if needed) to the right of the decimal point in the product is equal to the sum obtained in (2).

Example:

$$0.\overbrace{0123}^{4} \times 1.\overbrace{57}^{2} \times 1.\overbrace{1}^{1} = 0.\overbrace{0212421}^{4+2+1=7}$$

Do exercise 9.

COMPARING DECIMALS

Many times we need to compare numbers when they are written in decimal form. We need, therefore, to explore the ideas related to the ordering of decimals.

Do exercises 10 through 14.

From the results of exercises 10–13, you can see that adding on zeros on the right of a numeral for a decimal does not change the value of the number: 0.50 = 0.5, 3.00700 = 3.007, etc. We can use this fact to compare two decimals easily by writing both of them as decimals with the same number of digits to the right of the decimal point.

Example 1. Which is larger, 0.302 or 0.3019?

Solution: Write 0.302 as 0.3020. Then

$$0.3020 > 0.3019 \text{ because } \frac{3020}{10{,}000} > \frac{3019}{10{,}000} \text{ since } 3020 > 3019$$

Note: > is read as *greater than* and < is read as *less than.*

9. Without using your HHC, do the following multiplications:

0.2 × 0.4

0.02 × 0.04

0.003 × 0.2

21 × 0.03

0.0013 × 0.2

3.01 × 0.0004

STOP

In exercises 10-13 do the given calculations on your HHC.

10. 0.5 – 0.50 =

11. 0.37 – 0.370 =

12. 3.00700 – 3.007 =

13. 0.102 – 0.1020 =

14. Does the value of the decimal change as you add zeros to the right of the last digit to the right of the decimal place?

STOP

Do exercises 15 and 16.

15. Use your HHC to check that $0.302 > 0.3019$ by filling in the following blanks:

$0.302 - 0.3019 =$ ______

This means that

$0.3019 +$ ______ $= 0.302$

16. What happens if you had subtracted in the following manner?

$0.3019 - 0.302 =$ ______

What does this tell you?

STOP

Example 2: Arrange the numbers below in order of size with the largest number first:

0.06, 1.4, 0.19, and 0.388

Solution: Write each of the decimals as decimals with three places to the right of the decimal point.

$$0.06 = 0.060 \rightarrow 60 \text{ thousandths}$$

$$1.4 = 1.400 \rightarrow 1400 \text{ thousandths}$$

$$0.19 = 0.190 \rightarrow 190 \text{ thousandths}$$

$$0.388 = 0.388 \rightarrow 388 \text{ thousandths}$$

We now arrange the thousandths in order of size as follows:

1400, 388, 190, and 60

From this, we can now arrange the original numbers in order from largest to the smallest as follows:

1.4, 0.388, 0.19, and 0.06

Do exercises 17 through 19.

17. Arrange the following numbers in order from largest to smallest.

4.72, 4.93, 4.8, 4.65

18. Arrange the following numbers in order with the smallest first.

0.01, 0.001, 0.1, 0.0001

19. Arrange the following numbers in order with the largest first.

0.465, 0.4053, 0.47, 0.405

STOP

Exercises 2.2

Name ____________

Class ____________

In exercises 1–8 check the following calculations using your HHC. If the given answer is incorrect, give the correct answer.

1. $0.3724 + 104.15 + 21.417 + 100.42 = 224.9814$ — 1.

2. $200.172 + 14.105 + 12.1465 + 0.705 + 7.2 = 234.3825$ — 2.

3. $70.00017 - 7.000017 = 63.000150$ — 3.

4. $630.630 - 0.6304 = 628.9996$ — 4.

5. $0.3685 \times 0.042 = 0.15477$ — 5.

6. $1.205 \times 1.15 = 1.38675$ — 6.

7. $0.0003 \div 3.75 = 0.00008$ — 7.

8. $326 \div 0.25 = 1403$ — 8.

In exercises 9–16 select, without actually multiplying, the number in parentheses at the right that appears to be the correct product. Then check your selection by performing the computation on your HHC.

9. $0.05 \times 0.03 = (1.5)\,(0.15)\,(0.015)\,(0.0015)$ — 9.

10. $0.4 \times 0.004 = (0.16)\,(0.016)\,(0.0016)\,(0.00016)$ — 10.

11. $0.02 \times 0.007 = (0.14)\,(0.014)\,(0.0014)\,(0.00014)$ — 11.

12. $0.8 \times 3.0 = (24)\,(2.4)\,(0.24)\,(0.024)$ — 12.

13. $0.8 \times 0.09 = (7.2)\,(0.72)\,(0.072)\,(0.0072)$ — 13.

14. $0.007 \times 0.09 = (0.063)\,(0.0063)\,(0.00063)\,(0.000063)$ — 14.

15. $0.6 \times 9.0 = (540)\,(54)\,(5.4)\,(0.54)$ — 15.

16. $0.8 \times 0.006 = (0.48)\,(0.048)\,(0.0048)\,(0.00048)$ — 16.

	In exercises 17–26 arrange each set of decimals in order as indicated.
17.	17. 0.301, 0.29, 2.2, 0.31 (largest first)
18.	18. 0.565, 0.5053, 0.57, 5.5 (largest first)
19.	19. 0.01, 0.001, 0.1, 0.0001 (smallest first)
20.	20. 3.25, 0.353, 0.3485, 3.249 (smallest first)
21.	21. 0.61, 0.683, 0.70, 0.6126 (largest first)
22.	22. 0.04, 1.25, 0.156, 2.3 (largest first)
23.	23. 0.007, 6.02, 0.603, 0.5483 (smallest first)
24.	24. 0.76, 0.7, 0.076, 0.710 (smallest first)
25.	25. 0.89, 0.98, 0.983, 0.9 (largest first)
26.	26. 0.38, 0.308, 0.387, 0.0308 (smallest first)
27.	27. 12.3567, 12.3498, 12.35, 12.3, 12.4, (largest first)
28.	28. The 100-meter freestyle swimming record is 51.11 seconds, whereas the 100-meter backstroke record is 56.3 seconds. Which method of swimming is the faster and by how many seconds?
29.	29. Three steel plates having thicknesses of 0.826 inches, 0.508 inches, and 0.736 inches are to be riveted together. What will be their combined thickness?
30.	30. A factory sells a certain kind of scrap metal for 74.6 cents per pound. How much money would you receive for 145 pounds of this metal?
31.	31. What is the total thickness of 75 sheets of aluminum if one sheet has a thickness of 0.45 inches?
32.	32. If the large wheel of an antique bike is 14.36 feet in circumference, how many times (to the nearest whole number) will it turn in going one mile (5280 feet)?

33. What is the cost (to the nearest cent) of 17.8 gallons of gasoline at 63.9 cents a gallon? 33.

34. On the first day of his vacation, Mr. Knutson found he had driven 345.6 miles in 9 hours. How many miles per hour had he averaged? 34.

35. Mrs. Thompson's grocery bill for one week was $37.84. What was her average food cost per day (to the nearest cent)? 35.

36. The women's world speed skating record for 500 meters is 41.06 seconds. How fast (to the nearest hundredth of a second) is this in meters per second? 36.

37. If 8.7 gallons of gasoline cost $5.99, what was the cost of the gasoline per gallon to the nearest tenth of a cent? 37.

38. On a certain day the total receipts in a school cafeteria were $591.36. If 840 transactions were made that day, what was the average amount of money involved in each transaction to the nearest cent? 38.

39. Find the cost of three dozen pairs of socks at $.75 per pair. 39.

40. Tom wants to buy three items whose prices are $2.98, $1.37, and $.67. Can he buy all three items if he has only $5.00? If your answer is no, how much more does he need? If your answer is yes, what would be his change from a five-dollar bill? 40.

41. A farmer owned 1800 acres of land. He sold 139.75 acres to one neighbor, 186.5 acres to another neighbor, and 116.12 acres to a developer. How many acres did he still own? 41.

42. A machine part is to have a cylindrical hole 0.216 inches in diameter. On inspection, the hole can be no more than 0.0005 inches larger than the indicated dimension. Could the inspector pass the part if the hole measured 0.2164 inches? If your answer is yes, how much was there to spare? If your answer is no, how much too big or too small was the hole? 42.

43. From a rod 28.625 inches in length, a piece 9.874 inches long was cut. What is the length of the remaining piece if 0.125 inches was wasted in making the cut? 43.

44. Ten posts were used in building a fence 112.5 feet long. If the posts were evenly spaced, find the distance from the center of one post to the center of the next post. 44.

In exercises 45–51 refer to the following table:

MOTOR FUEL CONSUMPTION, BY USE: 1950 TO 1973

[Prior to 1960, excludes Alaska and Hawaii. Comprises all fuels under State motor fuel laws. Excludes exports and Federal purchases for military use. See *Historical Statistics, Colonial Times to 1970,* series Q 150-162]

Item	1950	1955	1960	1965	1969	1970	1971	1972	1973
Total consumption[1]–million gallons	40,280	53,116	63,711	71,104	88,122	92,328	97,547	105,061	110,473
Annual change[2]–percent	24.0	19.7	13.4	4.7	6.2	4.8	5.7	7.7	5.2
Passenger vehicles–million gallons	25,037	34,319	41,996	51,169	63,395	66,728	70,461	74,344	78,858
Cars[3]–million gallons	24,305	33,548	41,169	50,275	62,448	65,784	69,514	73,463	78,011
Buses[4]–million gallons	732	771	827	894	947	944	947	881	847
Trucks[5]–million gallons	10,566	13,308	15,882	19,935	24,727	25,600	27,086	30,718	31,615
Average gallons per vehicle	728	759	777	775	821	830	838	859	851
Cars[3]	603	644	661	656	700	722	723	730	736
Buses[4]	3,752	3,021	3,040	2,844	2,600	2,491	2,382	2,165	1,991
Trucks[5]	1,257	1,278	1,330	1,347	1,384	1,365	1,368	1,446	1,361
Average mileage per gallon	12.87	12.67	12.42	12.49	12.15	12.14	12.16	12.07	11.85
Cars[3]	14.95	14.53	14.28	14.15	13.75	13.58	13.73	13.67	13.29
Buses[4]	5.57	5.85	5.26	5.35	5.32	5.34	5.38	5.80	5.86
Trucks[5]	8.57	8.37	7.96	8.60	8.36	8.39	8.38	8.59	8.45

[1] Through 1960, includes nonhighway usage and losses allowed for evaporation, handling, etc.; thereafter, highway usage only.
[2] Change from immediately prior year; e.g., from 1949 to 1950, from 1954 to 1955, etc.
[3] Includes taxicabs and motorcycles.
[4] Includes school buses.
[5] Includes combinations.
Source: U.S. Federal Highway Administration, *Highway Statistics,* annual.

45.

45. Did the mileage per gallon for cars decrease or increase when comparing the years 1950 and 1973? By how much was the increase or decrease?

46.

46. Did the mileage per gallon for buses increase or decrease when comparing the years 1950 and 1973? By how much was the increase or decrease?

*47.

***47.** What is the increase in the number of cars in use in 1973 compared with 1950?

*48.

***48.** What is the increase in the number of trucks in use in 1970 compared with 1960?

49.

49. On the average, how many gallons of gas (to the nearest tenth) would it take a car to drive 100 miles in 1950? In 1973? What is the difference in gallons of gasoline used?

50. On the average, how many gallons of gas (to the nearest tenth) would it take a bus to drive 100 miles in 1960? In 1970? What is the difference in gallons of gasoline used? 50.

51. If a car tank holds 21 gallons of gas, how far could this car go on a tank of gas in 1950? In 1970? What is the difference (to the nearest mile) in the distance traveled in the two years? 51.

*52. Do the computation in parts (a), (b), and (c), then guess an answer for part (d). Check your guess with the calculator. *52.

(a) $(1 \times 1) \div 1$ (a)

(b) $(22 \times 22) \div (1 + 2 + 1)$ (b)

(c) $(333 \times 333) \div (1 + 2 + 3 + 2 + 1)$ (c)

(d) Guess, then compute:

$$(4444 \times 4444) \div (1 + 2 + 3 + 4 + 3 + 2 + 1)$$

*53. Without using the divide key, use your HHC to compute $544 \div 16$. *53.

*54. Solve this equation by trial and error. You must put the same number in each box. *54.

$$\square \times \square = 1{,}527{,}696$$

*55. Find three numbers which have the same sum as their product. *55.

*56. Do the computation and complete the following statement: *56.

$$(34.168 \times 3 - 12.345) \div 0.123$$

An ______________ is a long thin fish.

2.3 Combining Operations with Decimals

The mechanics of doing a problem such as

$$2.12 \times 3.743 + 10.17 \times 0.017$$

with an HHC are, except for using the decimal point key, no different from doing such problems with whole numbers on an HHC.

Check that the answer to the problem above is 8.10805. (Remember the My-Dear-Aunt-Sally rule and note that, unless your HHC has a memory key, you must write down the answer to one of the products.)

Do exercises 1 through 5 in the adjoining column.

In section 1.5 we discussed the use of estimation in computations involving whole numbers. An ability to estimate answers in computations with decimals is equally important: it will not only enable you to detect gross errors in computation caused, for example, by forgetting to press the decimal point key at some point, but will also be useful in many applications where only a rough idea of the size of the answer is needed.

There are no fixed rules for estimating answers, and different procedures can result in somewhat different estimates. The main point to keep in mind is that we want to get an estimate that will tell us whether or not we have made any major errors in the use of the HHC. (Note that we can assume the HHC itself will not make errors, so that if our estimate shows our HHC answer is "in the ball park," it is highly probable the answer given by the HHC is *exactly* correct. This is in contrast to paper-and-pencil calculation where our estimate may indicate that there are no *major* errors in the computations, and yet there may be a minor error caused, for example, by saying that 7 + 8 = 13 at some stage of the calculation.)

Consider the calculation with which we began this section. Using the symbol "≈" to mean "is approximately equal to," we have

$$2.12 \approx 2, \qquad 3.743 \approx 4, \qquad 10.17 \approx 10, \qquad 0.017 \approx 0.02$$

so that our approximation becomes

$$2 \times 4 + 10 \times 0.02 = 8 + 0.2 = 8.2$$

which is quite close to the actual answer of 8.10805.

Now let us make estimates to the answers to exercises 1 and 2.

1.
$$\begin{aligned} & 0.0234 \times 11.78 + 14.05 \times 0.0038 \\ \approx\ & 0.02 \times 12 + 14 \times 0.004 \\ =\ & 0.24 + 0.056 \approx 0.2 + 0.1 = 0.3 \end{aligned}$$

Notice how we have alternated a roundoff down with a roundoff up, e.g., 0.0234 is rounded down to 0.02 and then 11.78 is rounded up to 12; 14.05 is rounded down to 14 and then 0.0038 is rounded up to 0.004; 0.24 is rounded down to 0.2 and then 0.056 is rounded up to 0.1. Our estimate here compares quite closely to the HHC answer of 0.329042. (Here, as elsewhere, slightly different ways of rounding off will give slightly different estimates.)

OBJECTIVES

You should be able to:

(a) Perform calculations with decimals that involve more than one of the four basic operations of arithmetic.

(b) Estimate the answer to computations involving decimals.

(c) Apply computations with decimals to the solution of problems.

(d) Know that the answers to problems should often be rounded off.

****

In exercises 1–5, perform the given computations.

1. $0.0234 \times 11.78 + 14.05 \times 0.0038$

2. $13.94 \times 9.18 \times 0.0026 - 0.1432$

3. $15.749 \times 0.081 \div 49.82$

4. $0.254 \div 15.92 \times 0.0031$

5. $(524.27 - 418.23) \times 0.0003142$

STOP

2.

$$13.94 \times 9.18 \times 0.0026 - 0.1432$$
$$\approx 13 \times 10 \times 0.003 - 0.14$$
$$= 130 \times 0.003 - 0.14 = 0.39 - 0.14 = 0.25$$

Here we have rounded up 9.18 because multiplying by 10 is easier than multiplying by 9, and so we rounded down both 13.94 and 0.0026. Our estimate this time is not too close to the HHC answer of 0.1895199 but it does indicate that we have probably made no errors in using the HHC.

Do exercises 6 through 10.

In estimating the answer to a multiplication or an addition, we try to alternate up and down roundings, since rounding all factors or addends up would clearly produce an answer that is too large and, similarly, rounding all down would produce an answer that is too small. In estimating the answer to a division or subtraction problem, however, both of the numbers involved should be rounded up or both should be rounded down in order to balance the approximations.

We now use this procedure in estimating answers to exercises 3 through 5.

3.

$$(15.749 \times 0.081) \div 49.82$$
$$\approx (16 \times 0.1) \div 50 = 1.6 \div 50 \approx 1.5 \div 50$$
$$= (1.5 \div 5) \div 10$$
$$= 0.3 \div 10 = 0.03$$

Here our estimate is quite close to the HHC answer of 0.0256055. Note that 15.749, 0.081, and 49.82 have all been rounded up because a division is involved.

4.

$$(0.254 \div 15.92) \times 0.0031$$
$$\approx (0.2 \div 10) \times 0.003 = 0.02 \times 0.003$$
$$= 0.00006$$

Notice that we have rounded both 0.254 and 15.92 down because a division is involved. (We could also have rounded both of them up to, say 0.32 and 16, and still have a simple division to do, except that it is not quite as easy to place the decimal point properly in $0.32 \div 16 = 0.02$.)

Notice also that although our roundoffs are rather extreme (to simplify computations), our estimate is still fairly close to the HHC answer of 0.0000494.

In exercises 6–8, *first* make an estimate of the answer. Then perform the computation on your HHC and compare your result with your estimate. Recheck your HHC calculation and/or your estimate if the two results differ substantially.

6. $57.781 \times 0.94 + 0.079 \times 67.158$

Estimate:

HHC answer:

7. $9.8 \times 8.825 - 70.135 \times 0.853$

Estimate:

HHC answer:

8. $52.234 \times 0.375 - 2.623$

Estimate:

HHC answer:

In exercises 9–10, first decide what arithmetic calculations must be performed to answer the question. Next, estimate the answer to the calculation. Finally, perform the calculation, round off the answer in reasonable fashion, and compare it with your estimate.

9. A candy dealer mixed 2 pounds of \$0.88 per pound candy with 5 pounds of \$1.94 per pound candy and 3 pounds of \$1.12 per pound candy. What should he price the mixture per pound?

Method:

Estimate:

Answer:

Rounded answer:

5.
$$(524.27 - 418.23) \times 0.003142$$
$$\approx (500 - 400) \times 0.003 = 100 \times 0.003 = 0.3$$

Note that, because a subtraction is involved, we have rounded down both 524.27 and 418.23. Here our estimate of 0.3 is quite close to the HHC answer of 0.3331776.

Do exercises 11 through 13.

We will return to further work with decimals after we have considered computations with fractions in Chapter 4.

10. A car's odometer read 21,376.4 miles at the beginning of a trip, and 21,688.3 at the end of the trip. If 13.8 gallons of gasoline were used on the trip, how many miles to the gallon did the car average?

Method:

Estimate:

Answer:

Rounded answer:

STOP

In exercises 11–13, first estimate the answer and then perform the computation on your HHC.

11. $40.33 \times 0.84 + 24.7 \times 0.007$

Estimate:

HHC answer:

12. $(605.45 \div 3.22) - (27.055 \times 5.87)$

Estimate:

HHC answer:

13. From a piece of metal 96.35 centimeters long, how many pieces 2.25 centimeters long can be cut if each cut wastes 0.15 centimeters?

Method:

Estimate:

HHC answer:

Rounded answer:

STOP

Exercises 2.3

Name ____________________

Class ____________________

In exercises 1-20 perform the given computation, but first estimate your answer. If possible, do some of the computations mentally and/or rearrange the way the operations are done to simplify your computations.

1. $9.946 - 4.59 + 0.413$
2. $0.03535 + 0.6908 - 0.0575$
3. $(6.84 + 10.274) - (9.325 + 0.602)$
4. $(7.58 + 37.296) - (3.738 + 4.239)$
5. $(28.185 - 3.188) + (0.525 - 0.316)$
6. $(375.26 - 95.076) + (79.789 - 73.8)$
7. $27.44 + 8.579 - 6.975 \times 2.094$
8. $97.85 + 2.65 \times 32.7 - 51.513$
9. $47.475 + 48.621 \div 2.006 + 8.897$
10. $395.06 \div 20.45 - 4.92 + 6.235$
11. $(4.153 \times 3.0057) - (1.0375 \times 3.0057)$
12. $(5.307 \times 10.375) - (2.04 \times 10.375)$

1. Estimate:

   Solution:

2. Estimate:

   Solution:

3. Estimate:

   Solution:

4. Estimate:

   Solution:

5. Estimate:

   Solution:

6. Estimate:

   Solution:

7. Estimate:

   Solution:

8. Estimate:

   Solution:

9. Estimate:

   Solution:

10. Estimate:

    Solution:

11. Estimate:

    Solution:

12. Estimate:

    Solution:

13. Estimate:

Solution:

13. $(98.6 \times 7.76) + (38.9 \times 7.76)$

14. Estimate:

Solution:

14. $(912.82 \times 0.993) + (490.22 \times 0.993)$

15. Estimate:

Solution:

15. $(17.405 + 18.830) \div (9.186 + 3.003)$

16. Estimate:

Solution:

16. $(0.07629 + 0.01785) \div (0.0783 + 0.0965)$

17. Estimate:

Solution:

17. $(15.638 - 9.0114) \times (9.373 - 1.374)$

18. Estimate:

Solution:

18. $(70.177 - 49.175) \times (4.2113 - 2.160)$

19. Estimate:

Solution:

19. $(216.00 - 69.625) \div (21.944 - 2.328)$

20. Estimate:

Solution:

20. $0.0692 + 8.91 \times 9.6025 - 0.138 \div 5.025 + 6.704$

In exercises 21–40 first write down the method you would use to solve the problem; simplify the method, if possible, for easier computation; estimate an answer to the problem; and then calculate the answer (rounding it off if necessary).

Example: A grocer bought 12 cases of 24 cans each of peaches that cost her \$.32 a can and 12 cases of 24 cans of pears that cost her \$.37 a can. What was her total cost?

Solution:

Method: $(12 \times 24 \times 0.32) + (12 \times 24 \times 0.37)$

Simplification: $(0.32 + 0.37) \times 12 \times 24$

Estimated answer: $300 \times \$.70 = \$210.00$

Computed answer: \$198.72 (no roundoff needed)

21.

21. In a certain city, the average monthly rainfall is 1.37 inches. At this rate, what is the total rainfall per year?

22. What is the total thickness of 65 sheets of aluminum if the thickness of each sheet is 0.042 inches? 22.

23. If a kilometer is 0.6 mile, how many miles are there in 24.18 kilometers? 23.

24. If a butcher charged \$7.68 for a piece of meat priced at 96 cents per pound, how many pounds of meat were purchased? 24.

25. How many 1.5-pound boxes will 47.6 pounds of candy fill? 25.

26. If 15.2 gallons of gasoline costs \$5.00, what, to the nearest tenth of a cent, is the price per gallon? 26.

27. The hold of a freighter has a capacity of 47,869 cubic feet. How many bushels of grain can it contain? (1 bushel = 1.25 cubic feet) 27.

28. Joyce and Jill borrowed a $2\frac{1}{5}$ gallon (2.2 gallon) punch bowl for a party. How many full cups of punch can be served from a full bowl if each cup holds 6 ounces (0.0469 gallon)? 28.

29. The rainfall reported for five different years at one weather station was 39.5, 36.75, 42.12, 37.87, and 40.01 inches. What was the average rainfall at that station during this five-year period? 29.

30. For each can containing 4.75 fluid ounces of frozen concentrated juice, 3 cans (of the same capacity as the juice can) of water must be added to make a tasty mixture. Ralph used 8 cans of frozen juice concentrate. How many ounces of the tasty mixture did Ralph get? 30.

31. At the start of a trip, the odometer of a car read 13,243.7. At the end of the trip it read 13,536.2. If 15 gallons of gasoline were used on the trip, how many miles did the car travel per gallon of gasoline? 31.

32. A metal strip is 39.96 inches long. How many pieces, each 2.125 inches long, can be cut from it if 0.095 inches of waste is allowed for each piece? 32.

33. In a stack of sheet metal there are sheets of two different thicknesses, 0.035 and 0.050 inches. If there is the same number of sheets of both kinds and the pile is 21.25 inches high, what is the number of thinner sheets in the stack? 33.

34.

34. A neighborhood baseball team bought 15 balls at \$1.65 each or \$1.48 each when a complete box of a dozen was purchased, and 7 bats at \$3.50 each. If 9 boys shared in the cost equally, how much, to the nearest cent, was each boy's share?

35.

35. The rates at a certain laundry are \$2.50 for the first 25 pieces and \$0.08 for each additional piece. If the laundry bill was \$3.70, how many pieces were laundered?

36.

36. A handicraft booth at a student fair sold sand candles. During the day, John sold 8 candles at \$3.50 each, 12 candles at \$3.75 each, and 15 candles at \$1.75 each. If his cost for materials for the candles was \$21.72, what was his profit?

37.

37. A department store bought 50 television sets at a cost of \$156.32 each. They sold 42 of them at \$294.99, and then sold the rest in a clearance sale at \$224.99. How much profit did they have?

38.

38. A store sold pears at 3 for 35¢, oranges at 9 for 50¢, and apples at 3 for 44¢. How much would it cost you to buy one of each item? (How do stores round off prices?)

39.

39. A girls' club needs to wrap 48 Christmas packages for a children's hospital. If a roll of ribbon contains 37 feet of ribbon and 5.25 feet of ribbon are needed for each package, how many rolls of ribbon must be purchased?

***40.** A class in an elementary school saved \$85.25 through various projects. The class then voted to use this money to attend a concert. Tickets to the concert cost \$3.50, \$4.75, and \$5.75. If there were 23 children in the class, what price ticket and how many of each should they buy in order for all members of the class to attend the concert?

*40.

***41.** The world record for the 100-yard dash is 9.1 seconds. What is this speed in miles per hour to the nearest hundredth of a mile?

*41.

***42.** The nearest star (except for our sun) is 4.3 light years away. A light year is the distance light travels in one year (365.25 days) at the rate of 186,284 miles per second. How long (years and days) would it take a space ship traveling at 1,000,000 miles per hour to reach this star?

*42.

***43.** Perform the calculation below and use your calculator to complete the following statement:

$$(0.6362 + 0.7896) - (0.6675 + 0.1645)$$

The pretty maiden about to be tied to the railroad tracks by the villain (Boo!)

_______________ for her life.

*43.

Review Exercises 2

Name __________________

Class __________________

In exercises 1–5 write the given numeral in both forms of expanded notation.

1. 94.55 — 1.
2. 8.7546 — 2.
3. 0.4319 — 3.
4. 0.09119 — 4.
5. 0.0090209 — 5.

In exercises 6–10 read the numeral orally and then write out the number name.

6. 27.45 — 6.
7. 2.481 — 7.
8. 3.4408 — 8.
9. 0.08767 — 9.
10. 0.030701 — 10.

In exercises 11–15 write the decimal numeral for the given number.

11. Seventy and one hundred forty-six thousandths — 11.
12. Thirty-seven and eighty-three hundredths — 12.
13. Eight thousand nine hundred ninety-eight and thirty-seven thousandths — 13.

14.
15.
16.
17.
18.
19.
20.
21.
22.
23.
24.
25.
26.
27.
28.
29.
30.
31.
32.
33.
34.
35.

14. Six and eighty-nine ten thousandths

15. Twenty-five millionths

In exercises 16–20 give the place value of the "4" in the given numeral.

16. 224.9

17. 9469.05

18. 22.048

19. 2.5647

20. 512.0403

In exercises 21–25 give the value of the "7" in the given numeral.

21. 6196.2793

22. 46.0375

23. 0.0705

24. 0.02547

25. 9762.43

In exercises 26–35 round off the given number as indicated.

26. 64.51 to the nearest unit

27. 188.15 to the nearest ten

28. 9.619 to the nearest hundredth

29. 0.9638 to the nearest thousandth

30. 54.692 to the nearest tenth

31. 8.9654 to the nearest tenth

32. 8.2391 to the nearest hundredth

33. 2.0391 to the nearest tenth

34. 27.82 to the nearest ten

35. 29.00529 to the nearest thousandth

In exercises 36–45 select, without actually doing the calculation, the number in parentheses that appears to be closest to the correct answer. Check your selection by doing the calculation on your HHC.

36. 98.06 + 87.34 (1.854) (18.54) (185.4) — 36.

37. 0.0162 – 0.00459 (1.1161) (0.1161) (0.01161) — 37.

38. 44.9 × 0.889 (399.161) (39.9161) (3.99161) — 38.

39. 3.544 + 3.188 + 3.31 (100.42) (10.042) (1.0042) — 39.

40. 8.94 × 4.8 (4.2912) (42.912) (429.912) — 40.

41. 0.88 × 0.41 (3.608) (0.3608) (0.03608) — 41.

42. 8.8695 – 8.799 (0.00705) (0.0705) (0.705) — 42.

43. 2.92 + 19.18 × 98.9 (2185) (1899) (19,522) — 43.

44. (19.6 – 7.8) × 94.7 (1117) (758) (1848) — 44.

45. 211.66 + 55.08 × 99.4 – 18.4 (5668) (4672) (6712) — 45.

In exercises 46–60 perform the given computation, but first estimate your answer. If possible, do some of the computations mentally and/or rearrange the order in which the operations are done to simplify the computation.

46. 4.5 × (7.3 + 3.7)

46. Estimate:

Solution:

47. 16.2 + 97.34 × 0.08721

47. Estimate:

Solution:

48. (121.7 – 21.7) × 4.705

48. Estimate:

Solution:

49. 2.35 + (7.8 ÷ 73.2) – (0.89 ÷ 3.7)

49. Estimate:

Solution:

50. 647.01 – 4.07 × (99.01 + 0.99)

50. Estimate:

Solution:

51. (640.5 ÷ 88.159) – (9.6119 × 0.63)

51. Estimate:

Solution:

52. Estimate:

Solution:

52. 54.6 + 92.82 ÷ 39.67 – 23.287

53. Estimate:

Solution:

53. 88.159 × 96.119 × 0.0511 ÷ 6.19

54. Estimate:

Solution:

54. (10.449 + 5.24) – (5.0075 + 3.3)

55. Estimate:

Solution:

55. (33.8 + 45.86) – (5.102 + 6.19)

56. Estimate:

Solution:

56. (79.3 + 65.37) – (124.7 ÷ 55.3)

57. Estimate:

Solution:

57. (64.05 × 88.195) + (96.19 × 88.195)

58. Estimate:

Solution:

58. (153.07 × 110.7) – (153.07 × 10.7)

59. Estimate:

Solution:

59. (54.6 – 8.23) × (32.8 – 8.23)

60. Estimate:

Solution:

60. (29.5 + 35.9) ÷ (153.6 + 35.9)

In exercises 61–78 solve the given problem by first writing down the method you would use to solve the problem; simplify, if possible, for easier computation; estimate an answer to the problem; and then calculate your answer.

61.

61. Joe receives $0.037 for each item that he produces on his machine. If he averages 367 items an hour, how much does he make a year? (Assume 50 forty-hour weeks.)

62.

62. Jack made the following purchases during a sale at a local store: a three-piece oil change kit for $2.19 (regularly $2.99), an oil filter for $1.89 (regularly $2.59), a five-pound bag of oil absorbant for $.99 (regularly $1.39), and five quarts of oil for $.49 per quart (regularly $.75). How much change did he receive from a ten-dollar bill? How much did he save by buying the items on sale?

63. A piece of steel tubing 5.13 meters long has an outside diameter of 8.16 centimeters and an inside diameter of 7.37 centimeters. How thick is the tubing material? (1 meter = 100 centimeters) 63.

64. Gasoline sells for 54.9 cents a gallon. If 7 cents of this cost is state tax and the rest of the 12.5 cents in taxes is a federal tax, how much are the total taxes paid on 21.8 gallons of gasoline? How much of the cost of the 21.8 gallons goes to the state government and how much goes to the federal government? 64.

65. If a long-distance telephone call costs $2.15 for the first 3 minutes and $.07 for each additional minute, how much would a 25-minute call cost you? 65.

66. How many pieces of metal 6.67 centimeters in length can be cut from a 5-meter length of metal if 0.3 centimeters of material is wasted with each cut? (1 meter = 100 centimeters) 66.

67. A fishing boat uses 3.2 gallons of diesel oil each hour that it cruises. If the fuel capacity of the boat is 300 gallons and its average cruising speed is 8 miles per hour, what is the range of the boat? 67.

68. If the quality is the same, compare the cost per unit and determine the best buy between 2 pounds of peanut butter for $1.69 and 15 ounces of peanut butter for 98 cents. 68.

69. Mr. Ferris needs 4 pieces of bar steel in the following lengths: 8.25 feet, 9.09 feet, 7.00 feet, and 6.51 feet. What length of bar will he need if the stock bars are sold only in whole number of feet lengths? After each length is cut, how much will be left over? (Disregard any waste in cutting.) 69.

70.

70. In a mile relay race, four runners each ran a quarter mile. Their times for each quarter mile were 59.9 seconds, 60.3 seconds, 59.0 seconds, and 58.5 seconds, respectively. The meet record is 3 minutes, 59.9 seconds. By how many seconds did the team fail to match the record?

71.

71. John Bounce took a trip in which the car's odometer at the start read 3827.4 miles. At the end of the trip, the odometer read 4013.2. If the trip took him 6 hours and he used 11.2 gallons of gas, how far did he travel, what was his average speed, and how many miles per gallon did his car average?

72.

72. The rainfall in a certain town was 30.4 inches in one year. The rainfall for the first six months was as follows:

Jan.	1.39 inches	April	4.86 inches
Feb.	0.92 inches	May	3.49 inches
Mar.	3.14 inches	June	1.96 inches

(a)

(a) How many inches of rain fell during the last half of the year?

(b)

(b) How much more rainfall was recorded in April than in February?

73.

73. A chemistry student mixed 6.59 grams of substance A with 2.47 grams of substance B. If he placed an equal amount of the mixture in each of three test tubes, what was the weight of the material in each test tube?

74.

74. The "Royal Coach" is a car that weighs 3767 pounds, is 16.2 feet long, and gets 18.7 miles on a gallon of gas that costs 62.9 cents a gallon. What will it cost for gasoline to drive this car on a trip of 500 miles?

75.

75. A 6-inch bolt, 0.75 inches in diameter, weighs 0.625 pounds. How many bolts are there in a sack weighing 100 pounds?

76.

76. A merchant pays $2.38 a yard for an 83-yard bolt of material that is 54 inches wide. If she wishes to make a total profit of $125, what should she charge per yard for the material?

77. The cost of flying from Seattle to San Francisco and back is $174. The airline distance between the two cities is 678 miles, whereas the highway distance is 890 miles. What is the cost per mile to fly from Seattle to San Francisco and back?

77.

78. At the beginning of a trip the odometer on a car reads 29,272.5. After the trip, which took 8 hours, the odometer read 29,637.3 and the driver found that it took 11.2 gallons of gasoline to fill her tank.

78.

(a) If she had started with a full tank, how many miles to the gallon did she get on the trip?

(a)

(b) What was her average speed on the trip?

(b)

*79. A man has 5 pieces of chain, each with 3 links shown as follows.

*79.

If it costs $0.20 to cut a link and $0.35 to weld it back together, what is the cheapest way to make one chain of 15 links out of the 5 pieces?

*80. Find a four-digit number ABCD such that if a decimal point is put between B and C the resulting number is the average of the numbers AB and CD.

*80.

Just for Fun

A. PATTERNS AND SEQUENCES

Write the next five numbers in each sequence.

1. 1, 2, 4, 8, 16, ___, ___, ___, ___, ___.
2. 1, 2, 6, 24, ___, ___, ___, ______, ______.
3. 1, 2, 4, 7, 11, 16, ___, ___, ___, ___, ___.
4. 2, 6, 30, 210, ____, _____, _____, _______, ________.

B. MORE PATTERNS

For those items indicated by an asterisk (*) you are to examine the pattern indicated by the previous computations and then write what you think is the result.

5.

$9 \times 9 =$ ______

$99 \times 99 =$ ______

$999 \times 999 =$ ______

*$9999 \times 9999 =$ ______

*$99999 \times 99999 =$ ______

*$999999 \times 999999 =$ ______

6.

$9 \times 9 + 7 =$ ______

$9 \times 98 + 6 =$ ______

$9 \times 987 + 5 =$ ______

$9 \times 9876 + 4 =$ ______

*$9 \times 98765 + 3 =$ ______

*$9 \times 987654 + 2 =$ ______

*$9 \times 9876543 + 1 =$ ______

*$9 \times 98765432 + 0 =$ ______

*$9 \times 987654321 - 1 =$ ______

C. MISSING DIGITS

Find the missing digits in each subtraction problem.

7.

	__	__	0	__	6	__
(−)		1	__	0	__	8
		9	9	9	6	6

8.

	__	__	4	__	9
(−)		9	__	5	__
	4	8	__	1	5

D. TRY THESE

9. (a) If you count "one, two, three, . . . , naming one number per second, how many days would it take you to count to a million?

   (b) If you counted only 8 hours a day, 5 days a week, how many weeks would it take you to count to a million?

10. A rope ladder is attached vertically to the side of a ship so that 5 meters of the ladder is under water at low tide. If the tide rises at 60 centimeters an hour, how much of the ladder is under water at high tide 6 hours later?

11. Place +, −, ×, ÷, or ( ) between the digits so as to arrive at the answer of 10. (Do not change the order of the digits.) See how many ways you can do this.

$$1 \quad 2 \quad 3 \quad 4 \quad 5 \quad 6 \quad 7 \quad 8 \quad 9 = 10$$

E. MAGIC SQUARE

12. Complete the following magic square

72		36
		81
	63	18

F. CROSS PUZZLE

13. Place the numbers 1 through 9 (using each only once) in the circles so that the numbers in each line have the same sum.

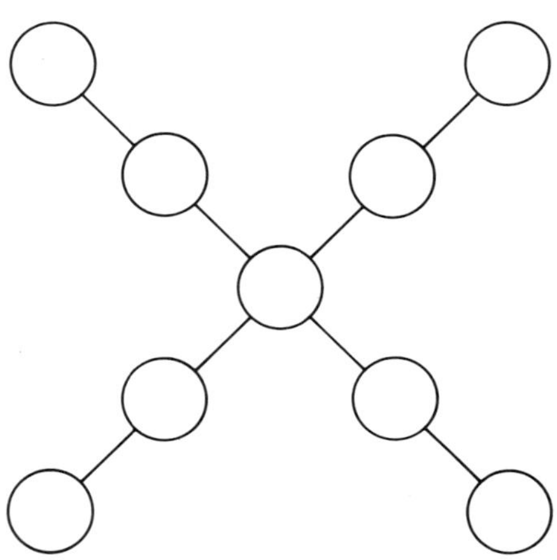

G. CALCULATOR GAME

14. *Subtraction Game:* One of two players picks a number and enters it in the calculator (say 21). The players than take turns subtracting 1, 2, or 3. The first player to produce zero or a negative number loses.

Sample game:				*Display shows:*
Player	A	enters	21	21
	B	subtracts	3	18
	A	subtracts	2	16
	B	subtracts	3	13
	A	subtracts	3	10
	B	subtracts	1	9
	A	subtracts	3	6
	B	subtracts	2	4
	A	subtracts	3	1

B loses: $1 - 1 = 0$, $1 - 2 = -1$, $1 - 3 = -2$

(a) Is there a winning strategy in this game? If so, what is it?

(b) Play the game with other initial values than 21 and allow other numbers for subtracting. Does the same strategy hold? Will the person who knows this strategy and plays second always win?

H. THE TALKING CALCULATOR

15. Do the calculations below and use your calculator to find the name of the sun god in Greek mythology.

$$([(65 + 79) \times 12 - (345 + 383)] \times 20 + 611) \times 24 + 7070$$

Mathematics and Nutrition

Use the table on page 144 to do the following exercises.

1. (a) How many calories and (b) how many grams of carbohydrate are there in a breakfast of $\frac{1}{2}$ grapefruit, 2 eggs, 2 slices of bacon, 2 slices of whole wheat toast, and a cup of black coffee?

2. (a) How many calories and (b) how many grams of carbohydrate are there in a lunch of a 3-oz. hamburger, 2 slices of whole wheat bread, 20 french fries, and a 12-oz. bottle of cola?

3. (a) How many calories and (b) how many grams of carbohydrate are there in a dinner of a baked potato, two pork chops, one cup of green peas, black coffee, and a piece of apple pie?

4. How many calories and how many grams of carbohydrate were obtained in the three meals of exercises 1, 2, and 3?

5. Design a menu for three meals that take care of the calorie needs for (a) a man 6′ 0″ tall with a medium frame; (b) a woman 5′ 4″ tall with a small frame.

6. Design a menu for someone who must limit her caloric intake to 1000 calories a day.

7. Design a menu for someone who must limit carbohydrate intake to (a) 100 grams a day; (b) 50 grams a day.

NOTE: To lose or gain a pound of body fat there needs to be about 3500 calories less or more in the diet than the body uses.

8. A 5′7″ woman with a medium frame weighs 150 pounds. If she eats 500 calories less than what is normal for a woman with her height and frame, runs 1 hour a day, walks (at 2.8 mph) for 2 hours a day, sits for 13 hours a day, and sleeps soundly for 8 hours a day, how many days will it take her to reach the mid-point of her normal weight range?

9. A man who is 5′11″, has a small frame, and weighs 130 pounds wishes to gain weight. What will he have to increase his daily caloric intake so that at the end of six months he will have reached his minimum normal weight?

10. Using the table of caloric energy requirements on page 144, estimate what your calorie needs would be for one day and then compare it with those listed in the chart.

11. According to the tables listed, how many calories per day should you add or reduce your caloric intake to maintain your normal weight?

CALORIES AND GRAMS OF CARBOHYDRATE IN SELECTED FOODS

Item	Calories	Gm Carb.
APPLE 1 medium	70	18
BACON 2 slices	90	TRACE
BANANA 1 medium	100	23
BEANS, KIDNEY 1 cup	230	37
BEEF, HAMBURGER 3 ounces	245	0
BEEF, STEW 1 cup	210	15
BREAD 1 slice whole wheat	65	10
COFFEE black	0	TRACE
COLA 12 ounces	145	39
CHICKEN, FRIED 6 ounces	155	27
EGG 1 medium	80	TRACE
GRAPEFRUIT ½	45	11
MACARONI AND CHEESE 1 cup	430	60
MILK, WHOLE 1 cup	160	11
MILK, SKIM 1 cup	60	13
ORANGE 1 medium	65	17
PIE, APPLE 4-inch wedge	350	42
PORK CHOP 3.5 ounces	260	0
PEAS, GREEN 1 cup	115	19
PIZZA, CHEESE 5-inch wedge	185	25
POTATO, BAKED 1 medium	90	28
POTATOES, FRIED 10 pieces	155	18

CALORIC ENERGY REQUIREMENT CHART

Activity Done	Calories Needed Per Lb., Per Hr.
Sleeping (soundly)	0.4
Sleeping (fitfully)	0.5
Sitting (no movement)	0.6
Standing (relaxed)	0.7
Sewing (by hand)	0.75
Dressing (or undressing)	0.8
Singing	0.85
Typewriting (average speed)	0.9
Washing Dishes (ironing or dusting)	0.95
Sweeping Floors (broom or vacuum)	1.0
Exercising (lightly)	1.25
Walking (approximately 2.8 mph)	1.5
Trade Work (carpentry or plumbing)	1.75
Exercising (active)	1.9
Walking Fast (4 mph)	2.0
Going Down Steps	2.25
Loading Heavy Objects (or unloading)	2.5
Exercising (heavy)	2.75
Active Sports (tennis or swimming)	3.25
Running (approximately 5.5 mph)	3.75
Exercising (very heavy)	4.0
Going Up Steps	7.0

WEIGHTS FOR WOMEN

Height (with shoes on) 2-inch heels	Small Frame	Medium Frame	Large Frame
4' 10"	92-98	96-107	104-119
4' 11"	94-101	98-110	106-122
5' 0"	96-104	101-113	109-125
5' 1"	99-107	104-116	112-128
5' 2"	102-110	107-119	115-131
5' 3"	105-113	110-122	118-134
5' 4"	108-116	113-126	121-138
5' 5"	111-119	116-130	125-142
5' 6"	114-123	120-135	129-146
5' 7"	118-127	124-139	133-150
5' 8"	122-131	128-143	137-154
5' 9"	126-135	132-147	141-158
5' 10"	130-140	136-151	145-163
5' 11"	134-144	140-155	149-168
6' 0"	138-148	144-159	153-173

WEIGHTS FOR MEN

Height (with shoes on) 1-inch heels	Small Frame	Medium Frame	Large Frame
5' 2"	112-120	118-129	126-141
5' 3"	115-123	121-133	129-144
5' 4"	118-126	124-136	132-148
5' 5"	121-129	127-139	135-152
5' 6"	124-133	130-143	138-156
5' 7"	128-137	134-147	142-161
5' 8"	132-141	138-152	147-166
5' 9"	136-145	142-156	151-170
5' 10"	140-150	146-160	155-174
5' 11"	144-154	150-165	159-179
6' 0"	148-158	154-170	164-184
6' 1"	152-162	158-175	168-189
6' 2"	156-167	162-180	173-194
6' 3"	160-171	167-185	178-199
6' 4"	164-175	172-190	182-204

DAILY CALORIE NEEDS FOR WOMEN

Height	Small Frame	Medium Frame	Large Frame
4' 11"	1635	1725	1845
5' 0"	1665	1770	1890
5' 1"	1695	1800	1935
5' 2"	1740	1845	1995
5' 3"	1785	1875	2040
5' 4"	1845	1950	2100
5' 5"	1890	1995	2145
5' 6"	1950	2040	2220
5' 7"	2010	2130	2280
5' 8"	2055	2175	2340
5' 9"	2115	2235	2400
5' 10"	2175	2295	2460
5' 11"	2220	2340	2505
6' 0"	2265	2415	2580

DAILY CALORIE NEEDS FOR MEN

Height	Small Frame	Medium Frame	Large Frame
5' 2"	1845	1965	2085
5' 3"	1890	2010	2130
5' 4"	1950	2070	2205
5' 5"	2010	2130	2265
5' 6"	2045	2175	2325
5' 7"	2115	2235	2400
5' 8"	2175	2310	2460
5' 9"	2235	2370	2520
5' 10"	2295	2430	2595
5' 11"	2355	2490	2670
6' 0"	2430	2565	2745
6' 1"	2505	2640	2820
6' 2"	2595	2730	2910
6' 3"	2670	2805	3000

Source: *Brand Name Carbohydrate Diet Guide*, 10th edition. Tenth printing 1976. ©1974. Published by Success Publications, Inc.

3

Factors, Multiples, and Divisibility

3.1 About Factors, Multiples, and Divisors

In this chapter, the word *number* will always refer to one of the numbers 1, 2, 3, 4, 5, ... Such numbers are called *counting* or *natural* numbers.

When a number n is written as a product of numbers, we call the numbers in the product *factors* of n. Thus since

$$6 = 1 \times 6 \qquad \text{and also} \qquad 6 = 2 \times 3$$

we see that 1, 2, 3, and 6 are all factors of 6. Are there any other factors of 6?

If a is a factor of n, then, by definition,

$$n = a \times b$$

for some number b and hence

$$n \div a = b$$

For example, because

$$6 = 1 \times 6, \qquad \text{we have} \qquad 6 \div 1 = 6 \qquad \text{and} \qquad 6 \div 6 = 1$$

Similarly, because

$$6 = 2 \times 3, \qquad \text{we have} \qquad 6 \div 2 = 3 \qquad \text{and} \qquad 6 \div 3 = 2$$

Thus we can test to see whether a number m is a factor of a number n by seeing whether $n \div m$ is a natural number. That is, a factor of n is also a *divisor* of n; 1, 2, 3, and 6 are divisors of 6 as well as factors of 6. For example, to see whether 23 is a divisor (factor) of 851 we calculate that $851 \div 23 = 37$. Hence 23 is a divisor (factor) of 851 and $851 = 23 \times 37$. On the other hand, $1484 \div 23 = 64.521739$ so that 23 is not a divisor (factor) of 1484.

Do exercises 1 through 5 in the adjoining column.

OBJECTIVES

You should:

(a) Know the meaning of factor, multiple, and divisor.

(b) Be able to test whether a number is a factor (divisor) of another number.

(c) Be able to use simple sight tests for divisibility by 2, 5, and 10.

****

In exercises 1–5 determine whether the second number is a factor of the first by dividing the first number by the second.

1. 6593; 37

2. 6048; 288

3. 32,832; 609

4. 892,155; 105

5. 60,264; 81

STOP

In exercises 6–10 write the first number as a multiple of the second.

6. 6104; 56

7. 72,186; 227

8. 5842; 5842

9. 10,756; 1

10. 51,320,034; 981

STOP

In exercises 11–14 use your HHC to decide whether 2, 5, or 10 are divisors of the given number.

11. 1320

12. 15,191

13. 20,192

14. 15,115

STOP

In exercises 15–18 determine by inspection whether the given number is divisible by 2, 5, or 10. Which are even numbers and which are odd?

15. 15,292

16. 182,195

17. 342,191

18. 3,147,280

STOP

We have seen that 1, 2, 3, and 6 are all factors (divisors) of 6; we also say that 6 is a *multiple* of 1, 2, 3, and 6 since

$$6 = 1 \times 6 \quad \text{and} \quad 6 = 2 \times 3$$

In general, a number n is said to be a multiple of a number m if there exists a number b such that

$$n = m \times b$$

Do exercises 6 through 10.

As we have seen, we can always find whether a given number is a divisor (factor) of another number by performing a division on an HHC. However, it is easy to see by inspection whether a number is divisible by 2, 5, or 10.

Do exercises 11 through 14.

The results of exercises 11–14 indicate that a number is divisible

(1) by 2 if and only if its last digit is 0, 2, 4, 6, or 8
(2) by 5 if and only if its last digit is 0 or 5
(3) by 10 if and only if its last digit is 0.

A number that is divisible by 2 is called an *even* number; a number that is not divisible by 2 is called an *odd* number. Thus 228 is an even number and 229 is an odd number. An even number can always be written as $2 \times n$ where n is a whole number. Thus 228 is an even number because

$$228 = 2 \times 114$$

and 0 is also an even number because

$$0 = 2 \times 0$$

Do exercises 15 through 18.

There are also easy tests for deciding whether a given number is divisible by 3, 4, 6, or 9. They can be summarized as follows:

A number is divisible by	If and only if this condition is met:
3	The sum of the digits is a multiple of 3.

Example: 642 is divisible by 3 because 6 + 4 + 2 = 12 which is a multiple of 3 (3 × 4 = 12 and 642 = 3 × 214).

4	The last two digits on the right name a number that is a multiple of 4.

Example: 724 is divisible by 4 because 24 is a multiple of 4 (4 × 6 = 24 and 724 = 4 × 181).

A number is divisible by	
6	The conditions for divisibility by both 2 and 3 are met.

Example: 642 is divisible by 2 (ends in the even number 2) and is divisible by 3 (6 + 4 + 2, the sum of the digits, is equal to 12 and hence is a multiple of 3). Hence 642 is divisible by 6 ($642 = 6 \times 107$).

9	The sum of the digits is a multiple of 9.

Example: 774 is divisible by 9 because $7 + 7 + 4 = 18$ and 18 is a multiple of 9 ($18 = 2 \times 9$ and $774 = 86 \times 9$).

Do exercises 19 through 21.

In exercises 19–21 apply the tests for divisibility by 3, 4, 6, and 9 to determine whether the given number is divisible by 3, 4, 6, or 9.

19. 5232

20. 1263

21. 2916

STOP

Exercises 3.1

Name ______________________

Class ______________________

In exercises 1-10, without actually doing the division, tell whether any of the numbers 2, 5, or 10 are factors (divisors) of the given number.

1. 113
2. 3186
3. 7020
4. 78,422
5. 12,960
6. 20,115
7. 5625
8. 7011
9. 9610
10. 157,465

In exercises 11-20, without actually doing the division, refer to the rules given in the text and tell which of the numbers 3, 4, 6, or 9 are factors (divisors) of the given number.

11. 234
12. 4224
13. 6252
14. 93,600
15. 3564
16. 9876
17. 1002
18. 125,092
19. 213
20. 3969

1.
2.
3.
4.
5.
6.
7.
8.
9.
10.

11.
12.
13.
14.
15.
16.
17.
18.
19.
20.

21.

22.

23.

24.

25.

26.

27.

28.

29.

30.

31. (a)

(b)

(c)

32.

33.

In exercises 21–30 list all the factors (divisors) of the given number.

Example: 6 is divisible only by 1, 2, 3, and 6; therefore 1, 2, 3, and 6 are all of the factors of 6.

42 is divisible only by 1, 2, 3, 6, 7, 14, 21, and 42; therefore, 1, 2, 3, 6, 7, 14, 21, and 42 are all of the factors of 42.

21. 8 **22.** 12

23. 15 **24.** 24

25. 32 **26.** 48

27. 51 **28.** 60

29. 67 **30.** 196

31. (a) Find all the factors of 4, of 9, of 14, of 16, of 18, and of 21.

(b) Find all the factors of 2, of 3, of 5, of 7, of 11, and of 13.

(c) Compare the number of factors of the numbers in (a) with the number of factors of the numbers in (b).

32. Write an eight-digit numeral for a number which is divisible by 3 but not by 9.

33. Write a six-digit numeral for a number which is divisible by both 5 and 9.

34. What is the smallest number divisible by 7, 11, and 13?

35. Write a six-digit numeral in which the same digit is repeated six times. Is the number divisible by 7? By 11? By 13?

36. What numbers less than 16 will divide 973,822?

37. What numbers less than 16 will divide 373,835?

38. The following numbers are each divisible by three numbers less than 16. What are they?

(a) 8365 (b) 543,907

39. Show, by giving at least two examples, that if two numbers have the same digits but in opposite order, then their difference is always divisible by 9.

40. Show, by giving at least two examples, that if one number has an even number of digits and a second number has the same digits but in the opposite order, then the sum of the two numbers is always divisible by 11.

41. Show, by giving at least two examples, that the number $n \times (n^2 + 1)$ is always even for any whole number n. (One of your examples should be for an odd number n.)

42. Show, by giving at least two examples, that the product of any three consecutive whole numbers is always divisible by 3.

43. Show, by giving at least two examples, that the product of any four consecutive whole numbers is always divisible by 4.

34.

35.

36.

37.

38. (a)

(b)

39.

40.

41.

42.

43.

44.

44. Considering your examples of exercises 42 and 43, do you think that the product of any n consecutive whole numbers is divisible by n? Try some examples.

45.

45. The number 1234 is not divisible by 11. How many numbers divisible by 11 can be formed by changing the order of its digits? What are they? (There are eight of them.)

Example: 1243 is divisible by 11.

46.

46. If a and b are two natural numbers and each of a and b is a multiple of 3, is $a + b$ divisible by 3? Give three examples to support your answer.

47.

47. If a and b are two natural numbers and only one of them is a multiple of 3, is $a + b$ divisible by 3? Give three examples to support your answer.

48.

48. If a and b are two natural numbers, what must be true if their product, $a \times b$, is to be divisible by 3?

49.

49. A girl was carrying a basket of eggs. A boy riding a bicycle hit the basket and broke all the eggs. He asked the girl how many eggs she had. The girl replied, "When I counted by two's, there was one left over; when I counted by three's, there was one left over; when I counted by four's, there was one left over; but when I counted by five's, no eggs were left over." What is the smallest number of eggs she could have had in her basket?

*50.

*50. A shepherd was asked how many sheep he had. He replied, "I do not know exactly but when I count by 2, 3, 4, 5, or 6, there is always one left over, but when I count by seven there is none left over." What is the smallest number of sheep the shepherd could have?

*51. Do some research and see if you can find a rule for divisibility by 7.

*51.

*52. Do some research and see if you can find a rule for divisibility by 11.

*52.

3.2 Prime and Composite Numbers

Do exercise 1 in the adjoining column.

Of the numbers listed in exercise 1, only 2, 3, 5, and 7 have exactly *two* divisors; 1 has just one divisor; 4 and 9 have three; 6, 8, and 10 have four.

Numbers that have exactly two divisors are called *prime* numbers. Thus 2, 3, 5, and 7 are the first four prime numbers.

Any number, other than 1, that is not a prime number is called a *composite* number; 1 is neither a prime number nor a composite number. Thus 4, 6, 8, 9, and 10 are the first five composite numbers.

Do exercises 2 and 3.

A long time ago, about 200 B.C., a Greek scholar and chief librarian in the city of Alexandria thought of a process which separates all natural numbers greater than 1 into two sets of numbers, prime and composite. His name was Eratosthenes and the process is known as the *sieve of Eratosthenes.*

We shall make a sieve for the numbers 2 through 50. To do this we need a chart listing the numbers 2 through 50. The numbers may be listed in rows and columns. For convenience we will list the natural numbers from 2 through 50 in rows as follows.

	2	3	4	5	6	7	8	9	10
11	12	13	14	15	16	17	18	19	20
21	22	23	24	25	26	27	28	29	30
31	32	33	34	35	36	37	38	39	40
41	42	43	44	45	46	47	48	49	50

To use the sieve of Eratosthenes follow the instructions below and mark the array of numbers from 2 through 50 above.

1. Make a circle around 2.
2. After 2, cross out every second number. You will then cross out 2, 4, 6, 8, and every other number that is a multiple of 2; that is, every even number on the list.
3. Make a circle around 3.
4. After 3, cross out every third number. You will then cross out 6, 9, 12, and all other multiples of 3. (You have already crossed out those which are multiples of 2.)
5. Is 4 crossed out? Is every fourth number after 4 or every multiple of 4 crossed out? Why?
6. Make a circle around 5.
7. After 5, cross out every fifth number. You will then cross out 10, 15, 20, and every number on the list which is a multiple of 5.
8. Is 6 crossed out? Is every multiple of 6 crossed out? Why?
9. Make a circle around 7. Cross out every seventh number after 7.
10. Look at the remaining numbers that have been neither circled nor crossed out and note that none of them can be multiples of any numbers in the array. Circle these remaining numbers.

OBJECTIVE

You should know how to determine whether a natural number greater than 1 is a prime number or a composite number.

****

1. List all of the divisors for each of the following numbers.

1:	6:
2:	7:
3:	8:
4:	9:
5:	10:

STOP

2. List the next five prime numbers after 7.

3. List the next five composite numbers after 10.

STOP

Your array should now appear as follows.

	2	3	~~4~~	5	~~6~~	7	~~8~~	~~9~~	~~10~~
11	~~12~~	13	~~14~~	~~15~~	~~16~~	17	~~18~~	19	~~20~~
~~21~~	~~22~~	23	~~24~~	~~25~~	~~26~~	~~27~~	~~28~~	29	~~30~~
31	~~32~~	~~33~~	~~34~~	~~35~~	~~36~~	37	~~38~~	~~39~~	~~40~~
41	~~42~~	43	~~44~~	~~45~~	~~46~~	47	~~48~~	~~49~~	~~50~~

What we have done is to "sift out" all composite numbers in the array and we have left only prime numbers, those numbers that are circled.

Do exercises 4 through 6.

4. List the prime numbers shown by the circled numbers in the array and verify that the first nine of them are the prime numbers that you identified in exercises 1 and 2.

5. Do you think that it is possible to find all the prime numbers in the entire set of natural numbers by the sieve of Eratosthenes?

6. Is it possible to find all the prime numbers less than any given natural number by the method of Eratosthenes?

STOP

A list of all the prime numbers less than 1000 is given below:

2	97	227	367	509	661	829
3	101	229	373	521	673	839
5	103	233	379	523	677	853
7	107	239	383	541	683	857
11	109	241	389	547	691	859
13	113	251	397	557	701	863
17	127	257	401	563	709	877
19	131	263	409	569	719	881
23	137	269	419	571	727	883
29	139	271	421	577	733	887
31	149	277	431	587	739	907
37	151	281	433	593	743	911
41	157	283	439	599	751	919
43	163	293	443	601	757	929
47	167	307	449	607	761	937
53	173	311	457	613	769	941
59	179	313	461	617	773	947
61	181	317	463	619	787	953
67	191	331	467	631	797	967
71	193	337	479	641	809	971
73	197	347	487	643	811	977
79	199	349	491	647	821	983
83	211	353	499	653	823	991
89	223	359	503	659	827	997

Exercises 3.2

Name ____________

Class ____________

1. On a separate piece of paper make a chart which shows the numbers from 2 through 100 in ten rows. By the method of Eratosthenes, find all the prime numbers less than 100.

2. How many prime numbers less than 50 are there? Are there twice as many prime numbers less than 100 as there are less than 50?

3. Are there more prime numbers or more composite numbers between 1 and 100?

4. What natural number is neither prime nor composite?

5. Defend the statement "All odd numbers are prime," or show that it is false.

6. What number is both even and prime?

7. What can you say about the statement "Except for the prime number 2, all prime numbers are odd numbers?" Is it true or false?

8. What are the prime numbers between 190 and 200?

9. What are the prime numbers between 290 and 300?

In exercises 10–30 indicate which numbers are prime (P) and which are composite (C). (You may find the divisibility tests of Section 3.1 useful.)

10. 19

11. 33

12. 47

13. 51

14. 57

15. 65

16. 89

17. 97

1.
2.
3.
4.
5.
6.
7.
8.
9.
10.
11.
12.
13.
14.
15.
16.
17.

18.
19.
20.
21.
22.
23.
24.
25.
26.
27.
28.
29.
30.
31. (a)
(b)
32.
33. (a)
(b)
34.
35.

18. 101

19. 127

20. 129

21. 173

22. 391

23. 781

24. 591

25. 9777

26. 111,111

27. 3,767,476

28. 131,313

29. 824,615

30. 923,913

31. Any prime number greater than 6 will have one of the following two properties:

(a) If 1 is added to the prime number, the sum is divisible by 6.

(b) If 1 is subtracted from the prime number, the difference is divisible by 6.

Find three prime numbers that have property (a) and three that have property (b).

32. An interesting fact about prime numbers is that if n is any natural number greater than 1, then there is at least one prime number between n and $2n$. Show that this is true for five different natural numbers.

33. There are at least four primes between squares of consecutive primes greater than 3.

(a) Find the primes between 5^2 and 7^2.

(b) Choose another pair of consecutive primes and find four prime numbers between their squares.

34. Give five examples that verify the statement "Every prime other than 2 and 3, when divided by 6 must leave a remainder of 1 or 5."

35. If a number is a prime greater than 5, what are the possible ones digits of its numeral?

36. Some primes can be written in the form $1 + n^2$ for some natural number n.

Examples: $5 = 1 + 2^2$, $17 = 1 + 4^2$.

Find five more primes of the form $1 + n^2$.

36.

37. Some primes can be written in the form $n^2 - 1$ for some natural number n.

Example: $3 = 2^2 - 1$

(a) Can n be an odd number?

(b) Are there any other primes less than 1000 that can be written this way? If so, what are they?

37. (a)

(b)

38. Find a prime that is of the form

(a) $4 \times k + 1$ (b) $4 \times k + 3$

(k is a natural number.)

38. (a)

(b)

39. Find five primes of the form $n^2 - 2$.

39.

40. Find three replacements for n such that $n^2 - 2$ is twice a prime.

40.

41. Find three primes p such that $q = 2 \times p + 1$ is also a prime.

41.

42.

42. It has been conjectured that every even natural number can be expressed as the difference of two consecutive primes in an infinite number of ways. Example: $6 = 29 - 23, 37 - 31, 53 - 47$, etc. Write 10 as the difference of two consecutive primes in three ways.

43.

43. Prime numbers, such as 5 and 7, whose difference is 2 are called *twin primes.* List all the twin primes between 1 and 100.

44.

44. Mathematicians believe that there is an infinite number of twin primes, but they become quite scarce as the numbers become larger. List the twin primes between 100 and 200; between 200 and 300. Compare the number of these with the number of twin primes less than 100.

45.

45. Over 200 years ago a mathematician by the name of Goldbach made two conjectures (guesses).

Conjecture 1: Every even number greater than 2 is the sum of two prime numbers.

Conjecture 2: Every odd number greater than 7 is the sum of three prime numbers.

Many mathematicians and other persons interested in numbers have tried to prove these conjectures. To date, no one has been able to do so. Show that Goldbach's conjecture 1 holds for five even numbers greater than 2 and that his conjecture 2 holds for five odd numbers greater than 7.

46.

46. A *Mersenne* prime (named after the sixteenth-century mathematician Marin Mersenne) is a prime of the form $2^p - 1$ where p is itself a prime. Show that $2^p - 1$ is a prime for $p = 2, 3, 5, 7$, and 13 but is not a prime if $p = 11$.

*47.

*47. What primes can be written as the sum of squares of two natural numbers? (Make a table of the prime numbers 1 through 47 that can be so written.)

Examples: $5 = 1^2 + 2^2$

$13 = 2^2 + 3^2$

*48. The product of two primes is 346. What are the prime numbers?

*48.

*49. Find five consecutive whole numbers, none of which is prime.

*49.

*50. In the 4 X 4 square shown below write each of the numbers 1 through 16 in the small squares so that any two squares with a common side have a sum that is a prime number. (There are five different possible ways to do this.)

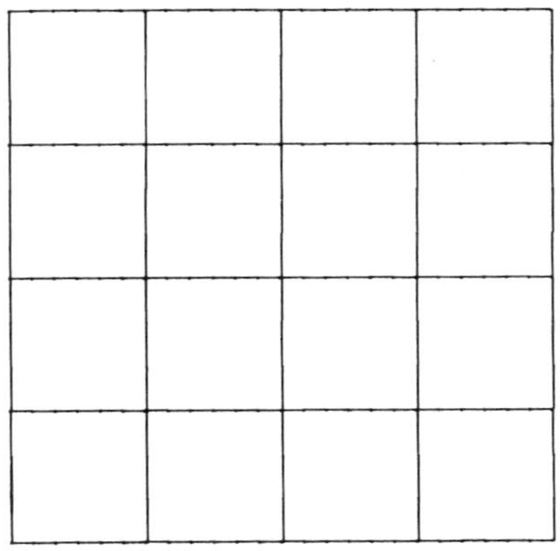

*50.

3.3 Composite Numbers and Prime Factorization

Consider the composite number 30. It has the prime number 2 as a divisor and we can write

$$30 = 2 \times 15$$

Now 15 is also a composite number and has the prime number 3 as a divisor so that

$$30 = 2 \times 3 \times 5$$

where 5 is a prime number. Similarly,

$$55 = 5 \times 11, \qquad 60 = 2 \times 2 \times 3 \times 5, \qquad \text{and} \qquad 125 = 5 \times 5 \times 5$$

That is, any composite number can be written as a product of prime numbers.

If any prime number appears more than once in such a product, we can use exponential notation and write, for example,

$$60 = 2^2 \times 3 \times 5, \qquad 125 = 5^3, \qquad \text{and} \qquad 72 = 2^3 \times 3^2$$

When we write a composite number n as the product of prime numbers we call this the *prime factorization* of n. Thus

$2^2 \times 3 \times 5$ is the prime factorization of 60

5^3 is the prime factorization of 125

and

$2^3 \times 3^2$ is the prime factorization of 72

Do exercises 1 through 7 in the adjoining column.

A direct procedure for finding the prime factorization of a number is to test the prime numbers in order (2, 3, 5, 7, . . .) to see whether they are divisors. Thus, for example, to find the prime factorization of 1287 we note first that 2 is not a divisor. (Why?) So we try 3 and find that

$$1287 = 3 \times 429$$

Then we try 3 again on the number 429 to get

$$1287 = 3 \times 3 \times 143$$

Now we find that 143 is not divisible by 3. It is certainly not divisible by 5 (why?) and either inspection or the use of the HHC shows that it is not divisible by 7. Trying 11 we obtain

$$1287 = 3 \times 3 \times 11 \times 13 = 3^2 \times 11 \times 13$$

OBJECTIVE

You should know how to find the prime factorizations of composite numbers.

****

In exercises 1–7 determine whether the given prime factorization is correct.

1. $6{,}740{,}636 = 2^2 \times 7^3 \times 17^3$

2. $240 = 2^2 \times 3^2 \times 5^2$

3. $1890 = 3^3 \times 5^2 \times 7^3$

4. $9{,}634{,}625 = 5^3 \times 7^2 \times 11^2 \times 13$

5. $6{,}154{,}544 = 2^4 \times 11^3 \times 17^2$

6. $29{,}712{,}375 = 2 \times 3^2 \times 5^3 \times 7^4 \times 11$

7. $70{,}981{,}625 = 5^3 \times 11^2 \times 13 \times 19^2$

STOP

In exercises 8–13 determine whether the given number is a prime. If it is not, find its prime factorization.

8. 389

9. 1302

10. 1319

11. 1558

12. 2405

13. 8303

STOP

Since 13 is also a prime we have found the prime factorization of 1287:

$$1287 = 3^2 \times 11 \times 13$$

This factorization of 1287 can also be found and shown by means of the following *factor tree:*

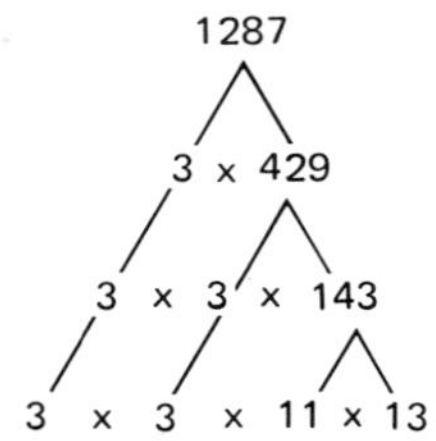

In this example, we know that 3, 11, and 13 are prime numbers, but how can we determine that a larger number such as 157 is a prime number? Again, we can proceed by trial. Clearly 2 and 5 are not divisors of 157 (why?) and we can easily check (with or without an HHC) that 3, 7, 11, and 13 are not divisors of 157. There are, however, many other primes less than 157: 17, 19, 23, etc. How many of these do we need to consider before concluding that 157 is a prime? To answer this question note that

$$13^2 = 169 > 157$$

Hence if we were to try 17 and were to find that it is a divisor, we would have to have

$$157 = 17 \times b$$

where b is a natural number *less* than 17. But we have already tried all of the primes less than 17 and found that they are not divisors. Conclusion:

To test whether a number n is a prime, we need only test all primes p as divisors such that $p^2 \leqslant n$. ($\leqslant$ is read as "less than or equal to.")

Do exercises 8 through 13.

When we are working with the prime factorization of numbers we should recognize a very important property of composite numbers—usually referred to as the *fundamental theorem of arithmetic:*

Every composite number has a unique prime factorization.

This theorem tells us that no matter how we proceed to factor a composite number, we will always get exactly the same prime factors. (Although, of course, the factors may appear in different orders; for example, $30 = 2 \times 3 \times 5 = 2 \times 5 \times 3 = 3 \times 5 \times 2 = 5 \times 2 \times 3 = 5 \times 3 \times 2$.)

We can use factor trees to show three different ways of factoring 24. In each instance, 24 is shown as the product of three 2's and a 3. The order is different but the same four prime factors are shown in each product: $24 = 2^3 \times 3$.

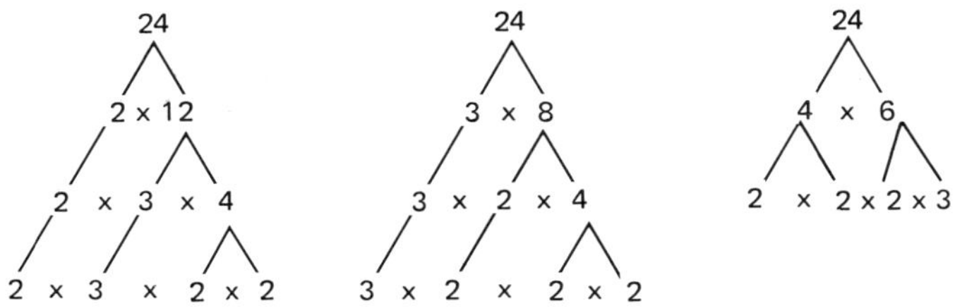

Do exercise 14.

14. Use a factor tree to find the prime factorization of 36.

STOP

Exercises 3.3

Name ______________________

Class ______________________

In exercises 1–10 the indicated product is the prime factorization of a number. Find this number.

1. $2^2 \times 5^2 \times 7$

2. $2 \times 3^3 \times 41$

3. $2^2 \times 3^3 \times 5^2$

4. $2 \times 5 \times 7 \times 11$

5. $2^4 \times 3 \times 5$

6. $2^3 \times 3^2 \times 43$

7. $2^2 \times 3^2 \times 7^2 \times 13^2$

8. $2^5 \times 3^4 \times 5^3 \times 7^2$

9. $2^3 \times 3^5 \times 5^4 \times 37$

10. $3 \times 5^2 \times 7^3 \times 17$

In exercises 11–16 use a factor tree to find the prime factorization of the given number.

11. 12

12. 18

13. 60

14. 72

15. 84

16. 128

1.
2.
3.
4.
5.
6.
7.
8.
9.
10.
11.
12.
13.
14.
15.
16.

17.

18.

19.

20.

21.

22.

23.

24.

25.

26.

27.

28.

29.

30.

In exercises 17–30 find the prime factorization of the given number.

17. 156

18. 188

19. 288

20. 301

21. 312

22. 323

23. 345

24. 365

25. 1688

26. 1728

27. 3455

28. 8395

29. 19,635

30. 48,204

In exercises 31–40 find the largest prime number that divides the given number.

31. 105

32. 252

33. 450

34. 589

35. 707

36. 998

37. 1182

38. 1828

39. 4325

40. 8080

31.
32.
33.
34.
35.
36.
37.
38.
39.
40.

In exercises 41–46 find all of the divisors of the given number.

41. 54

42. 75

43. 120

44. 192

45. 300

46. 162

41.
42.
43.
44.
45.
46.

47.

47. What are the divisors of a prime number p?

*48.

*48. How can we tell from the prime factorization of a number whether it is a perfect square? A perfect cube?

*49.

*49. A *Fermat* prime (named after a sixteenth-century mathematician, Pierre de Fermat) is a prime of the form $N = 2^n + 1$ where $n = 2^t$. Show that N is a prime number for $t = 1, 2, 3$, and 4. (Fermat thought that N was a prime for all values of t but, in 1739, Leonard Euler showed that it was composite for $t = 5$. Unfortunately, for $t = 5$ the value of N is too large to enter into an HHC.)

*50.

*50. Do you think that the formula $N^2 - N + 41$ will always give you a prime number when N is replaced by a natural number? (Example: When $N = 2$, $N^2 - N + 41$ becomes $2^2 - 2 + 41 = 43$ and 43 is a prime number.) Try $N = 1, 3, 4$, and 5. Then try to find a natural number N that will *not* give you a prime number. What is the smallest such number?

3.4 Greatest Common Divisor (GCD) and Least Common Multiple (LCM)

GREATEST COMMON DIVISOR

Consider the two numbers 30 and 45; 30 has the divisors

1, 2, 3, 5, 6, 10, 15, and 30

and 45 has the divisors

1, 3, 5, 9, 15, and 45

The *common* divisors of 30 and 45 (i.e., the numbers that are divisors of both 30 and 45) are

1, 3, 5, and 15

The largest of the common divisors is 15 and we say that 15 is the *greatest common divisor* (*GCD*) of 30 and 45. In general, we say that

If $a, b, c, d, \ldots$ are any numbers, their GCD is the largest number that divides each of $a, b, c, d, \ldots$.

One way of finding GCD's is to use the prime factorization of the given numbers. For example, suppose that we want to find the GCD of 12, 24, and 180. We first note that

$$12 = 2^2 \times 3, \qquad 24 = 2^3 \times 3, \qquad 180 = 2^2 \times 3^2 \times 5$$

Then, using the definition that $a^1 = a$ for all numbers a, we can rewrite these factorizations as

$$12 = 2^2 \times 3^1, \qquad 24 = 2^3 \times 3^1, \qquad 180 = 2^2 \times 3^2 \times 5^1$$

Now we know that the GCD of 12, 24, and 180 must divide each of 12, 24, and 180. Hence certainly no primes other than 2, 3, and 5 can be involved. However, 5 cannot be involved because 5 is not a factor of 12 or 24. What about 2? It is a factor of each of 12, 24, and 180 and, indeed, 2^2 is a factor of all three numbers but 2^3 is not. For 3, we see that 3 is a factor of all three numbers but 3^2 is not. Hence to find the GCD of 12, 24, and 180, we take the product of the prime numbers that occur in *all* of the factorizations, each to the *lowest* power that it occurs. Thus

$$\text{GCD of 12, 24, and 180 is } 2^2 \times 3^1 = 12$$

Do exercises 1 through 5 in the adjoining column.

If two numbers have no prime factors in common as is the case for

$$15 = 3 \times 5 \qquad \text{and} \qquad 26 = 2 \times 13$$

their GCD is 1, and we say that the two numbers are *relatively prime*.

OBJECTIVES

You should:

(a) Know the meaning of greatest common divisor, least common multiple, and relatively prime.

(b) Be able to calculate greatest common divisors and least common multiples.

****

1. Since
$55{,}125 = 3^2 \times 5^3 \times 7^2$
and
$282{,}975 = 3^1 \times 5^2 \times 7^3 \times 11^1$
we can find the GCD of 55,125 and 282,975 by first observing that the prime numbers which are factors of both 55,125 and 282,975 are
____, ____, and ____.
Then we observe that the lowest power of 3 that occurs is ____; of 5 is ____; of 7 is ____.
Hence the GCD of 55,125 and 282,975 is
$3^{\_\_} \times 5^{\_\_} \times 7^{\_\_}$
= __________

In exercises 2-5 use prime factorizations to find the GCD of the given set of numbers.

2. 12, 18

3. 16, 20

4. 15, 18, 24

5. 24, 36, 60, 96

STOP

In exercises 6–12 determine if the given pair of numbers are relatively prime.

6. 8, 15

7. 9, 24

8. 20, 28

9. 20, 27

10. 16, 51

11. 14, 36

12. 72, 35

In exercises 13–17 find the GCD of the given set of numbers.

13. 3, 6, 12

14. 5, 10, 8

15. 3, 5, 7

16. 7, 14, 28

17. 17, 34, 45

STOP

If any of the numbers we are considering is a prime, no prime factorization is possible, of course. Then, for example, to find the GCD of 7, 42, and 48, we write

$$7 = 7^1, \qquad 42 = 2^1 \times 3^1 \times 7^1, \qquad 48 = 2^4 \times 3^1$$

and find that the GCD of 7, 42, and 48 is 1.

Do exercises 6 through 17.

LEAST COMMON MULTIPLE

Now let us consider the multiples of 6:

$$6, 12, 18, 24, 30, 36, 42, 48, 54, \ldots$$

and the multiples of 9:

$$9, 18, 27, 36, 45, 54, 63, 72, \ldots$$

The *common* multiples of 6 and 9 (i.e., the numbers that are multiples of both 6 and 9) are

$$18, 36, 54, \ldots$$

The smallest of these is 18 and we say that 18 is the *least common multiple* (LCM) of 6 and 9. In general, we say that

If *a, b, c, d*, . . . are any numbers, their LCM is the smallest number that is a multiple of *a, b, c, d*, ...

Like GCD's, LCM's can be found by using prime factorizations. For example, to find the LCM of 12, 24, and 180, we again write

$$12 = 2^2 \times 3^1, \qquad 24 = 2^3 \times 3^1, \qquad 180 = 2^2 \times 3^2 \times 5^1$$

Then, since the LCM of 12, 24, and 180 must be divisible by 12, 24, and 180, it must be divisible by

2^3 because 2^3 is a factor of 24

3^2 because 3^2 is a factor of 180

and by

5^1 because 5^1 is a factor of 180

Thus the LCM of 12, 24, and 180 is the product of all the prime numbers occurring in any of the factorizations, each to the highest power that it occurs in any of the factorizations. Hence

$$\text{LCM of 12, 24, and 180 is } 2^3 \times 3^2 \times 5^1 = 360$$

As another example of finding an LCM, this one involving a prime number, consider the following problem: Find the LCM of 7, 12, 25, and 36.

***Solution*:**

$$7 = 7^1, \qquad 12 = 2^2 \times 3, \qquad 25 = 5^2, \qquad 36 = 2^2 \times 3^2$$

The LCM of 7, 12, 25, and 36 is

$$2^2 \times 3^2 \times 5^2 \times 7^1 = 6300$$

As you will see in the next chapter, GCD's are useful in reducing fractions to lowest terms and LCM's are useful in the addition of fractions.

Do exercises 18 through 24.

In exercises 18–22 find the LCM of the given set of numbers.

18. 12, 20, 18

19. 36, 24, 48

20. 27, 30, 42

21. 125, 45, 150

22. 13, 24, 76

23. What can you say about the LCM of a and 1 for any number a? About the GCD?

24. What can you say about the LCM of a and b if a is a multiple of b? About the GCD?

STOP

Exercises 3.4

Name ______________________

Class ______________________

In exercises 1–7 list the first ten multiples of the given number.

1. 6

2. 5

3. 7

4. 8

5. 12

6. 15

7. 20

In exercises 8–15 find the LCM of the given set of numbers by referring to the sets of multiples in exercises 1–7.

8. 6, 5

9. 6, 7

10. 5, 8

11. 6, 12

12. 12, 15

13. 8, 20

14. 6, 12, 20

15. 6, 5, 15

In exercises 16–24 determine the LCM for the set of numbers whose prime factorization is given. In your answer first list the numbers you get when you perform the multiplication and then give their LCM.

16. $2^3 \times 3^2$; $2^2 \times 3$

17. 2×3; $2^2 \times 3$

18. $2 \times 3^2 \times 5$; 3×5^2

19. 7×11; $3 \times 5 \times 7$

1.

2.

3.

4.

5.

6.

7.

8.

9.

10.

11.

12.

13.

14.

15.

16.

17.

18.

19.

20.
21.
22.
23.
24.
25.
26.
27.
28.
29.
30.
31.
32.
33.
34.
35.

20. $2^2 \times 5^2 \times 3;\ 2 \times 5^3 \times 3^2$

21. $3^2 \times 11;\ 3 \times 11^2$

22. $2^2 \times 5;\ 2^3;\ 2 \times 3 \times 5$

23. $2^3;\ 2^2 \times 3;\ 3 \times 5$

24. $5 \times 13^3;\ 5^2 \times 13;\ 2 \times 5 \times 13$

In exercises 25–35 use prime factorizations to find the LCM of the given set of numbers.

25. 56, 21

26. 42, 66

27. 22, 90

28. 15, 21

29. 60, 108

30. 16, 42

31. 252, 96

32. 20, 22, 12

33. 57, 209, 190

34. 94, 141, 188

35. 324, 306, 180, 144

In exercises 36–45 determine the GCD of the set of numbers whose prime factorization is given. In your answer first list the numbers you get when you perform the multiplications and then give their GCD.

36. $3 \times 5;\ 3 \times 7$

37. $2^2 \times 13;\ 2^2 \times 7$

38. $2 \times 3^2 \times 5;\ 3 \times 5^2$

39. $5^2 \times 19;\ 3 \times 5^2$

40. $7 \times 11;\ 3 \times 5$

41. $2^2 \times 5^2 \times 3;\ 2 \times 5^3 \times 3^2$

42. $2^2 \times 7 \times 23;\ 13^2 \times 17 \times 19;\ 3^2 \times 5^2 \times 31$

43. $2^3 \times 3^2 \times 5;\ 2^2 \times 3^2 \times 5;\ 2^2 \times 3 \times 5^2$

44. $2^2 \times 5;\ 2^3;\ 2 \times 3 \times 5$

45. $3 \times 5^2 \times 13;\ 2 \times 5^2 \times 13;\ 3^2 \times 5^2 \times 13^2$

In exercises 46–55 use prime factorizations to find the GCD of the given set of numbers.

46. 16, 15

47. 24, 30

48. 36, 35, 143

49. 65, 42, 66

50. 144, 72, 36

51. 28, 63, 42

52. 42, 96, 104, 18

53. 2244, 418

36.
37.
38.
39.
40.
41.
42.
43.
44.
45.
46.
47.
48.
49.
50.
51.
52.
53.

54.

55.

56. (a)

(b)

57.

58. (a)

(b)

59.

60.

61.

54. 220, 315

55. 735, 850

56. Ms. Able, Ms. Brown, and Ms. Chase are three traveling salespersons and it takes them 4, 15, and 18 working days to cover their respective territories. If they start from the sales office on the same day, **(a)** which two will return to the office together first and after how many working days? **(b)** after how many working days will all three be back in the office together?

57. Two interurban buses make round trips of their routes in 48 and 60 minutes respectively. If they start from the same place at 6:00 A.M., when will they meet again at the starting point?

58. A logger has three logs whose lengths are 63, 84, and 105 feet respectively. He wants to cut the longest pieces possible, all the same length, from each of the three logs. **(a)** What length will each piece be if we disregard waste in the cutting? **(b)** How many pieces will be cut?

59. A whistle from a certain buoy in a bay blows every 22 seconds and a light from this same buoy flashes every 12 seconds. If both events occur at 11:00 P.M., what is the next time that the light will flash and the whistle blow together?

60. Must two relatively prime numbers both be prime numbers? Explain your answer.

61. Is it possible for two composite numbers to be relatively prime? Explain your answer.

*62. What can you say about the LCM of a and b if a and b are both prime numbers and $a \neq b$? About their GCD? *62.

*63. Under what conditions will the LCM of two composite numbers a and b be equal to their product? *63.

*64. What is the LCM of a, b, and c if $a = c^2$ and $b = 7 \times c$? *64.

*65. What is the GCD of a, b, and c if $a = 5 \times c^2$ and $c = 3 \times b$? *65.

3.5 How Many Prime Numbers Are There?

Is there a largest prime number? Well, suppose that we knew just the first two prime numbers, 2 and 3. Then

$$2 \times 3 + 1 = 7$$

is another prime number greater than 3. Now suppose that we knew just the first three prime numbers, 2, 3, and 5. Then

$$2 \times 3 \times 5 + 1 = 31$$

is another prime number greater than 5.

Do exercises 1 and 2 in the adjoining column.

Does this process we have been using always give a prime number? Consider what happens if we use this process on the first six primes, 2, 3, 5, 7, 11, and 13. We get

$$2 \times 3 \times 5 \times 7 \times 11 \times 13 + 1 = 30{,}031$$

Now 30,031 is *not* a prime number. However,

$$30{,}031 = 59 \times 509$$

and 59 and 509 are both prime numbers greater than 13.

Do exercise 3.

From these examples, it appears that if we know the first n primes, we can always find a larger prime; that is, there is no largest prime and the list of primes is unending. Why is this so? Consider again

$$2 \times 3 \times 5 + 1 = 31$$

We have

$$31 = 2 \times (3 \times 5) + 1$$

so that

$$31 \div 2 = 3 \times 5 \text{ with a remainder of } 1$$

Similarly

$$31 = 3 \times (2 \times 5) + 1$$

so that

$$31 \div 3 = 2 \times 5 \text{ with a remainder of } 1$$

OBJECTIVES

You should:

(a) Know that there are an infinite number of primes.

(b) Be able to find, given a list of consecutive primes, a prime number greater than any in the list.

****

1. Suppose that we knew just the first four prime numbers, 2, 3, 5, and 7. Show that

$$2 \times 3 \times 5 \times 7 + 1$$

is a prime number greater than 7.

2. Suppose that we knew just the first five prime numbers, 2, 3, 5, 7, and 11. Show that

$$2 \times 3 \times 5 \times 7 \times 11 + 1$$

is a prime number greater than 11.

STOP

3. Show that

$$2 \times 3 \times 5 \times 7 \times 11 \times 13 \times 17 + 1$$

is not a prime number but that it has as a factor a prime number greater than 17 (but not much greater!).

STOP

4. Show that

$2 \times 3 \times 5 \times 7 \times 11 \times 13 \times 17 \times 19 + 1$

is not a prime number but that it has the factor of 347 and then show that 347 is a prime number.

STOP

5. Show that

$2 \times 3 \times 5 \times 7 \times 11 - 1$

is a prime number greater than 11.

6. Show that

$2 \times 3 \times 5 \times 7 \times 11 \times 13 \times 17 - 1$

is not a prime number but that it has the factor 61 and then show that 61 is a prime number.

STOP

and

$$31 = 5 \times (2 \times 3) + 1$$

so that

$$31 \div 5 = 2 \times 3 \text{ with a remainder of } 1$$

Thus if 31 were not a prime, it would not be divisible by 2, 3, or 5 and hence would have to be divisible by a prime greater than 5.

The same kind of argument applies in general: we either get a prime or a number that is the product of primes—and in each case the prime or primes that we obtain are greater than the primes that we began with.

Do exercise 4.

A similar conclusion can be reached by using 1 less than the product rather than 1 more. Thus, for example,

$$2 \times 3 - 1 = 5$$

is a prime greater than 3. Again, we do not always get a prime;

$$2 \times 3 \times 5 \times 7 - 1 = 209$$

is not a prime. But

$$209 = 11 \times 19$$

and 11 and 19 are both primes greater than 7.

Do exercises 5 and 6.

Review Exercises 3

Name ____________________

Class ____________________

In exercises 1–16 find the prime factorization of the given number.

1. 42
2. 56
3. 60
4. 90
5. 132
6. 121
7. 180
8. 210
9. 225
10. 450
11. 480
12. 720
13. 1024
14. 1080
15. 1776

In exercises 16–25 find the LCM of the given set of numbers.

16. 12, 16
17. 9, 12
18. 24, 30
19. 10, 15

1.
2.
3.
4.
5.
6.
7.
8.
9.
10.
11.
12.
13.
14.
15.
16.
17.
18.
19.

20.
21.
22.
23.
24.
25.
26.
27.
28.
29.
30.
31.
32.
33.
34.
35.

20. 28, 63

21. 96, 108

22. 32, 56

23. 39, 65

24. 24, 36, 48

25. 15, 18, 24

In exercises 26–35 find the GCD of the given set of numbers.

26. 24, 40

27. 36, 42

28. 65, 91

29. 32, 48

30. 64, 72

31. 48, 162

32. 294, 450

33. 192, 175

34. 24, 36, 108

35. 54, 120, 210

36. Anne, Bob, and Charles can jog around a track in 9, 12, and 14 minutes, respectively. If they start out at the same time, how many minutes will it be before they all meet again at the starting point?

36.

37. The director of a local track meet had some red, white, and blue ribbon in the lengths of 72 inches, 45 inches, and 108 inches respectively. If she wanted to cut award ribbons from this material and wanted them all to be the same length and to have no material left over, how long should she cut them and how many of each color would she have?

37.

38. The sides of an irregularly shaped field are of lengths 110, 154, 264, and 198 feet, respectively. **(a)** What is the greatest length of prefabricated wood fencing that could be used on all four sides? **(b)** How many pieces of this length would be needed to put a fence around this field?

38. (a)

(b)

39. A men's store buys ties at 3 for \$10 and sells them at 2 for \$10. If the store made a profit of \$100, how many ties did they buy?

39.

40. Two satellites revolve around the earth in orbits over the North Pole in 90 minutes and 105 minutes, respectively. If they both pass over the North Pole at noon, when will they next be over the North Pole at the same time?

40.

In exercises 41–54, determine if the statement is true or false. For those that are false, explain why it is false. (Recall that by "number" we mean a natural number, 1, 2, 3, ...)

41. Every number is either a prime or a composite number.

41.

42. 1 is a factor of every number.

42.

43.

44.

45.

46.

47.

48.

49.

50.

51.

52.

53.

43. If a is a factor of b and b is a factor of a, then $a = b$.

44. The sum of any two even numbers is even.

45. Every prime number is odd.

46. 1 is a prime number.

47. The sum of any two primes is a composite.

48. There exist infinitely many primes.

49. Any two consecutive numbers are relatively prime.

50. The LCM of two primes is their product.

51. If the LCM of a and b is 1, then $a = b = 1$.

52. If the LCM of a and b is b, then $a = 1$.

53. If the LCM of a and b is the product of a and b, then a and b are relatively prime.

54. The LCM of a pair of numbers is never less than the GCD of the same numbers. 54.

In exercises 55–58 complete the given sentence by filling in the blank.

55. Every prime number has exactly __________ factors. 55.

56. No prime number greater than 7 ends with one of the digits 0, 2, 4, 6, 8, or _______ . 56.

57. The numbers less than 10 that have exactly three factors are _______ and _______ . 57.

58. The smallest two-digit number that has exactly three factors is __________ . 58.

59. What is the smallest number that has exactly four different prime number factors? 59.

60. What is the smallest odd composite number with two different prime factors? 60.

61. The two prime numbers 17 and 71 have the same digits. Name three other pairs of prime numbers less than 100 that have the same digits. 61.

The ancient Greeks divided all natural numbers into three classes: perfect, deficient, and abundant. If the sum of all the factors of a number, excluding the number itself but including 1, is equal to the given number, the number is called a *perfect number*; if the sum of the factors (excluding the number) is less than the number, it is called a *deficient number*; if the sum of the factors (excluding the number) is greater than the number, it is called an *abundant number.* An example of a perfect number is 28 because

$$1 + 2 + 4 + 7 + 14 = 28$$

The number 4 is deficient because

$$1 + 2 = 3 < 4$$

and the number 12 is abundant because

$$1 + 2 + 3 + 4 + 6 = 16 > 12$$

In exercises 62–71 classify each number as deficient, perfect, or abundant.

62.

62. 6

63. 8

63.

64.

64. 10

65. 18

65.

66.

66. 16

67. 20

67.

68.

68. 24

69. 30

69.

70.

70. 32

71. 36

71.

*72.

*72. Show that 496 is a perfect number.

*73.

*73. There are only 21 abundant numbers less than 100 and they are all even. The first odd abundant number is 945. Prove that 945 is an abundant number.

Just for Fun

A. PATTERNS AND SEQUENCES

Write the next five numbers in each sequence.

1. 1.4, 2.8, 4.2, 5.6, ______, ______, ______, ______, ______
2. 5.5, 7, 5, 6.5, 4.5, ______, ______, ______, ______, ______
3. 0.5, 0.75, 1.125, 1.625, ______, ______, ______, ______, ______
4. 1.3, 3.9, 11.7, 35.1, ______, ______, ______, ______, ______

B. MORE PATTERNS

For those items indicated by an asterisk (*) you are to examine the pattern indicated by the previous computations and then write what you think is the result.

5.
- 9 × 9 = __________
- 99 × 89 = __________
- 999 × 889 = __________
- *9999 × 8889 = __________
- *99999 × 88889 = __________
- *999999 × 888889 = __________

6.
- 6 × 4 = __________
- 66 × 64 = __________
- 666 × 664 = __________
- *6666 × 6664 = __________
- *66666 × 66664 = __________
- *666666 × 666666 = __________

C. MISSING DIGITS

Find the missing digits in the following multiplication problems.

7.

```
            _ 8 5 _
      (X)     6 _ _
      -------------
        _ _ 8 2 _
  _ _ _ 1 8 _ _
  -------------
  _ _ 7 4 _ _ 4
```

8.

```
              _ 0 _ _
        (X)     8 _ 5
        -------------
          1 0 3 1 5
      _ _ _ 3
  _ _ _ _ _
  ---------------
  _ _ _ _ _ 4 5
```

D. TRY THESE

9. How long would you spend cutting a one-kilometer ribbon into 20-centimeter strips if one second is required for each cut?

10. Ed and Zeb are 30 miles apart and are riding bicycles toward each other, each at the rate of 15 miles per hour. At the start a fly on Ed's nose leaves, travels 12 miles an hour, lands on Zeb's nose, and promptly returns to Ed's nose, flying at the same speed. If the fly continues its trips back and forth between the noses until Ed and Zeb meet, how far does the fly fly?

11. Place +, −, ×, ÷, or ( ) between the digits below so as to arrive at the answer of 100. (Do not change the order of the digits.) See how many different ways you can do this.

$$1 \quad 2 \quad 3 \quad 4 \quad 5 \quad 6 \quad 7 \quad 8 \quad 9 \quad = \quad 100$$

E. MAGIC SQUARE

12. Complete the following magic square.

6.5	1.5		8.0
			2.5
6.0		3.5	4.5
0.5		7.5	2.0

F. MAGIC CIRCLE

13. There are three circles in each line. Use the numbers from 1 through 9 and place them in the circles so that the sum of each three numbers is the same. Each number is to be used only once.

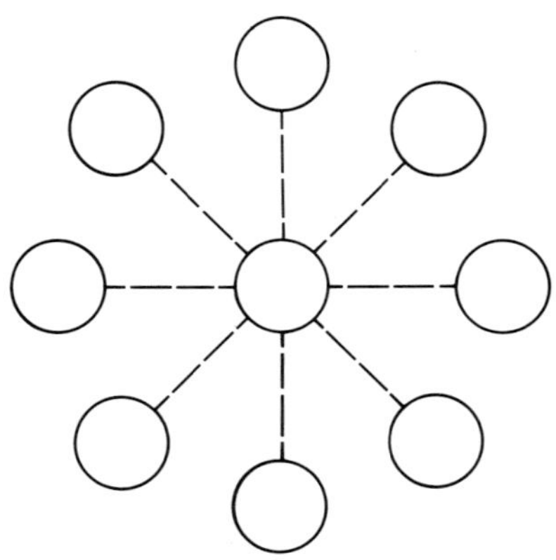

G. CALCULATOR GAME—"ZERO"

14. Play this game on your calculator.

Steps		*Sample Game:*		*Display Shows:*
1. Player A enters a number with three or more digits in the calculator.	A:	Chooses	4301	4301
2. Player B selects one of the non-zero digits of A's number, repeats it any number of times, and subtracts the resulting number from the number in the display.	B:	Chooses	3	
		Subtracts	3333	968
3. Player A repeats step 2 using the number that B obtains.	A:	Chooses	8	
		Subtracts	888	80
4. Play continues with the first player to produce "0" in the display the winner. (A player who produces a negative number in the display loses.)	B:	Chooses	8	
		Subtracts	8	72
	A:	Chooses	2	
		Subtracts	22	50
	B:	Chooses	5	
		Subtracts	5	45
	A:	Chooses	4	
		Subtracts	44	1
	B:	Chooses	1	
		Subtracts	1	0
	B WINS			

H. THE TALKING CALCULATOR

15. Perform the following calculation and use your calculator to find the name of the Egyptian goddess of fertility.

$$\left(\frac{150.912}{13.1} + 5.65\right) + (705.1 \times 0.6 - 405.89) + \left(\frac{18.5484}{1.23} + \frac{1.5675}{0.75}\right)$$

Mathematics and the Supermarket

Use the ads on page 194 to do the following exercises.

1. Calculate the unit cost for each of the items listed by completing the following tables.

Item: Bacon

	A	B	C
Amount (oz.)			
Cost (cents)			
Unit Cost			

Item: Cheese

	A	B	C
Amount (oz.)			
Cost (cents)			
Unit Cost			

Item: Tide

	A	B	C
Amount (oz.)			
Cost (cents)			
Unit Cost			

Item: Peanut Butter

	A	B	C
Amount (oz.)			
Cost (cents)			
Unit Cost			

Item: Lemonade

	A	B	C
Amount (oz.)			
Cost (cents)			
Unit Cost			

2. At what store was the unit cost of each item (a) the lowest? (b) the highest?

3. If your shopping list called for the following items:

    9 lbs bacon
    6 lbs cheese slices
    3 boxes of Tide
    $15\frac{3}{4}$ lbs peanut butter

   what would it cost to purchase the items on your shopping list at each store?

   Store A ____________ Store B ____________ Store C ____________

4. If you could shop at all three stores easily, what would you buy at each store so as to keep your cost the lowest? What would you then pay for the items on the list for question 3?

A. ________________
____
____

B. ________________
____
____

C. ________________
____
____

Total Cost ________________

5. Compare the cost of buying all of the items on the shopping list in question 3 at one store (exercise 3) with the cost of buying the items individually at the store where they cost least (exercise 4).

Cost of items individually ________________

Cost at Store A ________________ more by ________________

Cost at Store B ________________ more by ________________

Cost at Store C ________________ more by ________________

6. You are to make lemonade for 364 people expected at a summer outdoor wedding reception. Assuming 12 ounces of lemonade for each person and that 6 ounces of frozen lemonade will make 24 ounces of lemonade, what would it cost you to buy frozen lemonade at each store?

7. At what point (how much lemonade) is the cost the same at stores A and C?

A
WESTERN FAMILY
FROZEN LEMONADE
FIRST EIGHT 6 OZ. TINS
8 FOR $1
ADDITIONAL AT 6/1.00
SLICED BACON 12 OZ. PKG. 99¢
SLICED CHEESE AMERICAN PIMIENTO SWISS 12 OZ. PKG. 98¢
Famous Tide
SUPER SAVER
Heavy Duty Detergent
25¢ Off Label
84 oz. Box
$2.29
COUPON
HOODY, CREAMY or CHUNKY
PEANUT BUTTER $1.29
36-oz. JAR
EA.
$1.89 VALUE, WITH THIS COUPON
LIMIT 3 PER COUPON. ONE COUPON PER FAMILY
GOOD THRU 8/30/77
FOOD KiNG

B
Bonus Buy!
TIDE
Detergent 84-oz., 25c Off Label First One, Save 40c
1.98
Each
WESTERN FAMILY FROZEN
Lemon-ade
12-OZ. CAN
29¢
PEANUT BUTTER
18 oz. Creamy Or Super Chunky.
98¢
SAVE 31¢
Sliced Cheese 12-oz. $1.19
BACON ... POUND PKG. $1.69

C
SUNNY JIM
PEANUT BUTTER
Old Fashioned Creamy or Crunchy
28 oz.
$1.18
TIDE
Detergent
84 oz. BOX
$1.99
WITHOUT COUPON 2.35
LIMIT ONE PER COUPON
ONE COUPON PER FAMILY
EXP. 8-30-77
VALUABLE COUPON
BAR-S
SLICED BACON
12 oz. PKG.
98¢
CHEESE SLICES
• AMERICAN
• PIMENTO
• SWISS
12 oz. PKG.
$1.29
LEMONADE
SNOBOY, FROZEN
6 oz.
7 For $1

4

Computations with Fractions

4.1 Fractions and a Fundamental Principle

OBJECTIVES

You should know the:

(a) Meaning of a fraction.

(b) Meaning of numerator and denominator.

(c) Meaning of proper and improper fraction.

(d) Fundamental principle for fractions and the meaning of lowest terms.

****

Long ago, ancient man discovered that he needed more than the whole numbers, 0, 1, 2, 3, 4, 5, 6, . . . , to describe the world around him. He could not, for example, describe with whole numbers quantities whose measure came between two consecutive whole numbers. This realization of his inability to describe parts of things led him to the concept of *fractions.*

One of the best ways to picture the concept of a fraction is to use a horizontal *number line.* Choose a point on a line and label it 0. Next, choose another point to the right of 0 and label it 1; this determines the unit length from 0 to 1. Starting at 0, mark off this unit length repeatedly to the right to obtain the picture shown in Figure 4.1.

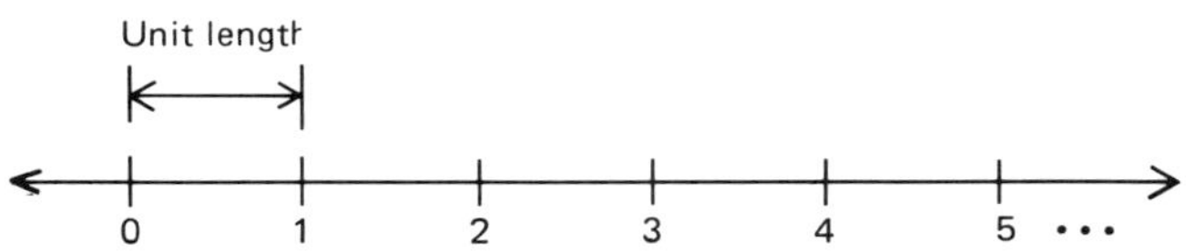

Figure 4.1

The fraction $\frac{1}{2}$ we associate with the point midway between 0 and 1. By marking off a segment of length one-half unit over and over again to the right of zero, we determine points that correspond to $\frac{1}{2}$, $\frac{2}{2}$, $\frac{3}{2}$, $\frac{4}{2}$, $\frac{5}{2}$, and so on as shown in Figure 4.2.

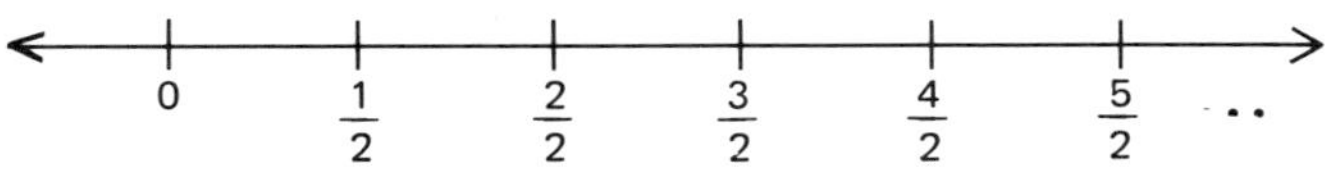

Figure 4.2

Given any fraction we can, by using other parts of a unit length, locate a point corresponding to this fraction. Some of these points are shown in Figure 4.3.

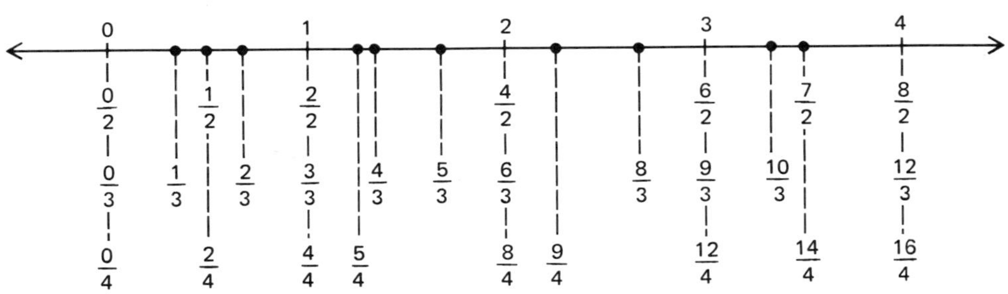

Figure 4.3

In exercises 1–4 give the fraction that the shaded region suggests.

1. 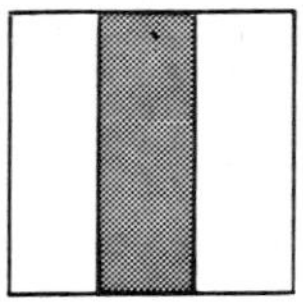

2. 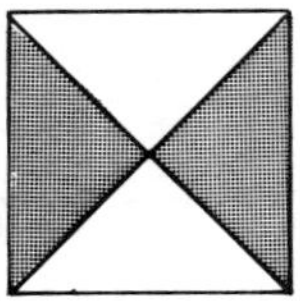

3. 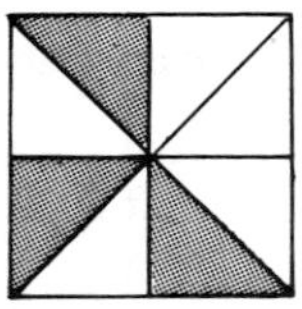

4. 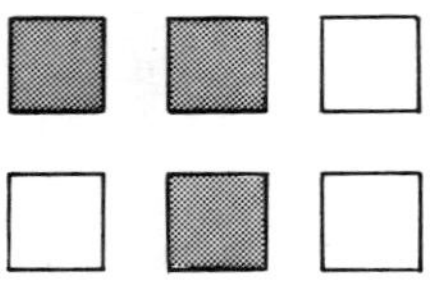

In exercises 5–8 give the fraction that the indicated point on the number line suggests.

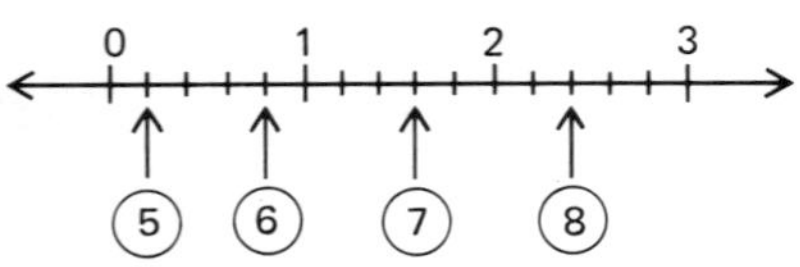

5.

6.

7.

8.

STOP

A fraction can also be used to describe other things that have been divided into parts. The rectangular region shown in Figure 4.4 has been divided into three parts, and two of them are shaded; thus the fraction $\frac{2}{3}$ could be used to describe the shaded portion.

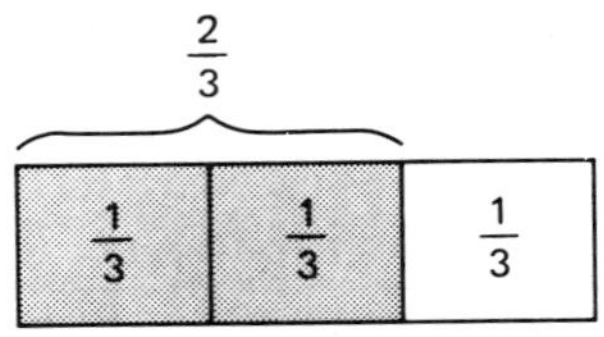

Figure 4.4

In Figure 4.5 we see that there are many fractions that could be used to show that $\frac{1}{2}$ of the square has been shaded. Such fractions are called *equivalent fractions.*

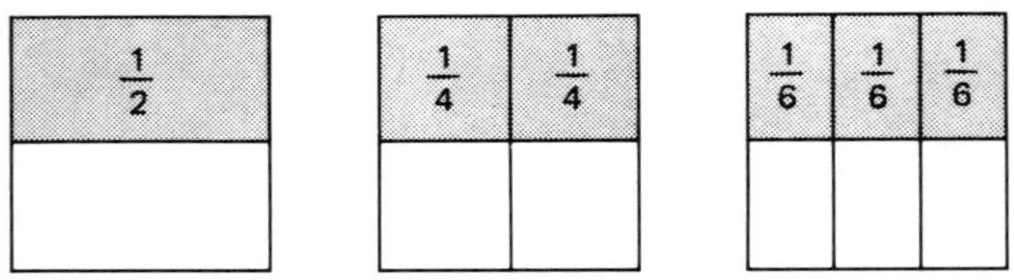

$\frac{1}{2}$ equivalent to $\frac{2}{4}$ equivalent to $\frac{3}{6}$

Figure 4.5

Do exercises 1 through 8 in the adjoining column.

In any fraction $\frac{a}{b}$, a is called the *numerator* and b the *denominator.* Thus in $\frac{3}{4}$, 3 is the numerator and 4 is the denominator.

A fraction such as $\frac{3}{4}$, in which the numerator is a smaller number than the denominator, is called a *proper fraction.* Such a fraction represents a number less than 1.

A fraction such as $\frac{5}{5}$ or $\frac{5}{4}$, in which the numerator is equal to or greater than the denominator, is called an *improper fraction.* Such a fraction represents a number equal to or greater than 1.

Do exercises 9 through 11.

Basic to operations with fractions is the *fundamental principle:*

The value of a fraction is unchanged if the numerator and the denominator are both multiplied by the same non-zero number.

Thus, for example,

$$\frac{1}{2} = \frac{1 \times 3}{2 \times 3} = \frac{3}{6} \quad \text{and} \quad \frac{15}{8} = \frac{15 \times 7}{8 \times 7} = \frac{105}{56}$$

The fractions $\frac{1}{2}$ and $\frac{3}{6}$ are equivalent fractions as are $\frac{15}{8}$ and $\frac{105}{56}$.

In general, corresponding to any fraction $\frac{a}{b}$ there is a set of fractions equivalent to it:

$$\frac{a}{b}, \frac{2 \times a}{2 \times b}, \frac{3 \times a}{3 \times b}, \frac{4 \times a}{4 \times b}, \ldots$$

Conversely, both numerator and denominator can be divided by the same non-zero number. Thus

$$\frac{75}{15} = \frac{75 \div 5}{15 \div 5} = \frac{15}{3} \quad \text{and} \quad \frac{186}{52} = \frac{186 \div 2}{52 \div 2} = \frac{93}{26}$$

We can verify such equalities with the HHC by using the fact that $\frac{a}{b} = a \div b$. Thus

$$\frac{1}{2} = 1 \div 2 = 0.5 \quad \text{and} \quad \frac{3}{6} = 3 \div 6 = 0.5$$

$$\frac{15}{8} = 15 \div 8 = 1.875 \quad \text{and} \quad \frac{105}{56} = 105 \div 56 = 1.875$$

$$\frac{75}{15} = 75 \div 15 = 5 \quad \text{and} \quad \frac{15}{3} = 15 \div 3 = 5$$

$$\frac{186}{52} = 186 \div 52 = 3.576923 \quad \text{and} \quad \frac{93}{26} = 93 \div 26 = 3.576923$$

In exercises 9–11 (a) list the numerator, (b) list the denominator, and (c) indicate whether the fraction is a proper or an improper fraction.

9. $\frac{5}{16}$

(a)

(b)

(c)

10. $\frac{13}{8}$

(a)

(b)

(c)

11. $\frac{15}{15}$

(a)

(b)

(c)

STOP

Do exercises 12 and 13.

12. **(a)** Find a fraction equivalent to $\frac{49}{61}$ by multiplying both numerator and denominator by 13.

(b) Verify that the fraction you found in (a) is equal to $\frac{49}{61}$ by performing divisions on your HHC.

13. **(a)** Find a fraction equivalent to $\frac{3933}{4674}$ by dividing both numerator and denominator by 57.

(b) Verify that the fraction you found in (a) is equal to $\frac{3933}{4674}$ by performing divisions on your HHC.

STOP

When a fraction is written in the form $\frac{a}{b}$, where a and b are relatively prime (i.e., when the GCD of a and b is 1), we say that the fraction $\frac{a}{b}$ is in *lowest terms.*

To write a fraction in lowest terms, then, is to divide both a and b by the GCD of a and b. For example, to write $\frac{312}{408}$ in lowest terms, we note that

$$\frac{312}{408} = \frac{2^3 \times 3 \times 13}{2^3 \times 3 \times 17} = \frac{13}{17}$$

where $2^3 \times 3 = 24$ is the GCD of 312 and 408.

Do exercises 14 through 16.

In exercises 14–16 write the given fraction in lowest terms.

14. $\frac{20}{24}$

15. $\frac{27}{64}$

16. $\frac{60}{45}$

STOP

Finally, note that fractions are sometimes written in the form a/b. Thus

$$2/3 = \frac{2}{3}, \qquad 51/62 = \frac{51}{62},$$

and so on.

Exercises 4.1

In exercises 1–4 write, for the given figure, a fraction that the shaded part of the figure suggests.

1. 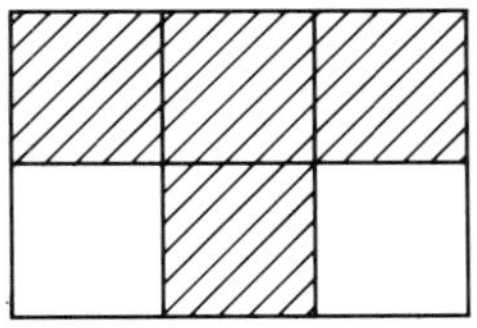

2. 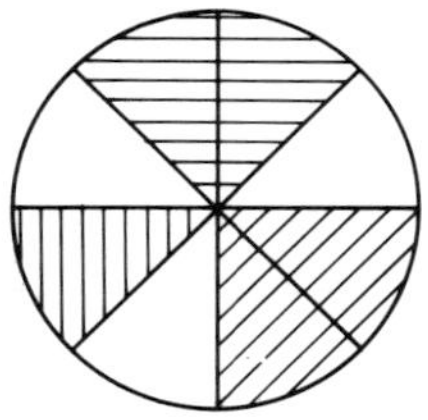

3. 

4. 

In exercises 5–8 use the figure to shade a region that represents the given fraction.

5. $\frac{5}{8}$ 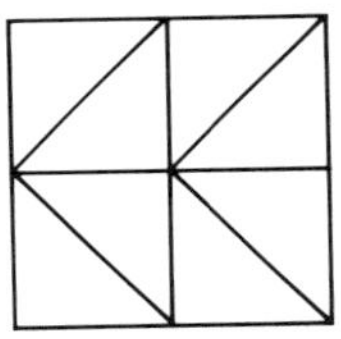

6. $\frac{6}{8}$ 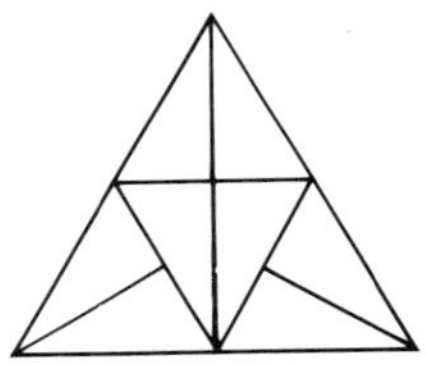

7. $\frac{3}{5}$ 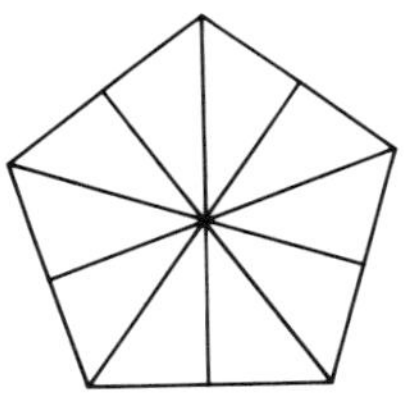

8. $\frac{1}{4}$ 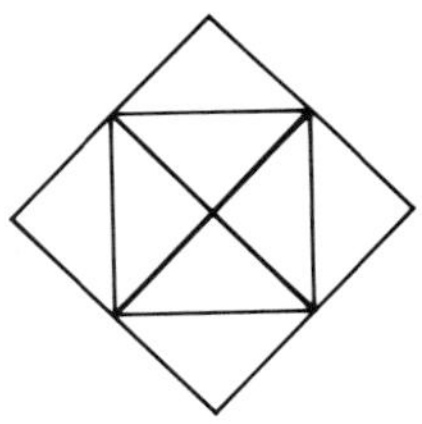

9. Use the square regions below to show $\frac{1}{4}$ in as many ways as you can think of.

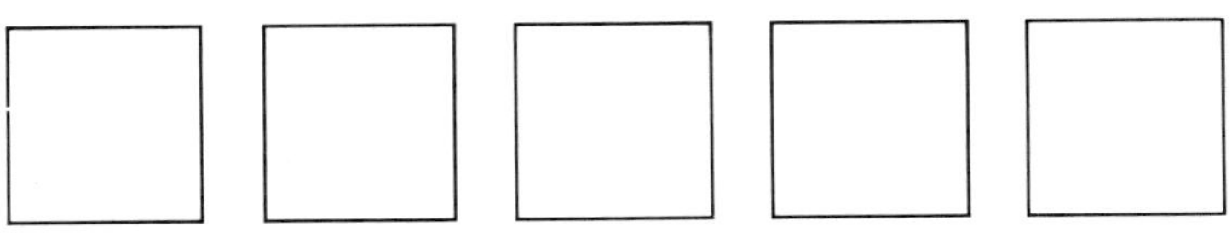

10. In a small mountain village, 19 of 26 homes were destroyed in a forest fire. Use a fraction to state (a) what part of the village was destroyed by fire; (b) what part of the village was not destroyed by fire.

Name ____________

Class ____________

1.

2.

3.

4.

10. (a)

(b)

11. (a) (b)

11. In a state park on a given weekend 90 campsites were reserved. If there were 75 tent spaces and 58 trailer spaces in the park, use a fraction to state (a) what part of the park campsites was reserved that weekend; (b) what part was not reserved.

12. (a) (b) (c) (d)

12. In a mathematics class there were 11 girls and 17 boys. Three of the girls and 4 of the boys were taking the class for the second time. Use a fraction to show (a) what part of the class was taking the course the second time; (b) what part of the class was taking the course for the first time; (c) what part of the boys was taking the class the first time; and (d) what part of the girls was taking the class the second time.

13. (a) (b) (c)

13. A family's monthly budget consisted of the following items:

Food	\$216	Medical	\$ 47
Housing	240	Recreation	50
Utilities	67	Gifts and contributions	100
Transportation	120	Taxes	108
Clothing	96	Insurance	72
Savings	48		

Use a fraction to show what part of the budget is used for (a) food, (b) housing, and (c) taxes.

14.

14. If the family in exercise 13 has an income of \$16,044 per year, use a fraction to show what part of their monthly income is not accounted for in their budget.

In exercises 15–20 state whether the given fraction is a proper fraction (P) or an improper fraction (I).

15. 16. 17. 18. 19. 20.

15. $\frac{9}{4}$ 16. $\frac{15}{33}$ 17. $\frac{22}{22}$

18. $\frac{86}{16}$ 19. $\frac{4}{62}$ 20. $\frac{11}{3}$

In exercises 21–25 fill in the missing numerators or denominators in each equivalent fraction.

21. 22. 23. 24. 25.

21. $\frac{1}{3} = \frac{\quad}{6} = \frac{\quad}{9} = \frac{\quad}{15} = \frac{\quad}{18} = \frac{\quad}{36}$

22. $\frac{4}{5} = \frac{\quad}{10} = \frac{\quad}{15} = \frac{\quad}{20} = \frac{\quad}{50} = \frac{\quad}{60}$

23. $\frac{5}{8} = \frac{\quad}{16} = \frac{\quad}{32} = \frac{\quad}{56} = \frac{\quad}{96} = \frac{\quad}{120}$

24. $\frac{7}{23} = \frac{14}{\quad} = \frac{21}{\quad} = \frac{\quad}{115} = \frac{49}{\quad} = \frac{\quad}{299}$

25. $\frac{13}{19} = \frac{\quad}{38} = \frac{65}{\quad} = \frac{\quad}{209} = \frac{169}{\quad} = \frac{\quad}{513}$

In exercises 26–30 give a set of five fractions equivalent to the given fraction.

26. $\frac{8}{15}$ 26.

27. $\frac{9}{6}$ 27.

28. $\frac{31}{35}$ 28.

29. $\frac{107}{103}$ 29.

30. $\frac{215}{17}$ 30.

In exercises 31–48 write the given fraction as an equivalent fraction in lowest terms.

31. $\frac{21}{27}$ 32. $\frac{45}{60}$ 31. 32.

33. $\frac{18}{32}$ 34. $\frac{36}{144}$ 33. 34.

35. $\frac{360}{72}$ 36. $\frac{53}{106}$ 35. 36.

37. $\frac{24}{32}$ 38. $\frac{62}{93}$ 37. 38.

39. $\frac{85}{68}$ 40. $\frac{42}{60}$ 39. 40.

41. $\frac{57}{76}$ 42. $\frac{135}{144}$ 41. 42.

43. $\frac{50}{6}$ 44. $\frac{100}{80}$ 43. 44.

45. $\frac{69}{72}$ 46. $\frac{34}{119}$ 45. 46.

47. $\frac{165}{678}$ 48. $\frac{567}{405}$ 47. 48.

*49.

*49. Keep track of how you spend a day by listing the time spent in school, eating, sleeping, studying, watching TV, etc. Round each time to the nearest hour and use a fraction (in lowest terms) to show the portion of each day spent on each item.

*50.

*50. Choose a particular television program and, using a stop watch or watch with a second hand, record the length of each commercial. Use a fraction to show what part of the program was devoted to commercials.

*51. (a)

(b)

(c)

(d)

(e)

*51. Using the four numbers 2, 5, 7, and 9

(a) Write as many fractions as you can. (You may use a number more than once.)

(b) Which fraction is the smallest? The largest?

(c) Which fractions are less than 1? Greater than 1? Equal to 1?

(d) Which fractions are equivalent?

(e) Write the fractions in order of size.

*52. Choose four different numbers from those in exercise 51 and answer each of the questions in exercise 51.

*52. (a)

(b)

(c)

(d)

(e)

*53. Depending upon the numbers you choose, the number of different fractions for any four numbers will be 13, 11, 9, 7, 5, 3, or 1. Try to find four numbers that will give each of these numbers of fractions. (Note: Two fractions that are equivalent are not to be regarded as different. Also note that the numbers in your set of four numbers need not be all different. Thus one answer for a set of four numbers giving only one fraction would be (1, 1, 1, 1).)

*53.

4.2 Adding Fractions

BASIC PRINCIPLES

It is certainly possible with an HHC to do arithmetic with whole numbers and decimals and yet not know how to do these operations without the use of an HHC! (Although, as you have seen, a knowledge of the *meaning* of the operations of arithmetic is important with or without the use of an HHC.)

Consider, however, the problem of calculating

$$\frac{2}{3} + \frac{4}{5}$$

Now there is actually at least one HHC on the market today on which it is possible to perform this computation directly, and have the answer of $1\frac{7}{15}$ displayed. However, almost all HHC's are designed to give displays that are whole numbers or decimals. Therefore, what we will do here is to reduce operations with fractions to operations with whole numbers or decimals.

First of all, note that it is easy to see how to add fractions that have the same denominator—we simply add numerators. Thus

$$\frac{592}{264} + \frac{127}{264} + \frac{91}{264} = \frac{591 + 127 + 91}{264} = \frac{809}{264}$$

Do exercises 1 through 5 in the adjoining column.

To add fractions with different denominators, we need only apply the fundamental principle to produce fractions with the same denominators. For example,

$$\frac{2}{3} + \frac{4}{5} = \frac{2 \times 5}{3 \times 5} + \frac{4 \times 3}{5 \times 3} = \frac{10}{15} + \frac{12}{15} = \frac{10 + 12}{15} = \frac{22}{15}$$

Similarly,

$$\frac{81}{79} + \frac{18}{107} = \frac{81 \times 107}{79 \times 107} + \frac{18 \times 79}{107 \times 79} = \frac{8667}{8453} + \frac{1422}{8453}$$

$$= \frac{8667 + 1422}{8453} = \frac{10{,}089}{8453}$$

OBJECTIVES

You should be able to:

(a) Change improper fractions to mixed numbers, and change mixed numbers to improper fractions.

(b) Add fractions.

(c) Apply addition of fractions to problems.

****

In exercises 1–5 perform the indicated addition and write your answer in lowest terms.

1. $\frac{5}{6} + \frac{5}{6}$

2. $\frac{3}{5} + \frac{4}{5}$

3. $\frac{7}{8} + \frac{5}{8}$

4. $\frac{13}{24} + \frac{17}{24}$

5. $\frac{47}{72} + \frac{43}{72}$

STOP

In exercises 6–12 perform the indicated addition. Write your answer in lowest terms.

6. $\frac{1}{4} + \frac{1}{8}$

7. $\frac{1}{6} + \frac{2}{3}$

8. $\frac{3}{4} + \frac{5}{6}$

9. $\frac{29}{7} + \frac{48}{13}$

10. $\frac{21}{12} + \frac{42}{14}$

11. $\frac{8}{9} + \frac{3}{4}$

12. $\frac{11}{12} + \frac{27}{32}$

STOP

Do exercises 6 through 12.

USING LEAST COMMON DENOMINATORS

Consider now the addition

$$\frac{51}{12} + \frac{19}{18}$$

Proceeding as before, we have

$$\frac{51 \times 18}{12 \times 18} + \frac{19 \times 12}{18 \times 12} = \frac{918}{216} + \frac{228}{216} = \frac{918 + 228}{216} = \frac{1146}{216}$$

Certainly $\frac{1146}{216}$ is not in lowest terms since the numerator and denominator are both even numbers and hence both divisible by 2. Indeed,

$$\frac{1146}{216} = \frac{2 \times 3 \times 191}{2 \times 3 \times 36} = \frac{191}{36}$$

is the answer in lowest terms since the only prime divisors of 36 are 2 and 3 and neither of these divide 191.

To increase our chances of having the answer to an addition problem in lowest terms, we can use the *least common denominator* (LCD). This is simply the LCM of the denominators (see section 3.4). In this case, since

$$12 = 2^2 \times 3 \quad \text{and} \quad 18 = 2 \times 3^2$$

the LCD is $2^2 \times 3^2 = 36$. Then, since $36 \div 12 = 3$ and $36 \div 18 = 2$, we have

$$\frac{51}{12} + \frac{19}{18} = \frac{51 \times 3}{12 \times 3} + \frac{19 \times 2}{18 \times 2} = \frac{153}{36} + \frac{38}{36} = \frac{153 + 38}{36} = \frac{191}{36}$$

and our answer is in lowest terms.

Do exercises 13 through 18.

A similar procedure can be used when more than two fractions are to be added. Consider, for example,

$$\frac{15}{216} + \frac{19}{48} + \frac{29}{54}$$

We have

$$216 = 2^3 \times 3^3, \qquad 48 = 2^4 \times 3, \qquad 54 = 2 \times 3^3$$

so that the LCD of our fractions is

$$2^4 \times 3^3 = 432$$

Then, since

$$432 \div 216 = 2, \qquad 432 \div 48 = 9, \qquad 432 \div 54 = 8$$

we write

$$\frac{15}{216} + \frac{19}{48} + \frac{29}{54} = \frac{15 \times 2}{216 \times 2} + \frac{19 \times 9}{48 \times 9} + \frac{29 \times 8}{54 \times 8}$$

$$\frac{30}{432} + \frac{171}{432} + \frac{232}{432} = \frac{30 + 171 + 232}{432} = \frac{433}{432}$$

Our answer is in lowest terms since the only prime factors of 432 are 2 and 3 and neither 2 nor 3 are factors of 433.

Using the LCD, however, does not guarantee that the answer will be in lowest terms—it just makes it more likely. This is illustrated in the following example:

$$\frac{3}{10} + \frac{8}{15} = \frac{3 \times 3}{10 \times 3} + \frac{8 \times 2}{15 \times 2} = \frac{9}{30} + \frac{16}{30} = \frac{25}{30}$$

Here the answer is not in lowest terms even though we have used 30, the LCD of the two fractions.

In exercises 13–18 perform the indicated addition by using the LCD of the two fractions. Write your answer in lowest terms.

13. $\frac{5}{6} + \frac{3}{4}$

14. $\frac{3}{10} + \frac{5}{8}$

15. $\frac{5}{8} + \frac{2}{3}$

16. $\frac{5}{16} + \frac{9}{20}$

17. $\frac{13}{12} + \frac{17}{32}$

18. $\frac{27}{10} + \frac{5}{32}$

STOP

In exercises 19–21 perform the indicated addition by using the LCD and write your answer in lowest terms.

19. $\frac{1}{2} + \frac{3}{4} + \frac{5}{8}$

20. $\frac{9}{16} + \frac{7}{8} + \frac{15}{32}$

21. $\frac{9}{28} + \frac{31}{49} + \frac{2}{7}$

In exercises 22–23 note that $a/1 = a$ for any whole number a and use this fact to perform the indicated addition, writing your answer as an improper fraction.

22. $3 + \frac{1}{2}$

23. $5 + \frac{7}{16}$

STOP

Do exercises 19 through 23.

WRITING IMPROPER FRACTIONS AS MIXED NUMBERS

In exercises 22–23 you found that

$$3 + \frac{1}{2} = \frac{7}{2} \quad \text{and} \quad 5 + \frac{7}{16} = \frac{87}{16}$$

We also write

$$3 + \frac{1}{2} \text{ as } 3\frac{1}{2} \quad \text{and} \quad 5 + \frac{7}{16} \text{ as } 5\frac{7}{16}$$

and call $3\frac{1}{2}$ and $5\frac{7}{16}$ *mixed numbers* or *mixed fractions.*

You have seen how to write any mixed number as an improper fraction. Conversely, any improper fraction can be written as a mixed number. To do so we can perform a division with a remainder. For example,

$$\frac{7}{2} = 7 \div 2, \qquad \begin{array}{r} 3 \\ 2\overline{)7} \\ \underline{6} \\ 1 \end{array}$$

Thus $7 \div 2$ gives 3 with a remainder of 1 and so

$$\frac{7}{2} = 3\frac{1}{2}$$

Similarly,

$$\frac{127}{8} = 127 \div 8, \qquad \begin{array}{r} 15 \\ 8\overline{)127} \\ \underline{80} \\ 47 \\ \underline{40} \\ 7 \end{array}$$

Thus $127 \div 8$ gives 15 with a remainder of 7 and so

$$\frac{127}{8} = 15\frac{7}{8}$$

The divisions above have been done by hand to obtain a remainder. To achieve this goal with an HHC, we can proceed as follows :

Enter	127	÷	8	−	15	×	8	=
Display	127.	127.	8.	15.875	15.	0.875	8.	7.

That is,

$$127 \div 8 = 15.875, \qquad 15.875 - 15 = 0.875,$$

$$\text{and } 0.875 \times 8 = 7, \text{ the remainder}$$

Do exercises 24 through 28.

ADDING MIXED NUMBERS

There are two ways to add mixed numbers. For example, consider the addition

$$3\frac{1}{4} + 5\frac{2}{3} + 2\frac{1}{8}$$

One way to do this addition is to write

$$\left(3 + \frac{1}{4}\right) + \left(5 + \frac{2}{3}\right) + \left(2 + \frac{1}{8}\right) =$$

$$(3 + 5 + 2) + \left(\frac{1}{4} + \frac{2}{3} + \frac{1}{8}\right) =$$

$$10 + \left(\frac{6}{24} + \frac{16}{24} + \frac{3}{24}\right) = 10 + \frac{25}{24} =$$

$$10 + \left(1 + \frac{1}{24}\right) = 11 + \frac{1}{24} = 11\frac{1}{24}$$

24. Change $\frac{6011}{23}$ to a mixed number by doing the following computations:

    (a) $6011 \div 23 =$ ________

    (b) $261.34782 - 261 =$ ________

    (c) $0.34782 \times 23 =$ ________

    (d) Rounding off 7.99986 to the nearest whole number gives _____.

    (e) Now $\frac{6011}{23} =$ ________ (mixed number)

In exercises 25–26 express the given improper fraction as a mixed number.

25. $\frac{43}{16}$

26. $\frac{248}{35}$

In exercises 27–28 perform the indicated addition and express your answer as a mixed number.

27. $\frac{5}{2} + \frac{9}{8}$

28. $\frac{31}{49} + \frac{32}{35}$

STOP

In exercises 29–31 perform the indicated addition in each of the two ways shown by the text example. Write your answers as mixed numbers.

29. $3\frac{5}{9} + 5\frac{7}{9}$

30. $9\frac{3}{7} + 5\frac{5}{7}$

31. $6\frac{7}{8} + 2\frac{3}{16}$

STOP

The other way is to write

$$3\frac{1}{4} + 5\frac{2}{3} + 2\frac{1}{8} = \frac{13}{4} + \frac{17}{3} + \frac{17}{8}$$

$$= \frac{13 \times 6}{4 \times 6} + \frac{17 \times 8}{3 \times 8} + \frac{17 \times 3}{8 \times 3}$$

$$= \frac{78}{24} + \frac{136}{24} + \frac{51}{24}$$

$$= \frac{265}{24} = 11\frac{1}{24}$$

Do exercises 29 through 31.

Exercises 4.2

Name ______________________

Class ______________________

In exercises 1–10 find the sum of the fractions. Write your answer in lowest terms.

1. $\frac{14}{17} + \frac{12}{17}$

2. $\frac{3}{8} + \frac{3}{4}$

3. $\frac{1}{12} + \frac{3}{8}$

4. $\frac{9}{10} + \frac{5}{6}$

5. $\frac{13}{29} + \frac{17}{87}$

6. $\frac{17}{42} + \frac{29}{144}$

7. $\frac{55}{72} + \frac{17}{96}$

8. $\frac{7}{15} + \frac{11}{12} + \frac{9}{10}$

9. $\frac{8}{9} + \frac{19}{36} + \frac{29}{42}$

10. $\frac{14}{39} + \frac{11}{28} + \frac{11}{63}$

In exercises 11–20 change the improper fraction to a mixed number. Write any fractions in your answer in lowest terms.

11. $\frac{35}{4}$

12. $\frac{100}{12}$

13. $\frac{125}{12}$

14. $\frac{425}{9}$

15. $\frac{620}{11}$

16. $\frac{533}{17}$

1.
2.
3.
4.
5.
6.
7.
8.
9.
10.
11. 12.
13. 14.
15. 16.

17. 18.

19. 20.

21. 22.

23. 24.

25. 26.

27. 28.

29. 30.

31.

32.

33.

34.

35.

36.

17. $\frac{626}{35}$

18. $\frac{849}{40}$

19. $\frac{758}{51}$

20. $\frac{8463}{629}$

In exercises 21–30 change the mixed number to an improper fraction.

21. $4\frac{5}{16}$

22. $8\frac{9}{16}$

23. $5\frac{11}{12}$

24. $15\frac{7}{9}$

25. $21\frac{12}{23}$

26. $28\frac{16}{63}$

27. $19\frac{17}{216}$

28. $15\frac{48}{61}$

29. $26\frac{19}{56}$

30. $29\frac{105}{124}$

In exercises 31–40 do the indicated addition and write your answer as a mixed number.

31. $5\frac{1}{15} + 1\frac{7}{15}$

32. $7\frac{5}{16} + 4\frac{9}{16}$

33. $8\frac{1}{2} + 7\frac{7}{12}$

34. $10\frac{5}{18} + 3\frac{7}{9}$

35. $9\frac{1}{8} + 8\frac{2}{3}$

36. $41\frac{2}{3} + 16\frac{3}{5}$

37. $73\frac{1}{6} + 59\frac{4}{15}$ 37.

38. $19\frac{5}{17} + 8\frac{13}{51} + 2\frac{2}{3}$ 38.

39. $8\frac{14}{39} + 7\frac{2}{3} + 5\frac{15}{78}$ 39.

40. $4\frac{11}{27} + 6\frac{24}{45} + 7\frac{13}{72}$ 40.

In exercises 41-50 solve the given problem by first writing a mathematical sentence that describes the computation involved in the problem and then performing the computation.

41. Jack usually rides his ten-speed bike five miles each day. This morning he rode to the park, $3\frac{1}{2}$ miles away, and back. In the afternoon he rode $1\frac{1}{2}$ miles to the shopping center and back. How many miles did he ride his bike today? 41.

42. Bob and Sue made $7\frac{1}{2}$ pounds of fudge and $4\frac{1}{2}$ pounds of divinity candy for the holidays. They boxed and took three packages to the post office to mail. The clerk said that the packages weighed $3\frac{3}{4}$ pounds, $2\frac{1}{2}$ pounds, and $4\frac{3}{8}$ pounds. What was the total weight of the packages mailed? 42.

43. Four standard precision blocks are needed to measure the width of a certain space. The widths of the blocks are $\frac{5}{16}$ inch, $\frac{3}{64}$ inch, $\frac{3}{8}$ inch, and $\frac{1}{32}$ inch. If the space is $6\frac{1}{2}$ inches long, how wide is the space? 43.

44. Janice bought three pieces of yardage to sew into dresses. One was $4\frac{1}{2}$ yards of 54-inch material, another $3\frac{3}{4}$ yards of 42-inch material, and the third was $4\frac{1}{3}$ yards of 48-inch material. What was the total length of the yardage that she purchased? 44.

45.

45. George is going to bake some things for a party. He needs $3\frac{3}{8}$ cups of flour for a cake, $1\frac{1}{4}$ cups of flour for cookies, and $\frac{2}{3}$ cup of flour for some lemon tarts. He also needs $1\frac{3}{4}$ cups of sugar and $2\frac{1}{2}$ cups of milk. How much flour does he need altogether?

46.

46. A piece of $\frac{5}{8}$-inch plywood is nailed to each side of a 12-foot beam that is $2\frac{1}{4}$ inches thick. What is the total thickness?

47.

47. Four circular washers were placed on a $4\frac{1}{2}$ by $\frac{3}{8}$ inch bolt. Two of the washers were $\frac{3}{64}$ inch thick and the other two were $\frac{7}{32}$ inch and $\frac{1}{8}$ inch thick, respectively. How much space did the four washers take up?

48.

48. The following lengths of ribbon were sold from a 10-yard length of ribbon: $\frac{1}{2}$ yard, $\frac{5}{6}$ yard, $\frac{3}{8}$ yard, $\frac{1}{4}$ yard, and $2\frac{3}{8}$ yards. How much ribbon was sold off the roll?

49.

49. Pat had a board 10 feet long. He cut off the following lengths, with $\frac{1}{8}$ inch wasted on each saw cut: $6\frac{1}{4}$ inches, $5\frac{3}{8}$ inches, $4\frac{1}{64}$ inches, $14\frac{1}{2}$ inches, and $16\frac{5}{8}$ inches. How much of the board was used in obtaining these five pieces?

50. (a)

(b)

50. A bolt has the dimensions shown below. (a) How long is the bolt? (b) How wide is the head of the bolt?

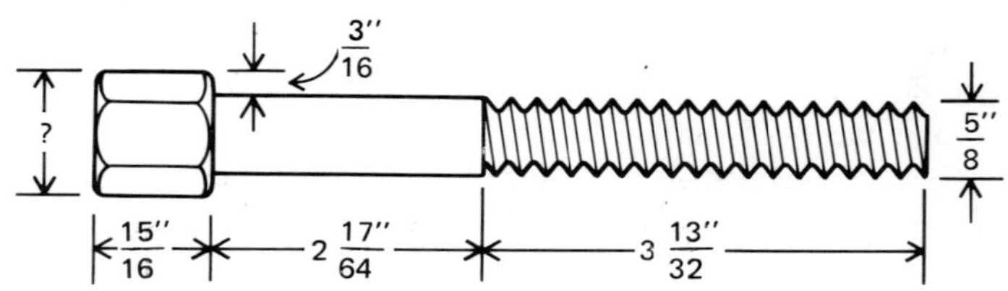

***51.** Arrange the numbers 5, 6, 7, and 8 into two fractions that will give the largest sum. (Do not use any number more than once.)

***52.** Arrange the numbers 5, 6, 7, and 8 into two fractions that will give the smallest sum. (Do not use any number more than once.)

***53.** Consider the following sequence of numbers:

$$1, \frac{1}{2}, \frac{1}{4}, \frac{1}{8}, \frac{1}{16}, \ldots$$

(a) Add the first two numbers in the sequence.

(b) Add the first three numbers in the sequence.

(c) Add the first four numbers in the sequence. Do you begin to see a pattern?

(d) Predict the sum of the first ten numbers in the sequence.

***54.** Using the sequence of numbers in exercise 53, the more numbers in the sequence you add together, you come closer to what number as the sum of all possible numbers in the sequence?

*51.

*52.

*53. (a)

(b)

(c)

(d)

*54.

4.3 Subtracting Fractions

No new principles are involved in the subtraction of fractions as compared to the addition of fractions. The following examples illustrate the procedures for the subtraction of fractions.

1. If the fractions have the same denominators, we simply need to subtract numerators. Thus

$$\frac{591}{264} - \frac{128}{264} = \frac{591 - 128}{264} = \frac{463}{264}$$

2. If the fractions have different denominators, we can proceed without finding the LCD, as in

$$\frac{81}{79} - \frac{18}{107} = \frac{81 \times 107}{79 \times 107} - \frac{18 \times 79}{107 \times 79} = \frac{8667}{8453} - \frac{1422}{8453} = \frac{7245}{8453}$$

3. However, as in the addition of fractions, we can increase the chances of our answer to a subtraction problem being in lowest terms by using the LCD, as in

$$\frac{5}{6} - \frac{4}{9} = \frac{5 \times 3}{6 \times 3} - \frac{4 \times 2}{9 \times 2} = \frac{15}{18} - \frac{8}{18} = \frac{7}{18}$$

where the LCD is 18.

Do exercises 1 through 6 in the adjoining column.

Subtraction of mixed numbers is illustrated by our concluding two examples.

4. $$7\frac{1}{2} - 5\frac{1}{3} = \left(7 + \frac{1}{2}\right) - \left(5 + \frac{1}{3}\right)$$

$$= (7 - 5) + \left(\frac{1}{2} - \frac{1}{3}\right)$$

$$= 2 + \left(\frac{3}{6} - \frac{2}{6}\right) = 2 + \frac{1}{6} = 2\frac{1}{6}$$

or

$$7\frac{1}{2} - 5\frac{1}{3} = \frac{15}{2} - \frac{16}{3} = \frac{45}{6} - \frac{32}{6} = \frac{13}{6} = 2\frac{1}{6}$$

OBJECTIVES

You should be able to:

(a) Subtract fractions.

(b) Apply subtraction of fractions to problems.

****

In exercises 1-6 perform the indicated subtraction and write your answer in lowest terms.

1. $\frac{8}{11} - \frac{5}{11}$

2. $\frac{13}{27} - \frac{7}{27}$

3. $\frac{15}{12} - \frac{2}{5}$

4. $\frac{5}{6} - \frac{3}{8}$

5. $\frac{7}{15} - \frac{5}{12}$

6. $\frac{9}{20} - \frac{11}{25}$

STOP

In exercises 7-11 perform the indicated subtraction and write your answer as a mixed number.

7. $6\frac{8}{9} - 2\frac{5}{9}$

8. $6\frac{1}{2} - 2\frac{3}{10}$

9. $3\frac{1}{2} - 2\frac{2}{9}$

10. $45\frac{5}{9} - 37\frac{5}{6}$

11. $96\frac{8}{15} - 37\frac{5}{6}$

STOP

5. For the subtraction

$$7\frac{1}{3} - 5\frac{1}{2}$$

we note that the fractional part, $\frac{1}{3}$, of $7\frac{1}{3}$ is less than the fractional part, $\frac{1}{2}$, of $5\frac{1}{2}$. So, to use the first method shown in example 4, we write

$$7\frac{1}{3} = 7 + \frac{1}{3} = (6+1) + \frac{1}{3} = 6 + \left(1 + \frac{1}{3}\right) = 6 + \frac{4}{3}$$

and then have

$$7\frac{1}{3} - 5\frac{1}{2} = \left(6 + \frac{4}{3}\right) - \left(5 + \frac{1}{2}\right)$$

$$= (6-5) + \left(\frac{4}{3} - \frac{1}{2}\right)$$

$$= 1 + \left(\frac{8}{6} - \frac{3}{6}\right) = 1 + \frac{5}{6}$$

$$= 1\frac{5}{6}$$

Or, using the second method,

$$7\frac{1}{3} - 5\frac{1}{2} = \frac{22}{3} - \frac{11}{2}$$

$$= \frac{44}{6} - \frac{33}{6} = \frac{11}{6}$$

$$= 1\frac{5}{6}$$

Do exercises 7 through 11.

Exercises 4.3

Name ______________________

Class ______________________

In exercises 1–25 perform the indicated subtraction. Express your answers in lowest terms.

1. $\frac{5}{7} - \frac{3}{7}$

2. $\frac{18}{19} - \frac{13}{19}$

3. $\frac{3}{5} - \frac{1}{10}$

4. $\frac{5}{6} - \frac{7}{12}$

5. $\frac{9}{10} - \frac{5}{6}$

6. $\frac{15}{24} - \frac{5}{82}$

7. $\frac{105}{124} - \frac{48}{61}$

8. $\frac{19}{54} - \frac{17}{216}$

9. $\frac{69}{168} - \frac{19}{56}$

10. $3\frac{2}{5} - 1\frac{1}{3}$

11. $6\frac{1}{2} - 2\frac{1}{8}$

12. $8\frac{1}{2} - 5\frac{5}{7}$

13. $10\frac{5}{11} - 3\frac{1}{10}$

1.

2.

3.

4.

5.

6.

7.

8.

9.

10.

11.

12.

13.

14.	14. $12\frac{4}{7} - 7\frac{5}{12}$
15.	15. $4\frac{1}{4} - 2\frac{3}{5}$
16.	16. $5\frac{2}{5} - 3\frac{7}{8}$
17.	17. $5\frac{1}{2} - 2\frac{5}{8}$
18.	18. $4\frac{2}{5} - 3\frac{3}{4}$
19.	19. $15\frac{2}{3} - 9\frac{1}{4}$
20.	20. $14\frac{7}{8} - 8\frac{19}{24}$
21.	21. $32\frac{5}{16} - 21\frac{11}{12}$
22.	22. $16\frac{25}{64} - 10\frac{95}{144}$
23.	23. $28\frac{16}{63} - 13\frac{11}{14}$
24.	24. $84\frac{21}{23} - 72\frac{15}{22}$
25.	25. $15\frac{48}{61} - 11\frac{4}{9}$

In exercises 26–35 perform the given combination of addition and subtraction operations with fractions.

26.	26. $4\frac{1}{3} + 5\frac{1}{2} - 2\frac{1}{4} - 1\frac{1}{8}$
27.	27. $16\frac{3}{4} - 2\frac{1}{2} - 4\frac{3}{16} + 2\frac{7}{16}$

28. $5\frac{3}{7} + 8\frac{7}{10} - 3\frac{1}{3} - 1\frac{1}{5}$ 28.

29. $8\frac{3}{4} - 1\frac{7}{16} + 4\frac{1}{8} - 2\frac{1}{3}$ 29.

30. $59\frac{2}{5} - \left(8\frac{3}{10} + 5\frac{7}{15} + 18\frac{4}{5}\right)$ 30.

31. $40\frac{3}{7} - \left(5\frac{2}{3} + 6\frac{4}{5} - 2\frac{5}{14}\right)$ 31.

32. $25 - \left(6\frac{5}{9} - 2\frac{2}{3} - 1\frac{6}{7}\right)$ 32.

33. $17\frac{5}{16} + 3\frac{7}{8} - \left(4\frac{2}{3} + 5\frac{3}{4} - 2\frac{5}{6}\right)$ 33.

34. $59\frac{2}{9} - \left(4\frac{1}{3} + 7\frac{2}{3} + 8\frac{21}{27}\right) + 6\frac{7}{9}$ 34.

35. $27\frac{4}{5} - 15\frac{3}{4} + 5\frac{3}{8} - 2\frac{7}{10} + 5\frac{31}{40}$ 35.

36. A saw makes a $\frac{3}{32}$-inch cut in a $6\frac{3}{4}$-inch by 18-inch board that is $\frac{3}{8}$ inch thick. What is the remaining thickness? 36.

37. A cylindrical piece of steel 6 inches long by $\frac{3}{4}$ inch in diameter is placed in a lathe and reduced until the diameter is $\frac{41}{64}$ inch. By how much was the diameter of the piece reduced? 37.

38.

38. What is the difference in thickness between two steel plates $4\frac{1}{2}$ inches by $5\frac{3}{4}$ inches if they have a thickness of $\frac{9}{16}$ inch and $\frac{7}{8}$ inch respectively?

39.

39. Dick picked the following from his garden: $1\frac{3}{4}$ bushels of tomatoes, $3\frac{1}{2}$ bushels of squash, $2\frac{1}{4}$ bushels of Delicious apples, and $3\frac{3}{4}$ bushels of Jonathan apples. How many more bushels of apples did he harvest than squash?

40.

40. At the opening of the New York Stock Exchange a certain stock listed for $35\frac{1}{2}$ and a second stock listed for $19\frac{3}{4}$. At the end of the first hour, the price of the first stock had risen $\frac{5}{8}$ of a point. During the next hour the price of the same stock rose $\frac{1}{4}$ of a point and the price remained the same the rest of the day until an hour before closing, when it fell $\frac{3}{8}$ of a point. What was the stock selling for at the close of the Exchange?

41.

41. Dorothy bought $6\frac{1}{3}$ yards of 48-inch nylon material and $3\frac{1}{2}$ yards of 36-inch cotton material. If she used $1\frac{5}{8}$ yards of the cotton material for a curtain, how much of that material did she have left? Does she have enough cotton material left to use $1\frac{3}{4}$ yards for a shirt?

42.

42. One tube of toothpaste is $6\frac{1}{2}$ inches long and weighs $2\frac{1}{8}$ ounces, while another tube is $7\frac{1}{4}$ inches long and weighs $1\frac{7}{8}$ ounces. Which tube is heavier and by how much?

43.

43. Mr. Thompson indicated to his wife that last year they spent $\frac{1}{4}$ of their income for rent, $\frac{2}{5}$ for food and recreation, $\frac{1}{10}$ for clothing, $\frac{1}{12}$ for insurance, and $\frac{1}{10}$ went into savings. Did Mr. Thompson account for all his money? If the answer is no, what part was not accounted for?

44.

44. Nominal 1-inch lumber now has an actual thickness of $\frac{25}{32}$ inch. It has been proposed that this be reduced to $\frac{3}{4}$ inch. What is the reduction in thickness that is being proposed?

45.

45. In the production of particle board, the board is soaked in water for a 24-hour period. During this time there is a change in thickness. Determine the increase in thickness if a sample was originally $\frac{3}{8}$ inch and the final thickness was $\frac{2}{5}$ inch.

46. An existing spur road $13\frac{1}{2}$ miles long in the Fort Smith Ranger District is to be rebuilt. $6\frac{1}{2}$ miles of the road requires extensive work. Of the remaining road, $4\frac{1}{3}$ miles requires two bladings and a ditch cleaning; the rest needs only one blading and a ditch cleaning. How much of the road needs only one blading and a ditch cleaning? 46.

47. Three pieces of sheet metal are to be riveted together. The thicknesses in inches are $\frac{1}{16}$, $\frac{3}{64}$, and $\frac{1}{32}$. What length of rivet should be used, allowing $\frac{1}{16}$ inch for heading over? 47.

48. A template is to be made $5\frac{15}{16}$ inches long with a tolerance of $\pm\frac{1}{32}$ inch. Between what two dimensions should its length be ground to meet the specifications? 48.

49. The front axle of an automobile is to be aligned with the rear axle. The rear axle to front axle measurement is 9 feet, $2\frac{3}{4}$ inches on one side and 9 feet, $3\frac{7}{16}$ inches on the other side. How much does one side need to be shifted to bring the axle into alignment? 49.

50. A garage owner has a piece of radiator hose 69 inches long. He cuts from it short pieces as follows: $15\frac{5}{8}$ inches, $11\frac{7}{16}$ inches, and $7\frac{3}{4}$ inches. How many inches of hose are left? 50.

*51. Arrange the numbers 9, 10, 11, and 12 into two fractions which will give the largest difference. (Use each number only once.) *51.

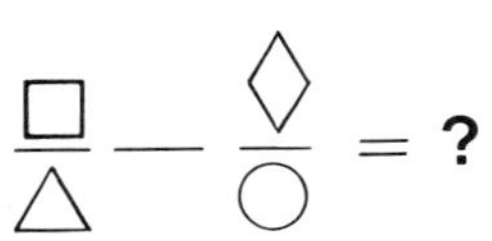

*52. Arrange the numbers 9, 10, 11, and 12 into two fractions which will give the smallest difference. (Use each number only once.) *52.

4.4 Using Decimals in Place of Fractions

With the increasing use of the metric system in the United States, there will be less and less need for all but such simple fractions as $\frac{1}{2}$, $\frac{1}{4}$, $\frac{2}{3}$, $\frac{3}{4}$, etc. In any event, it is often desirable to convert a fraction into a decimal by using the fact that

$$\frac{a}{b} = a \div b$$

Do exercises 1 through 4 in the adjoining column.

In exercises 1–4, all of the given fractions are exactly equal to a decimal. On the other hand, an HHC gives

$$\frac{1}{3} = 0.3333333$$

and

$$\frac{8}{7} = 1.1428571$$

where, if the display of the HHC allowed, still more digits would appear.

The decimals that appeared as answers to exercises 1–4 are all examples of what are called *terminating* decimals, whereas the decimals that we obtained for $\frac{1}{3}$ and $\frac{8}{7}$ are examples of *non-terminating* decimals.

Why are some fractions equal to terminating decimals and others equal to non-terminating decimals? The answer lies in the meaning of decimals, which represent fractions with denominators that are powers of 10: 10, 100, 1000, 10,000, etc. Thus

$$0.4 = \frac{4}{10}, \quad 0.042 = \frac{42}{1000}, \quad \text{and} \quad 1.3257 = \frac{13{,}257}{10{,}000}$$

Do exercises 5 through 8.

Conversely, to write a fraction as a terminating decimal, we must be able to write that fraction as a fraction with denominator a power of 10: 10, 100, 1000, etc. For example, we can use the fundamental principle to write

$$\frac{2}{5} = \frac{2 \times 2}{5 \times 2} = \frac{4}{10} = 0.4, \quad \frac{3}{50} = \frac{3 \times 2}{50 \times 2} = \frac{6}{100} = 0.06$$

and

$$\frac{9}{80} = \frac{9}{2^4 \times 5} = \frac{9 \times 5^3}{(2^4 \times 5) \times 5^3} = \frac{9 \times 125}{2^4 \times 5^4} = \frac{1125}{10{,}000} = 0.1125$$

OBJECTIVES

You should be able to:

(a) Convert a fraction into a terminating or repeating decimal.

(b) Understand why some fractions are equal to terminating decimals and others are equal to non-terminating but repeating decimals.

(c) Add and subtract fractions by using their decimal equivalents.

****

In exercises 1–4 write the given fraction as a decimal.

1. $\frac{2}{25}$

2. $\frac{7}{8}$

3. $\frac{21}{32}$

4. $\frac{19}{20}$

STOP

In exercises 5–8 write the given decimal as a fraction.

5. 0.7

6. 0.546

7. 2.431

8. 0.00023

STOP

In exercises 9–13 determine, without using your HHC, whether the given fraction can be expressed as a terminating decimal. Give a reason for your answer.

9. $\frac{3}{8}$

10. $\frac{7}{40}$

11. $\frac{4}{15}$

12. $\frac{5}{6}$

13. $\frac{7}{125}$

In exercises 14–16 first express the given fraction as a fraction with a denominator a power of 10 and then use this result to write the fraction as a terminating decimal. Finally, check your answer by performing a division with your HHC.

14. $\frac{3}{4}$

15. $\frac{7}{20}$

16. $\frac{3}{16}$

STOP

What about $\frac{1}{3}$? Is there any whole number m such that $3 \times m$ is 10, 100, 1000, 10,000 or any other power of 10? Certainly not, because 3 is not a divisor of 10, 100, 1000, 10,000, or any other power of 10. Indeed, the only prime divisors of any power of 10 are 2 and 5, and hence the denominator of any fraction in lowest terms that is equal to a terminating decimal must be such that its only prime factors are 2 and 5. Thus, for example, we know that

$$\frac{27}{280}$$

cannot be written as a terminating decimal because 280 has 7 as a factor and $\frac{27}{280}$ is in lowest terms.

Note that before applying this test, we must check that the fraction is in lowest terms. Thus, for example, $\frac{27}{24}$ has a denominator with a factor of 3. But

$$\frac{27}{24} = \frac{9 \times 3}{8 \times 3} = \frac{9}{8} = \frac{9 \times 125}{8 \times 125} = \frac{1125}{1000} = 1.125$$

Do exercises 9 through 16.

Now let us look more closely at those fractions that are equal to nonterminating decimals. Suppose that we write out the division involved in getting

$$\frac{8}{7} = 1.1428571$$

We have

```
     1.142857
  7 ) 8.0000000
      7
      10
       7
       30
       28
        20
        14
         60
         56
          40
          35
           50
           49
            1
```

and, at this point, it is clear that we begin to repeat the digits in the quotient since a remainder of 1 has occurred for the second time. Thus

$$\frac{8}{7} = 1.142857\ 142857\ 142857 \cdots$$

where the sequence of digits 1, 4, 2, 8, 5, and 7 repeats endlessly.

Why did this happen? Well, the possible remainders when dividing by 7 are 0 (but only if the division terminates), 1, 2, 3, 4, 5, and 6. (Why not 7, 8, 9, . . . ?) Thus at some point a remainder must be repeated and hence a sequence of digits begins to repeat.

A similar argument applies to any fraction a/b. If the division of a by b does not terminate, the possible remainders are 1, 2, 3, . . . , $b - 1$ and so, at some point, a remainder will be repeated. That is, any fraction is equal to either a terminating or a non-terminating but *repeating* decimal.

Do exercises 17 through 27.

In doing exercises 17–27, you will have noticed that the repetition of digits need not begin immediately. Thus, for example,

$$\frac{5}{6} = 0.8333333$$

where 8 is not a part of the repeating cycle.

A bar over the repeating cycle is often used to indicate the repetition. Thus we write

$$0.3333333\ldots = 0.\overline{3}, \qquad 0.8333333\ldots = 0.8\overline{3},$$

$$0.142857142857142857\ldots = 0.\overline{142857},$$

and

$$1.32557575757\ldots = 1.325\overline{57}$$

In exercises 17–26 use your HHC to find a repeating decimal equal to the given fraction.

17. $\frac{5}{6}$

18. $\frac{9}{11}$

19. $\frac{1}{7}$

20. $\frac{2}{3}$

21. $\frac{5}{12}$

22. $\frac{5}{37}$

23. $\frac{5}{9}$

24. $\frac{35}{99}$

25. $\frac{47}{99}$

26. $\frac{213}{999}$

27. Study your answers to exercises 23–26 to see if you can find a way to write the following fractions as repeating decimals without using your HHC.

$\frac{7}{9}$

$\frac{13}{99}$

$\frac{28}{99}$

Check by using your HHC

STOP

In exercises 28–32 find a repeating decimal equal to the given fraction. Write your answer using the bar notation.

28. $\frac{7}{9}$

29. $\frac{7}{30}$

30. $\frac{15}{11}$

31. $\frac{4}{33}$

32. $\frac{157}{999}$

33. Compute

142857 × 2

142857 × 3

142857 × 4

142857 × 5

142857 × 6

What interesting thing do you observe about these products? Now multiply 142857 by 7 and note that something quite different occurs. Now try to relate what has happened to the fact that

$$\frac{1}{7} = 0.\overline{142857},$$

$$\frac{2}{7} = 2 \times \frac{1}{7}$$

$$\frac{3}{7} = 3 \times \frac{1}{7}$$

...

$$1 = \frac{7}{7} = 7 \times \frac{1}{7}$$

STOP

Do exercises 28 through 33.

Of course, with an HHC it is not always possible to detect the repeating cycle directly. For example, with an HHC we get

$$\frac{1}{43} = 0.0232558$$

where no repetition of digits is evident. (In fact, when we calculate

$$\frac{1}{7} = 0.1428571$$

on an HHC we are certainly guessing when we conclude that the second 1 heralds the beginning of a repetition!) However, as you can check if you have the patience, it is true that

$$\frac{1}{43} = 0.\overline{023255813953488372093}$$

We have been writing, for example,

$$\frac{1}{3} = 0.3333333, \quad \frac{1}{7} = 0.1428571$$

etc. but, of course, this is really not correct. Instead we should write

$$\frac{1}{3} \approx 0.3333333, \quad \frac{1}{7} \approx 0.1428571$$

where "≈" means *approximately equal to.* It is, however, true that

$$\frac{1}{3} = 0.\overline{3} \quad \text{and} \quad \frac{1}{7} = 0.\overline{142857}$$

since the bar indicates that the digits are to continue on forever.

We can use decimal approximations to fractions to provide an alternate method of adding or subtracting fractions. For example,

$$(1) \qquad \frac{81}{125} + \frac{52}{627} \approx 0.648 + 0.0829346 = 0.7309346$$

This gives us an approximate decimal answer. To obtain the answer as a fraction, remember that $a/b = c$ means that $c \times b = a$; therefore, all we need to do is to multiply this answer by 125 × 627 to get

$$0.7309346 \times 125 \times 627 \approx 57{,}286.999$$

Then, recognizing that 57,286.999 is certainly a decimal approximation to 57,287, we have, since 125 × 627 = 78,375,

$$\frac{81}{125} + \frac{52}{627} = \frac{57{,}287}{78{,}375}$$

Note that

$$\frac{57{,}287}{78{,}375} \approx 0.7309346$$

which is equal to the decimal sum obtained in (1).

Do exercises 34 through 36.

In exercises 34–36 perform the indicated addition by using the decimal approximation procedure as shown in the text example.

34. $\frac{3}{4} + \frac{7}{10}$

35. $\frac{4}{7} + \frac{13}{15}$

36. $\frac{55}{97} + \frac{122}{843}$

STOP

Similarly, to calculate

$$\frac{81}{125} - \frac{52}{627}$$

we have

$$(2) \qquad \frac{81}{125} - \frac{52}{627} \approx 0.648 - 0.0829346 = 0.5650654$$

Then

$$0.5650654 \times 125 \times 627 = 44{,}287$$

and hence, since $125 \times 627 = 78{,}375$,

$$\frac{81}{125} - \frac{52}{627} = \frac{44{,}287}{78{,}375}$$

(Here there was no need to round off to get the numerator of 44,287.)
Note that

$$\frac{44{,}287}{78{,}375} \approx 0.5650653$$

which, except for the last decimal place, is the decimal answer obtained in (2).

Do exercises 37 through 39.

In exercises 37–39 perform the indicated subtraction by using the decimal approximation procedure shown in the text.

37. $\frac{3}{4} - \frac{2}{5}$

38. $\frac{14}{17} - \frac{15}{27}$

39. $\frac{38}{121} - \frac{201}{962}$

STOP

When mixed numbers are involved in additions or subtractions, we can again use decimals. For example, to find

$$7\frac{5}{19} - 2\frac{18}{15}$$

we can calculate

$$\frac{5}{19} \approx 0.2631578 \quad \text{and} \quad \frac{18}{35} \approx 0.5142857$$

so that we have

$$7\frac{5}{19} - 2\frac{18}{35} \approx 7.2631578 - 2.5142857 = 4.7488721$$

In exercises 40–41 perform the indicated calculation by using the decimal approximation procedure as shown in the text example. Express your answer as a mixed number.

40. $5\frac{3}{17} + 8\frac{9}{22}$

41. $6\frac{3}{28} - 4\frac{2}{21}$

STOP

42. Write the improper fraction $\frac{589}{14}$ as a mixed number by using the decimal approximation procedure shown in the text example.

STOP

Thus, since 19 × 35 = 665, we know that

$$7\frac{5}{19} - 2\frac{18}{35} = 4\frac{n}{665}$$

for some whole number n. Then to find n we have

$$n \approx 665 \times 0.7488721 \approx 497.99994$$

So, since n is a whole number, $n = 498$ and thus

$$7\frac{5}{19} - 2\frac{18}{45} = 4\frac{498}{664}$$

Do exercises 40 and 41.

Another use for this procedure of approximating fractions by decimals is, as discussed briefly in section 4.2, in converting an improper fraction to a mixed number. For example, an HHC gives

$$\frac{784}{137} \approx 5.7226277$$

so that we know that

$$\frac{784}{137} = 5\frac{n}{137}$$

where n is a whole number. Then to find n we can calculate

$$0.7226277 \times 137 \approx 98.999994$$

Thus, since n is a whole number, we have $n = 99$ and so

$$\frac{784}{137} = 5\frac{99}{137}$$

Check: $\frac{99}{137} \approx 0.7226277$

Do exercise 42.

Although in many computations with decimals we should use "≈" rather than "=" (as we have been doing), we will, from now on, often use "=" in place of "≈" (as is customary).

Exercises 4.4

Name ____________________

Class ____________________

In exercises 1–5 write the given fraction as a decimal.

1. $\frac{9}{10}$
2. $\frac{3}{4}$
3. $\frac{3}{8}$
4. $\frac{5}{16}$
5. $\frac{7}{40}$

In exercises 6–10 write the given decimal as a fraction.

6. 0.85
7. 0.481
8. 4.208
9. 0.10096
10. 0.000027

In exercises 11–15 determine (without using your HHC) whether the given fraction can be expressed as a terminating decimal and give a reason for your answer.

11. $\frac{5}{8}$
12. $\frac{7}{16}$
13. $\frac{8}{15}$
14. $\frac{17}{22}$
15. $\frac{59}{250}$

In exercises 16–20 express the given fraction as a fraction with a denominator a power of 10 and use your result to write the fraction as a terminating decimal. Then check your answer by performing a division with your HHC.

16. $\frac{3}{5}$
17. $\frac{7}{25}$
18. $\frac{5}{16}$
19. $\frac{13}{40}$

1.
2.
3.
4.
5.
6.
7.
8.
9.
10.
11.
12.
13.
14.
15.
16.
17.
18.
19.

Answers	Exercises	
20.	20. $\frac{23}{125}$	

In exercises 21–30 use your HHC to find a repeating decimal equal to the given fraction.

Answers	Exercises	
21.	21. $\frac{4}{9}$	22. $\frac{3}{7}$
22.		
23.	23. $\frac{5}{13}$	24. $\frac{4}{11}$
24.		
25.	25. $\frac{7}{11}$	26. $\frac{16}{22}$
26.		
27.	27. $\frac{26}{27}$	28. $\frac{5}{9}$
28.		
29.	29. $\frac{7}{15}$	30. $\frac{7}{60}$
30.		

In exercises 31–50 perform the indicated addition or subtraction by using decimal approximations. Write your answer as a proper fraction or a mixed number.

Answers	Exercises
31.	31. $\frac{3}{8} + \frac{4}{5}$
32.	32. $\frac{56}{11} + \frac{4}{9}$
33.	33. $\frac{7}{8} - \frac{3}{5}$
34.	34. $\frac{14}{15} - \frac{5}{6}$
35.	35. $\frac{31}{33} + \frac{17}{18}$
36.	36. $25\frac{1}{2} + 19\frac{11}{16}$
37.	37. $61\frac{11}{12} + 52\frac{2}{3}$
38.	38. $83\frac{5}{8} - 34\frac{3}{4}$

39. $48\frac{5}{6} - 26\frac{7}{8}$ 39.

40. $34\frac{1}{5} + 16\frac{7}{15} + 28\frac{2}{3}$ 40.

41. A bag of fertilizer is composed of $2\frac{3}{8}$ pounds of potassium, $1\frac{1}{2}$ pounds of nitrogen, $3\frac{5}{16}$ pounds of phosphate, and $\frac{1}{4}$ pound of other material. What is the total weight of the bag of fertilizer? 41.

42. A machinist is required to turn a shaft to a diameter of $1\frac{61}{64}$ inches. If the shaft is now $2\frac{1}{16}$ inches in diameter, by how much does she have to reduce the diameter? 42.

43. How much greater is a distance of $64\frac{3}{10}$ miles than a distance of $48\frac{3}{4}$ miles? 43.

44. Two boys ran the 100-yard dash in $13\frac{3}{4}$ seconds and $12\frac{1}{5}$ seconds respectively. How much faster was the second boy's time? 44.

45. The gasoline tank in a car holds $22\frac{1}{5}$ gallons. If it took $11\frac{7}{10}$ gallons to fill the tank, how many gallons were in the tank before it was filled? 45.

46. A company shipped three packages whose weights were $55\frac{1}{4}$ lb, $72\frac{3}{8}$ lb, and $25\frac{2}{3}$ lb. What was the total weight of the shipment? 46.

47. A piece of shafting $10\frac{5}{32}$ inches long is to be cut from a piece $36\frac{1}{2}$ inches long. Will there be enough of the original piece left to make a piece $26\frac{1}{4}$ inches long if $\frac{1}{16}$ inch of shafting is wasted in making a cut? 47.

48. Three pieces with the respective lengths of $3\frac{3}{4}$ inches, $1\frac{5}{16}$ inches, and $4\frac{5}{8}$ inches are to be cut from a rod $16\frac{1}{8}$ inches long. What will be the length of the remaining piece if you neglect the waste made in cutting? 48.

49. How many inches of trim are needed for an irregularly shaped piece of paneling whose sides are of lengths $30\frac{1}{2}$ inches, $62\frac{1}{8}$ inches, $70\frac{5}{8}$ inches, $37\frac{1}{16}$ inches, and $60\frac{13}{16}$ inches?

50. Find the total length of the connecting rod in the accompanying illustration.

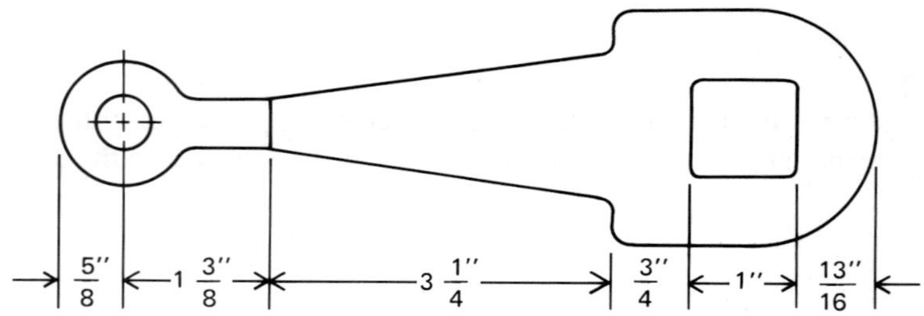

*51. (a) Name five fractions which are less than $\frac{1}{2}$ and greater than $\frac{1}{4}$.

(b) Name five decimals which are less than $\frac{1}{2}$ and greater than $\frac{1}{4}$.

(c) How many answers could be given to part (a)? To part (b)?

*52. Is there a smallest rational number (fraction or decimal) greater than zero?

*53. Is there a largest rational number (fraction or decimal)?

*54. Give all possible natural numbers that can be placed in the squares to make the statement below true.

$$\frac{\square}{5} < \frac{5}{\square}$$

*55. Show, by doing at least three examples, that the sum of two fractions $\frac{a}{b}$ and $\frac{c}{d}$ can be found on an HHC by the following method:

$$\frac{a}{b} + \frac{c}{d} = \left(\frac{a \times d}{b} + c\right) \div d$$

4.5 Multiplication and Division of Fractions

OBJECTIVES

You should be able to:

(a) Multiply and divide fractions.

(b) Apply multiplication and division of fractions to problems.

****

MULTIPLICATION

Consider the following figure which shows a square divided by five vertical lines into five rectangles of equal area and with $\frac{4}{5}$ of the area of the square hatched.

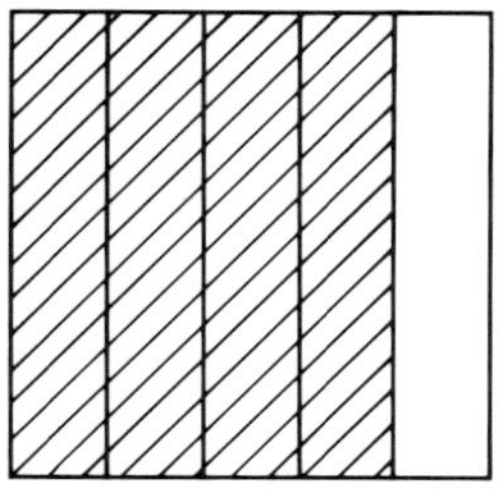

The next figure shows this hatched area further divided by three equally spaced horizontal lines with $\frac{2}{3}$ of this hatched area indicated by cross hatch.

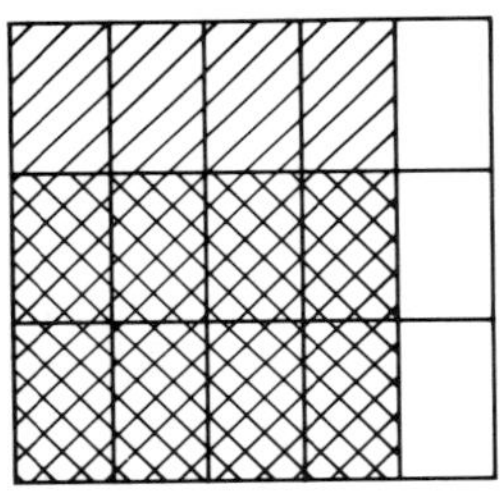

The square has now been divided into 15 rectangles of equal area of which 8 are cross hatched. This indicates that $\frac{2}{3}$ of $\frac{4}{5}$ is $\frac{8}{15}$, i.e., that

$$\frac{2}{3} \times \frac{4}{5} = \frac{2 \times 4}{3 \times 5} = \frac{8}{15}$$

Considerations like these lead us to the rule for multiplying fractions:

(1) $$\frac{a}{b} \times \frac{c}{d} = \frac{a \times c}{b \times d}$$

We can use an HHC to check this rule. For example, to check that

$$\frac{73}{61} \times \frac{95}{86} = \frac{73 \times 95}{61 \times 86}$$

use an HHC to get first

$$\frac{73}{61} = 73 \div 61 = 1.1967213$$

1. Check that

$$\frac{94}{27} \times \frac{67}{123} = \frac{94 \times 67}{27 \times 123}$$

by using your HHC as in the text example.

In exercises 2–6 use formula (1) to perform the indicated multiplication.

2. $\frac{16}{23} \times \frac{49}{31}$

3. $\frac{241}{19} \times \frac{72}{81}$

4. $\frac{97}{123} \times \frac{23}{219}$

5. $\frac{317}{92} \times \frac{119}{2314}$

6. $\frac{598}{227} \times \frac{973}{228} \times \frac{4}{27}$

STOP

and

$$\frac{95}{86} = 95 \div 86 = 1.1046511$$

so that

$$\frac{73}{61} \times \frac{95}{86} = 1.197213 \times 1.1046511 = 1.3219595$$

On the other hand, we also have

$$\frac{73 \times 95}{61 \times 86} = \frac{6935}{5246} = 1.3219595$$

Do exercises 1 through 6 in the adjoining column.

To multiply mixed numbers, the easiest procedure is to change to improper fractions. Thus

$$5\frac{2}{3} \times 7\frac{3}{5} = \frac{17}{3} \times \frac{38}{5} = \frac{646}{15} = 43\frac{1}{15}$$

Do exercises 7 through 11.

The direct application of (1) however, is not always the best procedure for multiplying fractions, especially since we usually want the answer in lowest terms. Consider, for example, the problem of calculating

$$\frac{399}{1027} \times \frac{325}{1026}$$

Direct use of (1) gives us

$$\frac{399}{1027} \times \frac{325}{1026} = \frac{129{,}675}{1{,}053{,}702}$$

It is rather a chore to reduce this answer to lowest terms. This chore can be avoided by first writing the prime factorizations (see Section 3.3) of the numbers involved in the multiplication. Doing this, we have

$$\frac{399}{1027} \times \frac{325}{1026} = \frac{7 \times 57}{79 \times 13} \times \frac{13 \times 5 \times 5}{2 \times 3 \times 3 \times 57}$$

$$= \frac{(7 \times 5 \times 5) \times (13 \times 57)}{(79 \times 2 \times 3 \times 3) \times (13 \times 57)}$$

Now we apply the fundamental principle to get

$$\frac{399}{1027} \times \frac{325}{1026} = \frac{7 \times 5 \times 5}{79 \times 2 \times 3 \times 3} = \frac{175}{1422}$$

an answer in lowest terms.

In exercises 7-11 perform the indicated multiplication and express your answer as a mixed number.

7. $8\frac{1}{3} \times 1\frac{1}{2}$

8. $4\frac{2}{5} \times 3\frac{3}{4}$

9. $1\frac{1}{3} \times 2\frac{7}{10}$

10. $5\frac{13}{17} \times 6\frac{4}{9}$

11. $8\frac{2}{13} \times 13\frac{9}{16} \times 6\frac{2}{3}$

STOP

Do exercises 12 through 15.

In exercises 12–15 perform the indicated multiplication by first converting any mixed numbers to improper fractions and then seeing if there are any factors in common in the numerators and denominators of the fractions. If there are, use the fundamental principle (as in the text example) to obtain an answer to the multiplication problem in lowest terms.

12. $\frac{49}{56} \times \frac{8}{24}$

13. $\frac{18}{27} \times \frac{15}{35}$

14. $2\frac{2}{9} \times 2\frac{4}{5}$

15. $3\frac{1}{5} \times 6\frac{3}{8}$

STOP

DIVISION

The usual rule used for division of fractions is

(2) $$\frac{a}{b} \div \frac{c}{d} = \frac{a}{b} \times \frac{d}{c}$$

(the invert-and-multiply rule).

We can use an HHC to check this rule. For example, to check that

$$\frac{27}{92} \div \frac{19}{24} = \frac{27}{92} \times \frac{24}{19}$$

we use the HHC to get

$$\frac{27}{92} \div \frac{19}{24} = 0.2934782 \div 0.7916666 = 0.3707093$$

On the other hand, we also have

$$\frac{27}{92} \times \frac{24}{19} = \frac{27 \times 24}{92 \times 19} = \frac{648}{1748} = 648 \div 1748 = 0.3707093$$

Again, as for multiplication, the easiest way to divide mixed numbers is to change to improper fractions. Thus

$$5\frac{2}{3} \div 7\frac{3}{5} = \frac{17}{3} \div \frac{38}{5} = \frac{17}{3} \times \frac{5}{38} = \frac{85}{114}$$

Do exercises 16 and 17.

Using (2) is not the only way to divide one fraction by another. We also have

$$\frac{a}{b} \div \frac{c}{d} = \frac{a \div c}{b \div d} \tag{3}$$

Checking (3) with our previous example, we have

$$\frac{27}{92} \div \frac{19}{24} = \frac{27 \div 19}{92 \div 24} \approx \frac{1.4210526}{3.8333333} \approx 0.3707093$$

Do exercise 18.

Of course, (3) doesn't, in our example, tell us what the answer is as a fraction. However, using (3) is an easy way to do some problems in division of fractions as, for example,

$$\frac{24}{39} \div \frac{6}{13} = \frac{24 \div 6}{39 \div 13} = \frac{4}{3}$$

Do exercises 19 through 22.

In exercises 16–17 perform the indicated division by using (2) and check your answer, as a decimal, by the procedure shown in the text.

16. $\frac{4}{5} \div \frac{8}{9}$

17. $5\frac{1}{3} \div 2\frac{2}{9}$

STOP

18. Check your answer to exercise 16 by using (3) to obtain a decimal answer.

STOP

In exercises 19–22 perform the division by using (3) to obtain a fraction as an answer.

19. $\frac{36}{45} \div \frac{9}{5}$

20. $\frac{390}{387} \div \frac{15}{9}$

21. $\frac{1349}{1269} \div \frac{71}{47}$

22. $\frac{2700}{2409} \div \frac{125}{219}$

STOP

Exercises 4.5

Name ______________

Class ______________

In exercises 1-30 perform the indicated computation. Write your answer as a proper fraction in lowest terms or as a mixed number.

1. $\frac{5}{11} \times \frac{7}{4}$

2. $\frac{3}{10} \times \frac{15}{4}$

3. $\frac{15}{16} \times \frac{24}{5}$

4. $\frac{4}{15} \times \frac{25}{18}$

5. $\frac{35}{6} \times \frac{14}{55}$

6. $\frac{12}{25} \times \frac{5}{6}$

7. $\frac{53}{55} \times \frac{35}{48}$

8. $3\frac{1}{2} \times 1\frac{1}{6}$

9. $2\frac{1}{4} \times 3\frac{2}{3}$

10. $2\frac{2}{9} \times 2\frac{4}{5}$

11. $3\frac{1}{3} \times 12\frac{3}{4}$

12. $3\frac{1}{5} \times 6\frac{3}{8}$

13. $2\frac{2}{5} \times 8\frac{1}{3}$

14. $\frac{18}{25} \times \frac{20}{27} \times \frac{45}{64}$

15. $\frac{7}{8} \times \frac{6}{49} \times \frac{10}{63}$

16. $1\frac{11}{24} \times 2\frac{8}{45} \times 1\frac{1}{35}$

17. $\frac{4}{49} \times \frac{28}{135} \times 2\frac{4}{25}$

18. $\frac{25}{12} \div \frac{35}{18}$

19. $\frac{63}{52} \div \frac{9}{13}$

20. $\frac{36}{85} \div \frac{78}{17}$

21. $2\frac{1}{4} \div \frac{3}{4}$

22. $2\frac{3}{5} \div \frac{4}{5}$

23. $3\frac{3}{4} \div 1\frac{1}{4}$

24. $12\frac{2}{3} \div 5\frac{8}{11}$

25. $\frac{40}{21} \div \left(\frac{4}{3} - \frac{7}{9}\right)$

26. $\frac{49}{25} \div \left(\frac{9}{8} \times \frac{28}{15}\right)$

1.	2.
3.	4.
5.	6.
7.	8.
9.	10.
11.	12.
13.	14.
15.	16.
17.	18.
19.	20.
21.	22.
23.	24.
25.	26.

27. 28.

27. $\left(\frac{1}{5} \times \frac{7}{12}\right) \times \left(\frac{2}{3} - \frac{4}{9}\right)$

28. $\left(\frac{5}{18} + \frac{7}{24}\right) \div \frac{82}{81}$

29. 30.

29. $\left(\frac{2}{3} + \frac{3}{4}\right) \div \left(\frac{5}{6} - \frac{4}{5}\right)$

30. $\left(\frac{7}{6} - \frac{5}{8}\right) \div \left(\frac{11}{9} + \frac{7}{12}\right)$

31.

31. Joan bought $\frac{3}{4}$ pound of pecans for cooking and used $\frac{2}{3}$ of them. What fraction of a pound did she use?

32.

32. At $8\frac{3}{4}$ miles a minute, how far will an airplane fly in 15 minutes?

33.

33. Grandma Basset had $\frac{7}{8}$ of a yard of ribbon. She used $\frac{1}{2}$ of it on a doll dress for her granddaughter. What fraction of a yard of ribbon did she use?

34.

34. The gas tank of Jerry's car holds 22 gallons of gas. If his gas gauge reads $\frac{3}{4}$ full, how many gallons of gas remain in the tank?

35.

35. A factory worker spends $2\frac{1}{2}$ hours at machine A and $4\frac{1}{2}$ hours at machine B. The rest of his 8-hour working day is spent at machine C. If his pay is $4.40 an hour, how much should he charge for the time spent at each machine?

36.

36. If a production worker takes $1\frac{1}{4}$ minutes for putting away a finished piece and replacing a piece of material in a press, how long will it take her to make 50 pieces if the press takes $3\frac{1}{2}$ minutes to form the material?

37.

37. If the length of a rod is to be $3\frac{3}{8}$ inches, how many such rods can be cut from a piece of stock rod 48 inches long? If there is any material left over, how long is it?

38.

38. If an iceberg floats with $\frac{1}{8}$ of its total height above water, how many feet below water will an iceberg extend if it stands 50 feet above the water?

39. A farmer used $12\frac{1}{2}$ bushels of seed corn to plant $16\frac{3}{4}$ acres. If the seed corn cost \$12 a bushel, how much did it cost to seed an acre? 39.

40. A tank is filled with 625 gallons of fuel oil. If $\frac{3}{5}$ of the fuel in the tank is used, how many gallons of oil are left in the tank? At a price of $41\frac{3}{4}$ cents per gallon, what would it cost to fill the tank? 40.

41. In an automobile electrical system the relationship between voltage, amperage, and resistance is expressed by the formula $E = I \times R$ (voltage equals amperage times the resistance in ohms). Find the resistance of the lighting system in a car if it draws $12\frac{1}{2}$ amperes at a battery voltage of $11\frac{3}{4}$ volts. 41.

42. In a 12-volt circuit, how many amperes of electricity are flowing through a wire which has a resistance of $1\frac{1}{2}$ ohms? (Use the formula in exercise 41.) 42.

43. If $\frac{2}{3}$ of an hour is the average time for greasing a car, how many cars can be greased in an 8-hour day? 43.

44. In grading lumber, a second-grade plank may have a split whose length is equal to $1\frac{1}{2}$ times the width of the board. What length of a split is allowed for a 2-inch by 8-inch plank that is 12 feet long? 44.

45. Aerial application of DDT mixed with diesel fuel has been used to control infestations of the tussock moth. Each gallon of the mixture contains $\frac{3}{4}$ pound of DDT and weighs $7\frac{13}{16}$ pounds. How many pounds of DDT are in a 20,000-pound plane load? 45.

46. A clearcut area of $19\frac{1}{2}$ acres is to be sprayed with atrazine to control a dense stand of grass before the area is reforested. Atrazine comes in bags of a mixture that contains $\frac{4}{5}$ atrazine. The mixture costs \$2 per pound and is applied at the rate of 4 pounds of active ingredient per acre. If it costs \$4 an acre to do the spraying, how much will the grass control cost? 46.

47.

47. How many arbors, each $9\frac{3}{8}$ inches long, can be cut from a piece of $1\frac{1}{4}$-inch diameter stock shafting that is $7\frac{1}{2}$ feet long, if $\frac{1}{16}$ inch of shafting is wasted in making a cut?

48.

48. A finishing tool on a shaper feeds $\frac{3}{32}$ inch on each stroke. How many strokes does the finishing tool make in finishing a surface $7\frac{1}{2}$ inches wide?

49.

49. Five holes are drilled in a straight line and the same distance apart. The distance from the center of the first hole to the center of the last hole is $13\frac{1}{2}$ inches. What will be the distance between the centers of any two adjacent holes?

50.

50. A garage stocks a 50-foot roll of $\frac{5}{8}$-inch heater hose. If an average of $12\frac{1}{2}$ feet of hose are required for each car, how many cars can have their heater hose replaced from the roll?

*51.

*51. If the reciprocal of $\frac{a}{b}$ is $\frac{b}{a}$ (reciprocal of $\frac{3}{4}$ is $\frac{4}{3}$), what number is equal to its reciprocal?

*52.

*52. What number is nine times its reciprocal?

*53.

*53. Simplify: $$1 + \cfrac{1}{\cfrac{1}{1+2}}$$

*54.

*54. Simplify: $$1 + \cfrac{1}{1 + \cfrac{1}{1 + \cfrac{1}{1+1}}}$$

*55.

*55. Jack gave Jill half his cookies and half a cookie. Jack had 9 cookies left. How many did Jill get?

Review Exercises 4

In exercises 1–4 write a fraction that the shaded part of the given figure represents.

1. 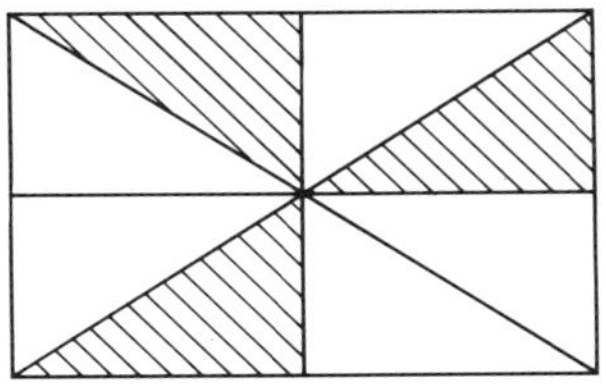

2. 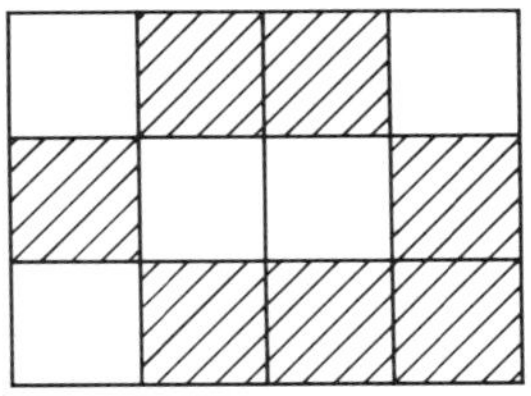

3. 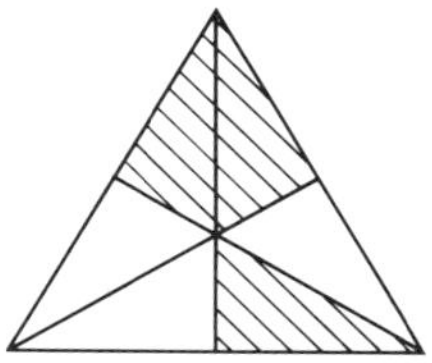

4. 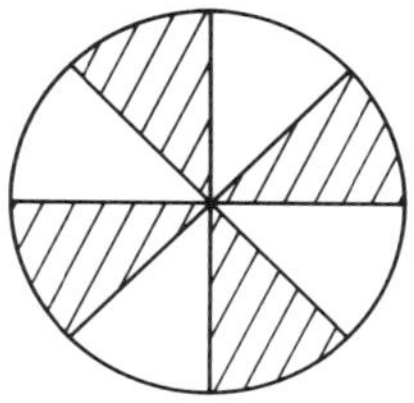

In exercises 5 and 6 write a fraction for each indicated point on the given number line.

5. 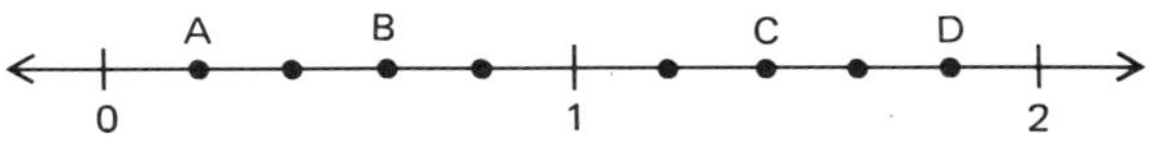

6. 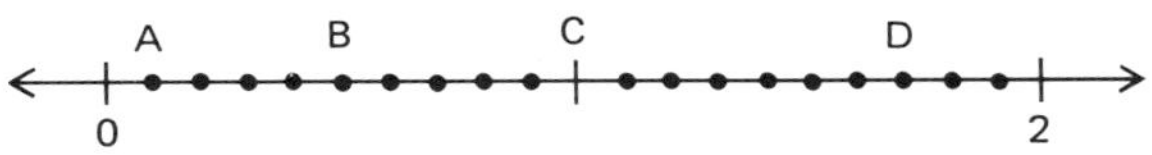

In exercises 7–12 indicate whether the given fraction is a proper fraction (P) or an improper fraction (I).

7. $\frac{5}{6}$

8. $\frac{5}{5}$

9. $\frac{8}{7}$

10. $\frac{15}{16}$

11. $\frac{29}{35}$

12. $\frac{13}{10}$

In exercises 13–20 list four fractions equivalent to the given fraction.

13. $\frac{3}{8}$

14. $\frac{6}{4}$

Name ____________________

Class ____________________

1. 2.

3. 4.

5. A B
C D

6. A B
C D

7. 8. 9.

10. 11. 12.

13.

14.

15.

16.

17.

18.

19.

20.

21. 22.

23. 24.

25.

26. 27.

28. 29.

30. 31.

32. 33.

34. 35.

15. $\frac{5}{6}$

16. $\frac{6}{9}$

17. $\frac{7}{25}$

18. $\frac{13}{19}$

19. $\frac{21}{17}$

20. $\frac{55}{64}$

In exercises 21–25 find the missing numerator for the given pair of equivalent fractions.

21. $\frac{2}{3} = \frac{?}{12}$

22. $\frac{7}{8} = \frac{?}{32}$

23. $\frac{5}{6} = \frac{?}{42}$

24. $\frac{7}{13} = \frac{?}{195}$

25. $\frac{?}{12} = \frac{35}{84}$

In exercises 26–35 write each fraction as an equivalent fraction in lowest terms.

26. $\frac{9}{12}$

27. $\frac{15}{20}$

28. $\frac{36}{24}$

29. $\frac{91}{14}$

30. $\frac{84}{96}$

31. $\frac{62}{26}$

32. $\frac{50}{125}$

33. $\frac{45}{51}$

34. $\frac{315}{135}$

35. $\frac{90}{138}$

In exercises 36–43 write each mixed number as an improper fraction.

36. $6\frac{7}{8}$

37. $18\frac{3}{4}$

38. $12\frac{2}{3}$

39. $22\frac{3}{8}$

40. $9\frac{8}{15}$

41. $21\frac{2}{9}$

42. $13\frac{7}{21}$

43. $45\frac{9}{17}$

In exercises 44–51 write each improper fraction as a mixed number.

44. $\frac{107}{6}$

45. $\frac{139}{8}$

46. $\frac{341}{16}$

47. $\frac{67}{4}$

48. $\frac{91}{16}$

49. $\frac{128}{9}$

50. $\frac{412}{15}$

51. $\frac{299}{12}$

In exercises 52–60 find the indicated sum.

52. $\frac{3}{4} + \frac{1}{6}$

53. $\frac{3}{8} + \frac{5}{6}$

54. $\frac{5}{12} + \frac{2}{15}$

55. $\frac{3}{8} + \frac{9}{20}$

56. $1\frac{2}{5} + 2\frac{3}{8}$

57. $3\frac{3}{4} + 3\frac{1}{6}$

36.	37.
38.	39.
40.	41.
42.	43.
44.	45.
46.	47.
48.	49.
50.	51.
52.	
53.	
54.	
55.	
56.	
57.	

58.

58. $4\frac{3}{20} + 5\frac{7}{12}$

59.

59. $7\frac{7}{10} + 9\frac{9}{14}$

60.

60. $8\frac{3}{8} + 6\frac{7}{12}$

In exercises 61–70 perform the indicated subtraction.

61.

61. $\frac{5}{12} - \frac{1}{3}$

62.

62. $\frac{3}{4} - \frac{2}{3}$

63.

63. $\frac{5}{6} - \frac{3}{4}$

64.

64. $15 - 8\frac{4}{5}$

65.

65. $2\frac{1}{4} - 1\frac{3}{4}$

66.

66. $10\frac{2}{3} - 3\frac{3}{5}$

67.

67. $28\frac{2}{3} - 5\frac{3}{8}$

68.

68. $15\frac{7}{13} - 9\frac{8}{25}$

69.

69. $29\frac{3}{65} - 15\frac{5}{17}$

70.

70. $245\frac{7}{9} - 137\frac{13}{17}$

In exercises 71–80 perform the indicated multiplication and write your answer as a proper fraction in lowest terms, a whole number, or a mixed number.

71.

71. $\frac{3}{8} \times \frac{5}{6}$

72. $\frac{4}{5} \times \frac{3}{2}$ 72.

73. $\frac{7}{9} \times \frac{14}{35}$ 73.

74. $\frac{2}{5} \times \frac{2}{3}$ 74.

75. $18 \times 1\frac{3}{4}$ 75.

76. $21 \times 3\frac{3}{7}$ 76.

77. $5\frac{3}{5} \times 2\frac{1}{3}$ 77.

78. $2\frac{1}{3} \times 2\frac{1}{2}$ 78.

79. $42\frac{1}{7} \times 36\frac{2}{3}$ 79.

80. $15\frac{3}{4} \times 20\frac{2}{3}$ 80.

In exercises 81–90 perform the indicated division and write your answer according to the instructions for exercises 71–80.

81. $\frac{2}{7} \div \frac{3}{5}$ 81.

82. $\frac{7}{8} \div \frac{3}{4}$ 82.

83. $\frac{2}{3} \div \frac{5}{6}$ 83.

84. $\frac{5}{8} \div \frac{3}{16}$ 84.

85. $\frac{3}{10} \div \frac{9}{20}$ 85.

86. $4 \div \frac{3}{4}$ 86.

87.

87. $\frac{1}{2} \div 100$

88.

88. $3\frac{1}{2} \div 2\frac{1}{4}$

89.

89. $6\frac{3}{4} \div 21\frac{1}{6}$

90.

90. $95\frac{1}{4} \div 21\frac{1}{6}$

In exercises 91 and 92 list those fractions that are equal to a repeating decimal.

91.

91. $\frac{3}{8}, \frac{2}{3}, \frac{5}{12}, \frac{9}{16}, \frac{7}{11}, \frac{132}{125}$

92.

92. $\frac{7}{20}, \frac{7}{8}, \frac{37}{25}, \frac{5}{9}, \frac{3}{7}, \frac{15}{32}$

In exercises 93 and 94 list those fractions that are equal to a terminating decimal.

93.

93. $\frac{3}{8}, \frac{2}{3}, \frac{5}{12}, \frac{9}{16}, \frac{7}{11}, \frac{132}{125}$

94.

94. $\frac{7}{20}, \frac{7}{8}, \frac{37}{25}, \frac{5}{9}, \frac{3}{7}, \frac{15}{32}$

In exercises 95 and 96 list those fractions that are equal to neither a terminating decimal nor a repeating decimal.

95.

95. $\frac{3}{8}, \frac{2}{3}, \frac{5}{12}, \frac{9}{16}, \frac{7}{11}, \frac{132}{125}$

96.

96. $\frac{7}{20}, \frac{7}{8}, \frac{37}{25}, \frac{5}{9}, \frac{3}{7}, \frac{15}{32}$

97.

97. Mr. Larson needs two strands of wire for each of three rows of pole beans in his garden. If the rows are 50 feet long and if he allows $\frac{1}{4}$ foot at each end of a strand for fastening, how many feet of wire will he need?

98.

98. Joe wears a size $6\frac{3}{4}$ hat. His Uncle Jim wears a size $7\frac{3}{8}$. If hat sizes increase by one-eighths, how many sizes smaller is Joe's hat than his uncle's?

99. You work for the Smiths doing odd jobs and yard work for \$2.50 an hour. Last month you worked four days for the following times: $2\frac{1}{4}$ hours, $1\frac{3}{4}$ hours, $2\frac{1}{2}$ hours, and 3 hours. How much money should you receive from the Smiths in payment for your work? 99.

100. Mary says that $\frac{3}{8}$ of the stamps in her collection are from the United States, $\frac{1}{4}$ from Europe, $\frac{1}{4}$ from South America, and the rest from Africa. What part of her stamp collection contains stamps from Africa? 100.

101. How far apart should the centers of the $\frac{3}{4}$-inch diameter holes be placed in the steel plate shown in the accompanying illustration, if the holes are to be equally spaced? 101.

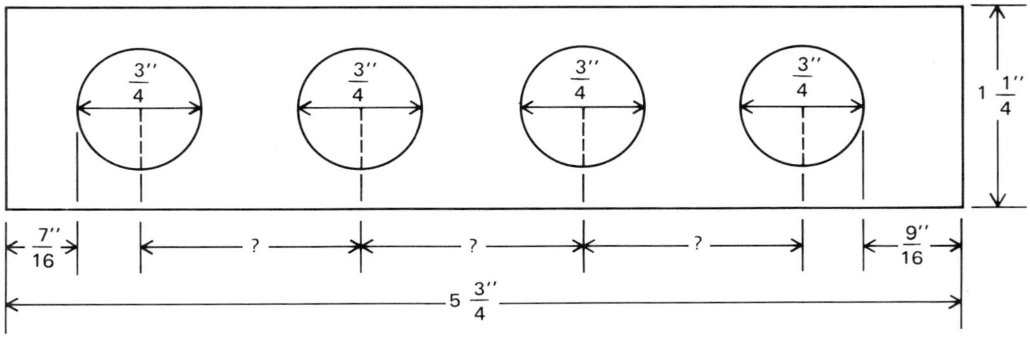

102. Ms. Dailey has $\frac{1}{2}$ of her money invested in real estate, $\frac{1}{10}$ in stocks, $\frac{1}{20}$ in heavy equipment, and the remainder, which is \$35,000, in savings accounts in two banks. How much is Ms. Dailey worth? 102.

103. After taxes, your family's budget of \$8400 is spent as follows: $\frac{1}{4}$, food; $\frac{1}{8}$, clothing; $\frac{1}{5}$, savings for a house; $\frac{3}{10}$, shelter; and the remainder for miscellaneous expenses. How much is spent for each item? 103.

104. (a)

(b)

105.

106.

107. (a)

(b)

108. (a)

(b)

109.

104. It is $3\frac{3}{4}$ miles from the parking area to Crystal Lake. There is a fine view point $\frac{2}{3}$ of the distance from the parking area to the lake. (a) How many miles is it from the parking area to the viewpoint? (b) Once you reach the viewpoint, how much farther is it to the lake?

105. You have a recipe that makes $2\frac{1}{2}$ dozen cookies. If you make $1\frac{1}{2}$ times the recipe, how many cookies could you expect to make?

106. Your two small dogs together eat $\frac{3}{4}$ of a can of dog food each day. How many cans of dog food does each dog eat in one month (30 days)?

107. On January 1, Mary weighed $98\frac{3}{4}$ pounds and was 4 feet, 10 inches tall. On July 1 she weighed $108\frac{1}{2}$ pounds and was 5 feet, $\frac{1}{2}$ inch tall. (a) How much weight did she gain in those six months? (b) How much did she gain in weight on the average per month?

108. Your brother lives $10\frac{1}{2}$ blocks from school and if he hurries he can walk 1 block in $1\frac{1}{2}$ minutes. (a) At that rate, how long should it take him to walk to school? (b) If 20 blocks equal a mile, how fast does he walk in miles per hour?

109. An iron ring has the dimensions shown in the following figure. What is the thickness of the ring?

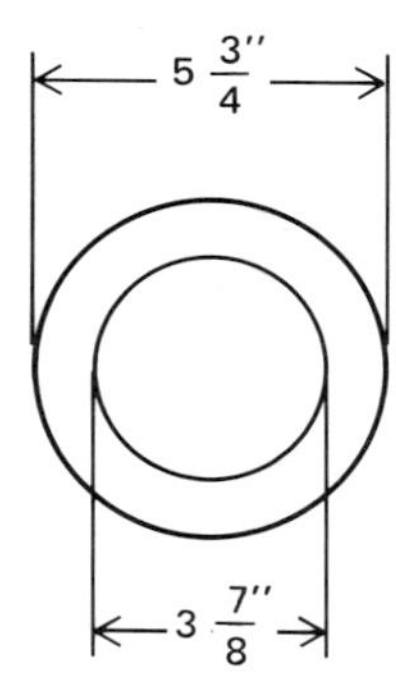

110. A piece of lumber is 18 feet long. You cut off 5 feet, 10 inches for a shelf. How many feet are left in the piece? 110.

111. In Canada, gas is sold in imperial gallons, each of which is equal to $1\frac{1}{5}$ U.S. gallons. If you purchased $12\frac{1}{2}$ gallons of gas in Canada, how many U.S. gallons would this equal? 111.

112. Alice typed 240 words in $7\frac{1}{2}$ minutes. How many words did she average per minute? 112.

113. How long a board must you have if you wish to cut five pieces 2 feet, $7\frac{1}{2}$ inches long and each saw cut wastes $\frac{1}{16}$ of an inch? 113.

114. Phyllis bought a remnant of velvet which is 5 yards long and 60 inches wide. If her pattern calls for $3\frac{7}{8}$ yards of 60-inch material, how much will she have left over? 114.

115. A pattern maker is required to allow $\frac{1}{8}$ inch on all surfaces of a pattern for machining and final finishing. If you are to make a pattern for a rectangular block which will measure $1\frac{3}{16}$ inches by $3\frac{1}{4}$ inches by $6\frac{3}{4}$ inches when finished, what must be the dimensions of the pattern? 115.

116. A ball of yarn weighs $\frac{1}{2}$ ounce and sells for $1.29. Find the cost of $1\frac{3}{8}$ pounds of this yarn. 116.

117. To find the amount of elastic needed for a skirt, Peggy must take her waist measure, which is $24\frac{3}{8}$ inches, subtract the length of the yoke, which is $11\frac{3}{4}$ inches, and add $\frac{1}{2}$ inch for fastening. How much elastic will Peggy need for her skirt? 117.

118.

118. The size of nails is measured not as so many inches but as so many *pennies* (d). A 2d (2-penny) nail is one inch long, and each additional penny adds $\frac{1}{4}$ inch to the length. How long are 4d, 6d, 10d, and 16d nails?

119.

119. A body falls $16\frac{1}{12}$ feet the first second, and in each succeeding second $32\frac{1}{6}$ feet more than the previous second. How far does it fall in three seconds?

120.

120. A dealer bought a number of bales of silk, each bale containing 135 yards, at $\$1\frac{2}{3}$ a yard, and sold it at $\$2\frac{1}{5}$ a yard, making a profit of \$792. How many bales did she buy?

121.

121. In determining hauling costs for a timber sale in a national forest, the following per load information is available:

Round trip hauling time	$1\frac{1}{3}$ hours
Loading time	$\frac{1}{2}$ hour
Weighing station time	$\frac{1}{12}$ hour
Unloading time	$\frac{1}{4}$ hour
Net operating time available	$11\frac{1}{2}$ hrs/day

Find the number of round trips possible per day.

*122.

*122. If 17 horses consume $8\frac{1}{3}$ bushels of oats in $3\frac{3}{4}$ days, how many bushels will 12 horses consume in $6\frac{4}{5}$ days?

*123.

*123. A pole stands $\frac{2}{7}$ in the mud, $\frac{7}{16}$ in the water, and the remainder, $12\frac{2}{5}$ feet, above the water. Find the length of the pole.

*124. Two pendulums beat once in $\frac{33}{35}$ of a second and once in $\frac{88}{91}$ of a second, respectively. If at some time the beats occur together, after how many seconds will they again occur together?

*124.

*125. A rabbit saw a carrot a certain distance away. The rabbit hopped $\frac{1}{2}$ of the distance to the carrot. Then it grew cautious and hopped only $\frac{1}{2}$ of the remaining distance. It continued this way, covering just $\frac{1}{2}$ of the remaining distance with each hop. What fraction of the distance remained after

(a) the first hop?

(b) the second hop?

(c) the third hop?

(d) the fourth hop?

(e) the tenth hop?

(f) The fractions in the answers to questions (a) through (e) are getting smaller and smaller. If you continued, is it possible to get a fraction equal to zero? Why?

*125.

(a)

(b)

(c)

(d)

(e)

(f)

*126. Ask someone to name the smallest fraction they can think of which is still greater than zero. Then try to name a fraction, greater than zero, that is smaller than your friend's.

*126.

Just for Fun

A. PATTERNS AND SEQUENCES

Write the next four or five terms of each sequence.

1. 7, 4, 11, 8, ___, ___, ___, ___, ___.
2. 100, 150, 250, 400, 600, ___, ___, ___, ___.
3. 25, 75, 50, 100, 75, 125, ___, ___, ___, ___.
4. Z, X, V, T, ___, ___, ___, ___, ___.

B. MORE PATTERNS

For those items indicated by an asterisk (*) you are to examine the pattern indicated by the previous computations and then write what you think is the result.

5.
$55 \times 33 =$
$555 \times 33 =$
$5555 \times 33 =$
$*55555 \times 33 =$
$*555555 \times 33 =$
$*5555555 \times 33 =$

6.
$78 \times 77 =$
$78 \times 777 =$
$78 \times 7777 =$
$*78 \times 77777 =$
$*78 \times 777777 =$
$*78 \times 7777777 =$

C. MISSING DIGITS

Find the missing digits in the following division problems.

7.
```
        2 2 _ _ _  R 7
  _ ) 2 _ 3 6 _ 2
```

8.
```
            4 _
  3 _ ) 1 _ _ 4
          _ _ 8
          -----
            6 _
            _ _
          -----
```

D. TRY THESE

9. Find four different numbers in the twenties which add to 100. Try this for tens or thirties. Is it possible?

10. A woman travels 90 miles and averages 45 miles per hour. On the return trip she averages 30 miles per hour. Find her average speed for the 180-mile trip.

11. Place +, −, ×, ÷, or ( ) between the following digits so as to arrive at the answer of 99. (Do not change the order of the digits.) In how many different ways can you do this?

$$1 \quad 2 \quad 3 \quad 4 \quad 5 \quad 6 \quad 7 \quad 8 \quad 9 \quad = \quad 99$$

E. MAGIC SQUARE

12. Complete the following magic square.

1000	200		136
350		500	
176	890	310	400
	610		

F. MAGIC CIRCLES

13. Place the nine numbers from 25 through 33 (using each one only once) in the circles so that the sum of the three numbers in each line is 87.

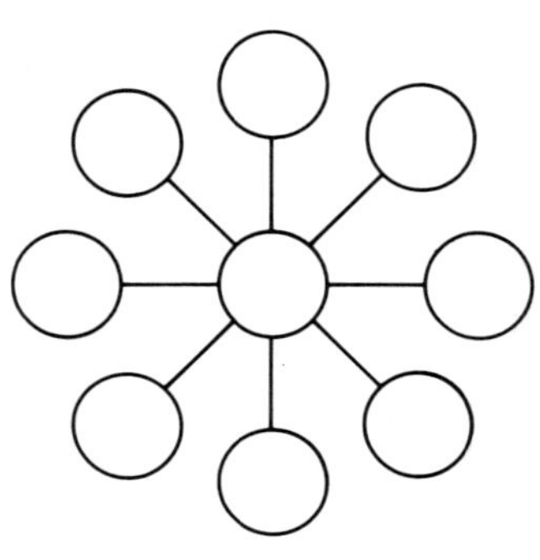

G. CALCULATOR GAME—"SCORE"

14. Play this game on your calculator.

Step 1: Players select one of the numbers 6, 7, 8, 9, or 11 to be a divisor of the winning number.

Players choose 8. — Display

Step 2: A begins play by entering a one-digit number in the calculator.

A enters 7. — 7

Step 3: B now plays by entering a one-digit number that is adjacent to the one used by A (but not the same number) on the HHC keyboard. The object is to produce a number divisible by the previously agreed upon number.

B chooses from 4, 5, or 8 which are adjacent to 7.

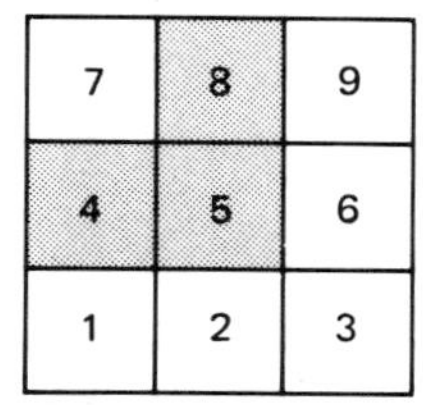

B chooses 4 and enters it on the HHC. — 74

Step 4: Play continues until one player scores a winning number.

A chooses from 1, 2, 5, 8 or 7.

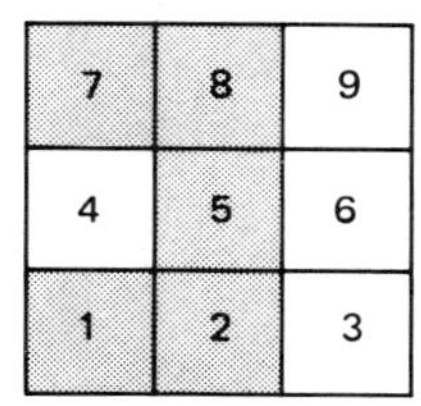

A chooses 8 and enters it on the HHC. — 748

B chooses from 7, 4, 5, 6, 9.

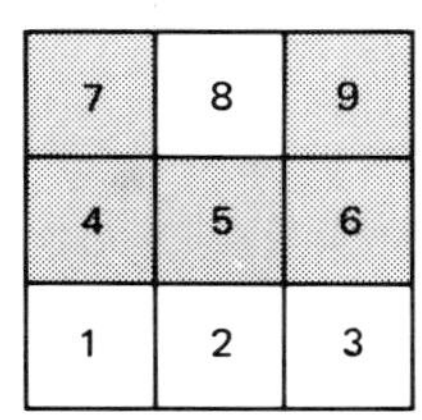

B chooses 4 and enters it on the HHC. — 7484

A chooses from 7, 8, 5, 1, or 2.

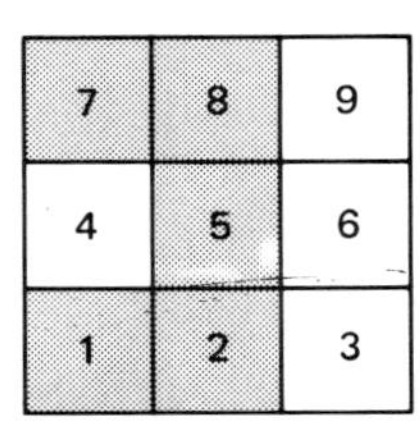

A chooses 8 and enters it on the HHC. — 74848

A wins, because

$$74848 \div 8 = 9356$$

H. THE TALKING CALCULATOR

15. Do the following calculation to find the name of the capitol city of the state of Idaho.

$$[(4065 \times 27) - (586 \times 27)] - [(142 \times 161) - (129 \times 161)] - [(13{,}975{,}164 \div 325) + (4{,}462{,}736 \div 325)]$$

Mathematics and Energy

Use the information on pages 264–265 to answer the following questions.

1. What will it cost to burn a 100-watt light bulb continuously for one year (365 days)?

2. What will it cost to operate a color TV set four hours a day, six days a week, for one year?

3. How many pounds of coal will it take to burn a 60-watt light bulb continuously for one year?

4. In 1974, the existing nuclear power plant capacity was 30,425 megawatt hours of energy. This is equivalent to (a) how many tons of coal, (b) how many cubic feet of natural gas, and (c) how many barrels of crude oil?

5. If an acre of strip-mined coal produces 2300 tons of coal, how many 100-watt light bulbs could be burned continuously for one year by using the energy produced by this coal?

6. A typical family in the USA uses 255 gallons of water a day and 25,000,000 BTU's of energy each year to heat some of this water for washing everything from hands to dishes. Calculate the number of (a) gallons of heating oil, (b) cubic feet of natural gas, and (c) kilowatt hours of electricity in heating water each year for a typical family.

7. What is the cost per gallon of #2 heating oil according to the data on page 264?

8. According to the information on page 265, how many miles to the gallon does (a) a small car get at 45 mph, (b) a large car get at 70 mph?

9. Heating a home uses up just over one quarter of an average family's budget. If the family budget amounts to $10,000 per year, what do heating costs amount to in terms of (a) kwh of electricity, (b) cubic feet of natural gas, or (c) gallons of #2 heating oil?

10. Calculate the unit cost of an electric bill for $36.35 for a home whose meter reading was 00924 kwh at the beginning of the month and 02424 kwh at the end of the month.

11. Using your own light bill and/or gas bill, calculate your unit cost of electricity or gas.

12. Using your own rate or the rate in exercise 10, calculate the yearly cost of operating the following electrical appliances:

(a) refrigerator-freezer combination

(b) color TV

(c) electric blanket

(d) water heater

(e) coffee maker

13. Read your electric and/or gas meter at the same time each day for a week. Calculate the amount of energy used each day. On which day was the most energy used? On which day was the least energy used? Can you account for the variation or lack of variation in daily use?

14. The following formula can be used to calculate the number of BTU's used in heating a house for a year (temperature in Fahrenheit):

$$\frac{(\text{Heat loss per hour}) \times (\text{Hours used})}{(\text{Efficiency})} \times \frac{(\text{Average inside temp.} - \text{Average outside temp.})}{70}$$

Find the cost of heating a home in Oregon (a) by #2 heating oil, (b) by natural gas, and (c) by electricity if the heat loss for the house is 35,000 BTU's per hour, the furnace is on from October 1 to June 1 and is used 16 hours a day, and if the average outside temperature is 40° F and the average inside temperature is 68° F. Assume that oil and gas heat have an efficiency of 0.80 (80%) and that electrical heat has an efficiency of 1 (100%).

15. For the house of exercise 14, if better insulation would reduce the heat loss by 5250 BTU's per hour, how much of a saving for each form of heating would this provide?

READING ELECTRICITY METERS

On most electricity meters the dials are organized from right to left: the right-most dial indicates kilowatt hours, the next dial tens of kilowatt hours, the next dial hundreds of kilowatt hours, and so on. Read the dials from left to right. On each dial read the lowest number to which the pointer points.

Write down the numbers in the order read—from left to right. Most meters have five dials but if your meter has only four, read in a similar manner. The face of the meter will indicate whether you should multiply the reading by 10.

READING GAS METERS

Since the dials on a gas meter are organized much the same as the electric meter, the reading is made the same way. However, there are usually only four dials. The smallest quantity of gas measured by most meters is 100 cubic feet. The markings on the right-most dial each represent 100 cubic-feet; thus, the reading must be multiplied by 100 to obtain total cubic feet of gas consumed.

COMMON ENERGY MEASUREMENTS

British Thermal Unit (BTU)–Energy required to increase the temperature of one pound of water by one degree Fahrenheit

Kilowatt hour (Kwh)–1000 watt-hours. (A 100-watt bulb burning for 10 hours will consume one kilowatt-hour of energy, or enough to lift a 150-pound person 17,700 feet in the air. One barrel of crude oil contains, on the average, 1700 Kwh.)

Watt hour–3.4128 BTU's.

Kilowatt (Kw)–1000 watts, or 1.341 horsepower.

Megawatt (Mw)–One million watts, or 1000 kilowatts.

Barrels (Bbls)–1 barrel equals 42 gallons.

Therm–100 cubic feet of natural gas. Since 1 cubic foot of natural gas averages 1031 BTU's, a therm contains 103,100 BTU's or roughly 100,000 BTU's.

ENERGY CONVERSIONS

To Change	To	Multiply by
Kilowatt hour (Kwh)	BTU	3,412.8
1 barrel (42 gallons crude oil)	BTU	5,600,000
1 barrel no. 5 residual oil	BTU	6,290,000
1 gallon no. 2 heating oil (home heating and diesel)	BTU	138,800
1 gallon gasoline	BTU	125,000
1 ton coal (bituminous)	BTU	26,200,000
1 therm natural gas	BTU	100,000
1 cubic foot natural gas	BTU	1,031

COMPARISON OF RETAIL PRICE TO RESIDENTIAL CONSUMER OF VARIOUS FORMS OF ENERGY*

Form of Energy	Retail price per million BTU's
Natural gas	$2.82
No. 2 fuel oil	2.92
Gasoline (regular)	4.64
Electricity	6.90**

*Oregon average, August 1975.

**Oregon electricity rates are lower than the national average due to readily available hydroelectric power.

ENERGY REQUIRED BY TYPICAL CONSUMERS DURING ONE HOUR OF NORMAL OPERATION OR ACTIVITY

	Kwh	BTU's
Human being	0.1	341
100-watt light bulb	0.1	341
Color TV	0.2	682
Toaster	1.1	3,750
Room air conditioner	3.0	10,240
Electric clothes dryer	4.0	13,600
Oven (baking a turkey)	4.5	15,400
Small car at 45 mph	60.0	205,000
Large car at 70 mph	270.0	921,500

APPROXIMATE AVERAGE MONTHLY KWH USE

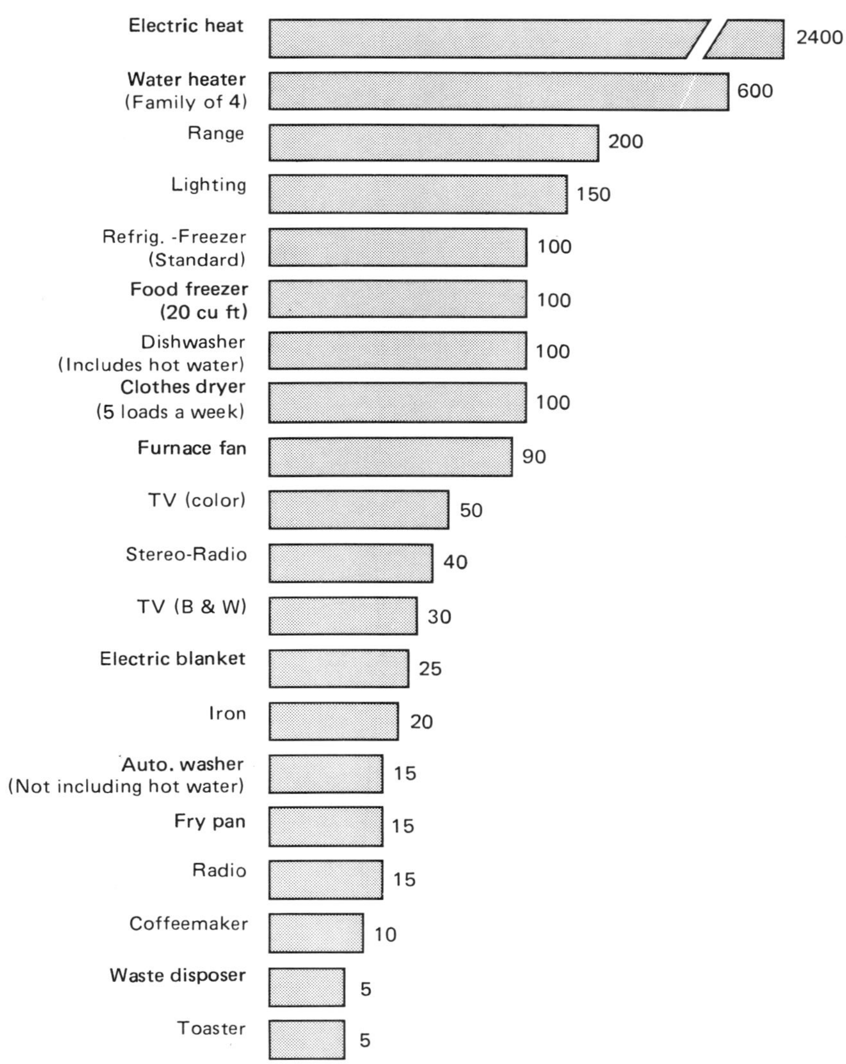

5

Ratio, Proportion, and Percent

5.1 Ratios

If a bookshelf contains 12 whodunits and 9 westerns, we can say that the *ratio* of whodunits to westerns is 12 to 9 or, since $\frac{12}{9} = \frac{4}{3}$, 4 to 3. We often write 12 to 9 as 12:9 or $\frac{12}{9}$ and, in general, for a and b positive numbers, write $a:b$ or $\frac{a}{b}$ for the ratio of a to b.

Batting averages in baseball are examples of ratios. Thus a batting average of .300 for a player means that this player got, on the average, 3 hits in every 10 times at bat. That is, the ratio of hits to times at bat is

$$3:10$$

since

$$0.300 = \frac{300}{1000} = \frac{3}{10}$$

Similarly, if the odds on a horse in a race are 3:5, this means that, if the horse wins, a \$3 bet on it will bring a return of \$5.

Note that the order of the two numbers in a ratio is important: If the odds were 5:3, then a bet of \$5 would be needed to win \$3.

Do exercises 1 through 6 in the adjoining column.

Just as fractions can be reduced to lowest terms (see section 4.1), so can ratios. We say that a ratio $a:b$ is in *lowest terms* if a and b are whole numbers whose only common divisor is 1. Thus 3:5 is a ratio in lowest terms because the only common divisor of 3 and 5 is 1. On the other hand, 4:6 is not in lowest terms since 2 is a divisor of both 4 and 6. We have, upon division of 4 and 6 by 2,

$$4:6 = 2:3$$

just as

$$\frac{4}{6} = \frac{4 \div 2}{6 \div 2} = \frac{2}{3}$$

Since any ratio can be written as a fraction and since fractions are familiar to you from the work of the preceding chapter, we will from now on use mainly fractions in discussing ratios.

OBJECTIVES

You should be able to:

(a) Use the language of ratio.

(b) Express ratios in lowest terms.

****

In exercises 1–6, write the given comparison as a ratio.

1. 12 rock records to 5 jazz records
2. 20 girls to 17 boys
3. 25 games won to 16 games lost
4. 15 candy bars to 32 popsicles
5. 9 pages read to 45 pages to read
6. \$7 interest for \$100 borrowed

STOP

In exercises 7–12 write the comparison as a ratio in lowest terms both in the form $a{:}b$ and in the form $\frac{a}{b}$.

7. 28 jazz records to 16 rock records

8. 20 boys to 15 girls

9. 6 dozen apples to 4 dozen oranges

10. 8 dollars interest to 100 dollars invested

11. 12 black cats to 18 white cats

12. 135 baseball cards to 140 baseball cards

STOP

In exercises 13–22 write the given comparison as a ratio $\frac{a}{b}$ in lowest terms, using the same units for both a and b in exercises 13–18.

13. 1 yard to 2 feet

14. 8 hours to 1 day

15. 5 pounds to 12 ounces

16. 6 quarts to 3 pints

Do exercises 7 through 12.

In a ratio $\frac{a}{b}$, the units of measurement of a and b should, whenever possible, be chosen to be the same to avoid confusion.

Suppose, for example, a picture frame is 15 inches wide and 2 feet long. If we were to say that the ratio of width to length was 15:2, we would certainly give a wrong impression of the shape of the frame! We could say that the ratio is

$$\frac{15 \text{ inches}}{2 \text{ feet}}$$

but it would be better to observe that 2 feet is 24 inches and write the ratio as

$$\frac{15}{24}$$

or, in lowest terms, $\frac{5}{8}$.

When a ratio between two different kinds of quantities is required, the word *per* is generally used, and the ratio is generally referred to as a *rate*. For example, in the statement, "Gayle traveled from Chicago to St. Louis, a distance of 300 miles, in 6 hours,"

(a) The ratio of the number of miles traveled to the number of hours traveled is

$$\frac{300 \text{ miles}}{6 \text{ hours}}$$

(b) Because $\frac{300}{6} = \frac{50}{1}$, we also say that Gayle's rate of travel was 50 miles per hour.

Do exercises 13 through 22.

Most ratios are given in a form involving only whole numbers, but this is not always the case. We could, for example, write the ratio between the width and length of the 15-inch by 2-foot picture frame as

$$\frac{5}{4} : 2$$

since 15 inches is $\frac{15}{12}$ of a foot and $\frac{15}{12} = \frac{5}{4}$. Observe, however, that

$$\frac{\frac{5}{4}}{2} = \frac{5}{4} \div 2 = \frac{5}{4} \div \frac{2}{1} = \frac{5}{4} \times \frac{1}{2} = \frac{5}{8}$$

so that

$$\frac{5}{4} : 2 = 5{:}8$$

—the same ratio that we obtained before.

Similarly,

$$2\frac{1}{2} : \frac{1}{2} = 5:1$$

since

$$\frac{2\frac{1}{2}}{\frac{1}{2}} = \frac{\frac{5}{2}}{\frac{1}{2}} = \frac{5}{2} \div \frac{1}{2} = \frac{5}{2} \times \frac{2}{1} = \frac{5}{1}$$

Likewise

$$\frac{17}{32} : \frac{69}{19} = 323:2208$$

since

$$\frac{\frac{17}{32}}{\frac{69}{19}} = \frac{17}{32} \div \frac{69}{19} = \frac{17}{32} \times \frac{19}{69} = \frac{323}{2208}$$

Also,

$$0.34:0.58 = 34:58 = 17:29$$

since

$$\frac{0.34}{0.58} = \frac{0.34 \times 100}{0.58 \times 100} = \frac{34}{58} = \frac{2 \times 17}{2 \times 29} = \frac{17}{29}$$

In general, any ratio involving numbers that are not whole numbers (fractions or decimals) can be rewritten as one that involves only whole numbers and is in lowest terms.

Do exercises 23 through 28.

17. 25 feet to 25 inches

18. 48 cents to \$2.52

19. 180 miles to 4 hours

20. Express the ratio in exercise 19 as a rate in miles per hour.

21. 6 gallons of paint per 1800 square feet

22. Express the ratio in exercise 21 as a rate of 1 gallon per so many square feet.

STOP

In exercises 23–28 write the given ratio in the form $\frac{a}{b}$ in lowest terms where a and b are whole numbers.

23. $\frac{5}{2}$ inches to 2 inches

24. $\frac{1}{6}$ to $\frac{5}{12}$

25. $3\frac{1}{2}$ to $1\frac{1}{4}$

26. $66\frac{2}{3}$ to $16\frac{2}{3}$

27. 0.36 to 0.408

28. 3.75 to 12.45

STOP

Exercises 5.1

Name ______________

Class ______________

In exercises 1–20 write the ratio of the first number to the second number in lowest terms.

1. 6, 3
2. 4, 16
3. 5, 35
4. 8, 12
5. 40, 70
6. 51, 30
7. 405, 567
8. 180, 216
9. 140, 125
10. 224, 700
11. $10, 2\frac{1}{2}$
12. $\frac{1}{4}, 1\frac{1}{4}$
13. $\frac{2}{3}, \frac{3}{5}$
14. $\frac{14}{16}, \frac{4}{8}$
15. $2\frac{3}{5}, \frac{41}{10}$
16. 7.8, 15.98
17. 0.77, 0.56
18. 0.386, 6.42
19. 0.035, 0.5
20. 500.6, 400.008

1. 2.
3. 4.
5. 6.
7. 8.
9. 10.
11. 12.
13. 14.
15. 16.
17. 18.
19. 20.

In exercises 21–40 express the given ratio in lowest terms. Use the same units for both *a* and *b*.

21. 5 black cats to 12 white cats
22. 2 dozen oranges to 3 dozen apples
23. 18 orange gumdrops to 5 dozen black gumdrops
24. 14 sport cars to 8 station wagons

21.

22.

23.

24

25.

26.

27.

28.

29.

30.

31.

32.

33.

34.

35.

36.

37.

38.

39.

40.

41.

(a)

(b)

(c)

25. 500 feet to 1 mile

26. 880 yards to 2 miles

27. 3 feet to 10 inches

28. 45 centimeters to 3 meters (1 meter = 100 centimeters)

29. 12 pounds to 24 ounces

30. 3 yards to 2 feet

31. 6 hours to 1 day

32. 6 quarts to 3 pints

33. 250 cubic centimeters to 1.5 liters (1 liter = 1000 cubic centimeters)

34. 2600 miles to 1000 miles

35. 20 girls to 15 boys

36. 1200 pounds to 1 ton

37. 66 feet to 3 inches

38. 75¢ to $2.25

39. 2 miles to 600 feet

40. 440 yards to 18 inches

41. A basketball team won 28 games and lost 16.

(a) What is the ratio of the number of games won to the number of games lost?

(b) What is the ratio of the number of games won to the number of games played?

(c) What is the ratio of the number of games lost to the number of games played?

42. The scale on a map is 1 inch to 20 miles. Express this scale as a ratio. 42.

43. A quart of canned orange juice sells for \$1.08 and a gallon of the same kind of juice sells for \$3.96. Find the ratio of the selling price (in cents) to the number of pints in each situation. 43.

44. Bus fare 30 years ago in Toonerville was 15¢; now it is 60¢. 44.

(a) Find the ratio of the present fare to the former fare. (a)

(b) Find the ratio of the increase in fare to the original fare. (b)

45. During a sale, shoes that normally sold for \$24 were priced at \$18. 45.

(a) Find the ratio of the decrease in cost to the original price. (a)

(b) Find the ratio of the sale price to the original price. (b)

46. A ream of $8\frac{1}{2}$ by 11-inch paper contains 500 sheets. What is the ratio of the length of the paper to the width? 46.

*47. On a given street corner record the kinds of vehicles that pass in five minutes. *47.

(a) What is the ratio of the number of cars to the number of trucks? (a)

(b) What is the ratio of the number of compact cars to the number of full-sized cars? (b)

*48. In a section of the school cafeteria, record the number of men and women smokers and non-smokers. *48.

(a) What is the ratio of smokers to non-smokers? (a)

(b) What is the ratio of men who smoke to women who smoke? (b)

*49. The sides of two squares are in a ratio of 3:1. *49.

(a) What is the ratio of their perimeters? (a)

(b) What is the ratio of their areas? (b)

*50.

(a)

(b)

(c)

(d)

*51.

*52.

*50. Imagine that all of the faces of a 4-inch cube were painted red and then imagine that the cube was sawed into 1-inch cubes as suggested by the accompanying figure.

(a) What is the ratio of the number of cubes that have paint on at least one face to the total number of cubes?

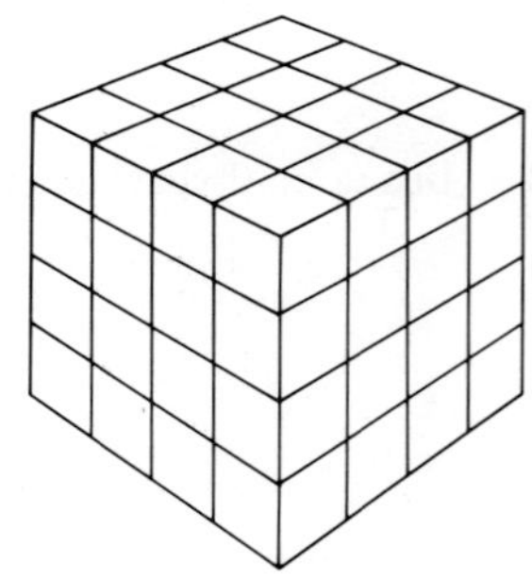

(b) What is the ratio of the number of cubes that have paint on them to the number of those that have no paint?

(c) What is the ratio of the number of cubes that have paint on exactly one face to the number of cubes that have paint on exactly two faces?

(d) What is the ratio of the number of those cubes that have no paint on them to the total number of cubes?

*51. Find the neck and waist measurements of several people. Compare the neck to waist ratios of each person. Can you draw any conclusions?

*52. Find the length (from knuckles to the tip of the longest finger) and the width (across the knuckles) of the hands of several people. Compare the ratio of these two measures for each person. Can you draw any conclusions?

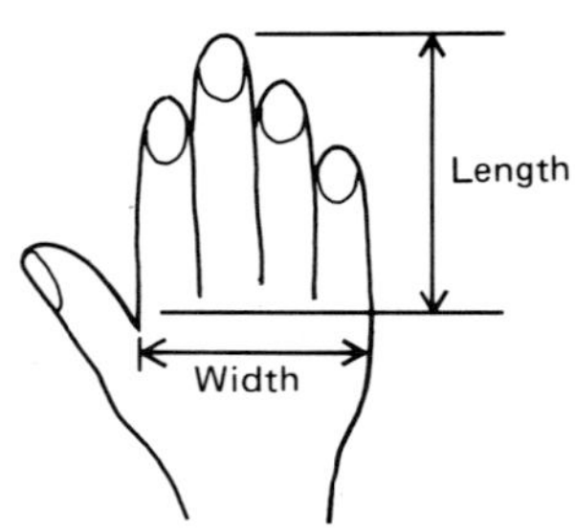

5.2 Proportions

An equality between two ratios is called a *proportion*. Thus

$$12:9 = 4:3$$

is a proportion. We read 12:9 = 4:3 as "twelve is to nine as four is to three."

Although there is a technical difference between a ratio and a fraction, in practice we do translate a statement of a proportion into a statement of equality of fractions. Thus we will write

$$12:9 = 4:3$$

as

$$\frac{12}{9} = \frac{4}{3}$$

Do exercises 1 through 4 in the adjoining column.

Consider the proportion

$$\frac{18}{21} = \frac{24}{28}$$

and note that

$$18 \times 28 = 504 = 21 \times 24$$

That is, "cross-multiplying,"

$$\frac{18}{21} \times \frac{24}{28} \quad \text{gives us} \quad 18 \times 28 = 21 \times 24$$

On the other hand, consider

$$\frac{18}{21} \text{ and } \frac{23}{29}$$

We have

$$\frac{18}{21} \approx 0.8571428 \text{ and } \frac{23}{29} \approx 0.7931034$$

so that

$$\frac{18}{21} \neq \frac{23}{29}$$

Here cross-multiplying gives us

$$18 \times 29 = 522 \neq 21 \times 23 = 483$$

OBJECTIVES

You should:

(a) Know the definition of a proportion.

(b) Be able to translate a statement about a proportion into a statement about equality of fractions and vice versa.

(c) Know how to test whether $\frac{a}{b} = \frac{c}{d}$ by using cross-multiplication.

(d) Be able to find the unknown term in a proportion.

(e) Be able to apply proportions to solve problems.

****

In exercises 1–2 write the given proportion in terms of the equality of fractions.

1. 2:3 = 10:15

2. 7:3 = 21:9

In exercises 3–4 write the given equality of fractions in the form $a:b = c:d$.

3. $\frac{16}{9} = \frac{48}{27}$

4. $\frac{18}{21} = \frac{24}{28}$

STOP

These two examples are special cases of the general rule that

$$\frac{a}{b} = \frac{c}{d} \quad \text{if} \quad a \times d = b \times c$$

and

$$\frac{a}{b} \neq \frac{c}{d} \quad \text{if} \quad a \times d \neq b \times c.$$

Do exercises 5 through 8.

In exercises 5–8 use cross-multiplication to determine whether "=" or "≠" should replace the "?."

5. $\frac{23}{29} \ ? \ \frac{322}{406}$

6. $\frac{49}{53} \ ? \ \frac{931}{1007}$

7. $\frac{58}{13} \ ? \ \frac{1334}{300}$

8. $\frac{67}{82} \ ? \ \frac{1742}{2132}$

STOP

We can use cross-multiplication to find the unknown term in a proportion. For example, if

$$N:28 = 425:17$$

so that

$$\frac{N}{28} = \frac{425}{17}$$

we have

$$17 \times N = 28 \times 425$$

Then, dividing both sides of the equality by 17, we have

$$N = \frac{28 \times 425}{17} = 700$$

Similarly, if

$$\frac{117}{N} = \frac{491}{8838}$$

then

$$491 \times N = 117 \times 8838$$

so that

$$N = \frac{117 \times 8838}{491} = 2106$$

Do exercises 9 through 14.

So far N has turned out to be a whole number. This, of course, doesn't always happen. If

$$\frac{N}{26} = \frac{15}{92}$$

we have

$$92 \times N = 26 \times 15$$

so that

$$N = \frac{26 \times 15}{92} = 4.2391304$$

to the limits of an eight-digit display. We can then arrive at the fractional form of N by applying the procedures discussed at the end of section 4.4. That is, we multiply 0.2391304 by 92 to get

$$0.2391304 \times 92 \approx 21.9999996 \approx 22$$

and hence conclude that

$$N = 4\frac{22}{92} = 4\frac{11}{46}$$

or, since

$$4\frac{11}{46} = 4 + \frac{11}{46} = \frac{4 \times 46}{46} + \frac{11}{46} = \frac{195}{46}$$

we also have

$$N = \frac{195}{46}$$

In exercises 9-14 find the unknown term in the given proportion.

9. $\frac{N}{8} = \frac{9}{24}$

10. $\frac{12}{N} = \frac{48}{8}$

11. $24:32 = N:48$

12. $\frac{360}{900} = \frac{270}{N}$

13. $96:144 = N:360$

14. $\frac{36}{N} = \frac{432}{1728}$

In exercises 15–18 find the unknown term in the given proportion. Express your answer as a fraction, a mixed number, and as an approximate decimal.

15. $\frac{N}{5} = \frac{7}{6}$

16. $\frac{N}{43} = \frac{15}{76}$

17. $\frac{38}{13} = \frac{N}{47}$

18. $756:495 = N:1365$

STOP

Do exercises 15 through 18.

Many word problems can be done by the use of proportions. Here are three examples:

Example 1: If six shirts sell for \$37.50, what would be the cost of eight shirts?

Solution: We have

$$\frac{6}{8} = \frac{37.50}{N}$$

where N is the price, in dollars, of eight shirts. Then we have

$$6 \times N = 8 \times 37.50$$

$$N = \frac{8 \times 37.50}{6} = 50$$

and so eight shirts would cost \$50.

Note that the ratio 6:8 is that of two quantities whereas the ratio 37.50:N is that of two prices in dollars.

Example 2: If flower seeds sell at a price of eight packets for \$2.80, how many packets will \$4.90 buy?

Solution: We have

$$\frac{8}{N} = \frac{2.80}{4.90}$$

where N is the number of packets that \$4.90 will buy. Then

$$2.80 \times N = 8 \times 4.90$$

$$N = \frac{8 \times 4.90}{2.80} = 14$$

so that \$4.90 will buy 14 packets.

Example 3: If the property tax on a car valued at \$4200 is \$28, what would be the property tax on a car valued at \$5400?

Solution: We have

$$\frac{N}{5400} = \frac{28}{4200}$$

where N is the tax on the \$5400 car. Thus

$$4200 \times N = 28 \times 5400$$

$$N = \frac{28 \times 5400}{4200} = 36$$

and so the tax would be \$36.

Do exercises 19 through 21.

In exercises 19–21 use a proportion to answer the question.

19. If soup sells at 3 cans for 59¢, what would be the cost of 15 cans?

20. If the county tax rate is \$3 per hundred dollars of assessed valuation, what would be the tax on a house assessed at \$23,000?

21. A motorist finds that he used 6 gallons of gas in going 92 miles. If he plans on a trip of 150 miles, how many gallons of gas might he expect to use?

STOP

Exercises 5.2

Name ____________

Class ____________

In exercises 1–16 determine if the two ratios are equal. If they are not, change the numerator of the first ratio so as to make them equal.

1. $\frac{1}{4}, \frac{8}{32}$

2. $\frac{10}{5}, \frac{15}{10}$

3. $\frac{48}{16}, \frac{42}{14}$

4. $\frac{7}{9}, \frac{20}{36}$

5. $\frac{5}{8}, \frac{35}{56}$

6. $\frac{0.75}{6}, \frac{3}{2}$

7. $\frac{1.6}{0.25}, \frac{0.25}{3}$

8. $\frac{20}{0.5}, \frac{40}{0.25}$

9. $\frac{3}{0.67}, \frac{15}{2}$

10. $\frac{15}{13}, \frac{45}{39}$

11. $\frac{3.2}{10}, \frac{1.6}{5}$

12. $\frac{30}{21}, \frac{10}{7}$

13. $\frac{15}{104}, \frac{7.5}{52}$

14. $\frac{42}{336}, \frac{126}{1008}$

15. $\frac{70}{105}, \frac{8}{12}$

16. $\frac{1460}{635}, \frac{291}{127}$

In exercises 17–32 find the unknown term in the given proportion. Express your answer as a whole number or a mixed number.

17. $\frac{N}{42} = \frac{5}{6}$

18. $\frac{N}{1000} = \frac{1}{40}$

19. $\frac{50}{160} = \frac{N}{16}$

20. $\frac{12}{15} = \frac{16}{N}$

1. 2.

3. 4.

5. 6.

7. 8.

9. 10.

11. 12.

13. 14.

15. 16.

17. 18.

19. 20.

21. 22.

23. 24.

25. 26.

27. 28.

29. 30.

31. 32.

33.

34.

35.

21. $\frac{16}{28} = \frac{N}{7}$

22. $\frac{23}{N} = \frac{4}{11}$

23. $\frac{12}{N} = \frac{10}{3}$

24. $\frac{65}{80} = \frac{91}{N}$

25. $\frac{25}{30} = \frac{30}{N}$

26. $\frac{N}{90} = \frac{14}{60}$

27. $\frac{24}{40} = \frac{N}{75}$

28. $\frac{4}{7} = \frac{N}{11}$

29. $\frac{603}{946} = \frac{N}{357}$

30. $\frac{339}{N} = \frac{4261}{292}$

31. $\frac{N}{1918} = \frac{2831}{698}$

32. $\frac{985}{73} = \frac{53}{N}$

In exercises 33-56 solve the given problem by using a proportion.

33. A recipe calls for 2 cups of sugar for every 3 cups of flour. If you need to use 24 cups of flour, how many cups of sugar will you need?

34. Gary can type 16 cards in 20 minutes. At this rate, how many cards can he type in 60 minutes?

35. Joyce has a picture 9 centimeters wide and 13 centimeters long. She wants an enlargement that will be 20 centimeters wide. How long will the enlargement be?

36. A farmer harvested 904 bushels of potatoes from 8 acres of land. At the same rate, how many bushels could she expect from 30 acres of land? 36.

37. Ms. Jones was paid $140 for a job that took 35 hours of work. At the same rate, how much should she charge for a job that takes 60 hours? 37.

38. In typing a 22-page manuscript which contained statistical tables, Ronald completed the first 3 pages in 1 hour and 20 minutes. At this same rate, how long, to the nearest minute, will it take him to finish typing the manuscript? 38.

39. If Sidney made 45 out of his last 56 free throws in league basketball games, how many free throws would you expect him to make if he had 10 chances in the next game? 39.

40. An electronic computer takes 4 microseconds to perform 7 operations (one microsecond is 0.000001 seconds). How long would it take the computer to do 273,000 of these operations? 40.

41. A hula hoop salesman keeps $37.50 out of every $125 of his sales and sends the rest to his employer. What were his total sales if he sent $1000 to his employer? 41.

42. If 35 onaks are worth 17 adnarks, how many adnarks would 595 onaks be worth? 42.

43. If dog food sells at 3 cans for 85¢, how many cans can be bought for $5? 43.

44. A wholesaler has a mixture of peanuts and cashews in which there are 9 kilograms of cashews and 15 kilograms of peanuts. How many kilograms of peanuts would she need to fill an order for 120 kilograms of such a mixture? 44.

45.

45. A motor scooter can go 30 miles per hour. This rate is equivalent to how many feet per second?

46.

46. What is the cost of 9750 bricks if they cost $35 per thousand?

47.

47. A water tower casts a shadow 75 feet long. At the same time, a 6-foot man casts a shadow 4 feet long. How high is the water tower?

48.

48. The effect of a medicine is sometimes dependent upon a person's weight. If the ratio is the same, how much medicine should a doctor prescribe for a 30-pound child if a 150-pound adult requires 100 milligrams?

49.

49. With good brakes and on dry level pavement, an automobile traveling at 20 miles per hour will move 22 feet before the driver can apply the brakes once he decides to stop. If this ratio holds, how far will a car, traveling at 70 miles per hour, travel before a driver, having decided to stop, applies the brakes?

50.

50. If in every 100 grams of a chemical mixture there is $\frac{3}{4}$ gram of impurities, how many grams of pure chemical can be expected in 3.75 kilograms of the mixture? (1 kilogram = 1000 grams)

51.

51. If it takes $3\frac{1}{2}$ boxes of berries and $1\frac{1}{2}$ pounds of sugar to make a quart of jam, how much jam can be made from 16 boxes of berries?

52.

52. In a manufacturing process an inspection procedure showed that 3 items out of 87 were defective. How many defective items could be expected if 3575 items were produced?

53. A contractor found it took 36 days to complete $2\frac{1}{4}$ miles of new road. If his contract calls for 15 miles of that road, how long will it take to do the whole job if he works at the same rate throughout?

53.

*54. A flea weighing 0.00001 ounces can jump $7\frac{3}{4}$ inches. If a man could jump in this ratio, how high, to the nearest mile, could a 175-pound man jump?

*54.

*55. A group of 357 soldiers had food for 112 days but after reinforcements arrived, the food lasted only 98 days. How many reinforcements were there?

*55.

*56. A 150-pound adult received a prescription for 10 grams of medicine divided into 50 capsules, calling for one capsule twice a day. What would be the number of capsules that could be made from 10 grams of medicine if the prescription were for a 25-pound child? How long would the prescription last?

*56.

5.3 The Language of Percent

The language of percent is all around us:

"Save 20% at our big sale!"
"Earn $5\frac{1}{4}$% on your savings."
"First National Bank has just reduced its prime lending rate from 7% to $6\frac{3}{4}$%.
"Prices have increased 120% since 1934."
"$33\frac{1}{3}$% off list prices."

The word *percent* comes from the Latin *per centum* meaning *per hundred.* So percent is simply the ratio of some number to 100 and is thus another way of naming fractions with a denominator of 100. That is,

$$20\% = \frac{20}{100} = 0.20 = 0.2$$

$$5\frac{1}{4}\% = 5.25\% = \frac{5.25}{100} = 0.0525$$

$$7\% = \frac{7}{100} = 0.07$$

$$6\frac{3}{4}\% = 6.75\% = \frac{6.75}{100} = 0.0675$$

$$120\% = \frac{120}{100} = 1.2$$

and

$$33\frac{1}{3}\% = \frac{33\frac{1}{3}}{100} = \frac{\frac{100}{3}}{100} = \frac{100}{3} \div 100$$

$$= \frac{100}{3} \times \frac{1}{100} = \frac{1}{3} = 0.\overline{3}$$

(where $0.\overline{3}$ is the repeating decimal 0.3333 . . . —see section 4.4)

Do exercises 1 through 10 in the adjoining column.

OBJECTIVES

You should:

(a) Understand the language of percent.

(b) Be able to write percents as decimals.

(c) Be able to write decimals of fractions as percents.

****

In exercises 1–10 write the given percent first as a fraction with denominator 100 and then as a decimal.

1. 5%

2. 40%

3. 24%

4. 154%

5. 300%

6. $2\frac{1}{2}$%

7. $\frac{1}{2}$%

8. 0.6%

9. 0.003%

10. 1%

STOP

In exercises 11–18 write the given decimal as a percent.

11. 0.24

12. 0.75

13. 0.0734

14. 0.5034

15. 0.003

16. 0.00005

17. 2.0

18. 5.38

STOP

In exercises 19–24 write the given fraction as a percent.

19. $\frac{2}{25}$

20. $\frac{4}{5}$

21. $\frac{1}{50}$

22. $\frac{1}{2}$

23. $\frac{3}{4}$

24. $\frac{7}{10}$

STOP

Just as percents can be expressed as decimals or fractions, so can decimals and fractions be expressed as percents. For example,

$$0.05 = 0.05 \times \frac{100}{100} = \frac{0.05 \times 100}{100} = \frac{5}{100} = 5\%$$

Changing a decimal to a percent simply involves multiplying the decimal by 100. Thus

$$0.05 = (0.05 \times 100)\% = 5\%$$

$$0.0625 = (0.0625 \times 100)\% = 6.25\%$$

$$0.27 = (0.27 \times 100)\% = 27\%$$

$$1.3 = (1.3 \times 100)\% = 130\%$$

Do exercises 11 through 18.

Changing a fraction to a percent involves finding an equal fraction with a denominator of 100; its numerator will then indicate the percent. Thus

$$\frac{1}{4} = \frac{1 \times 25}{4 \times 25} = \frac{25}{100} = 25\%$$

$$\frac{7}{5} = \frac{7 \times 20}{5 \times 20} = \frac{140}{100} = 140\%$$

$$4 = \frac{4}{1} = \frac{4 \times 100}{1 \times 100} = \frac{400}{100} = 400\%$$

Do exercises 19 through 24.

In the three examples given above, it is easy to see what to multiply by to get the required denominator of 100. If this is not the case, we can use proportions to get the answer. The following three examples illustrate this procedure.

Example 1: Change $\frac{1}{3}$ to a percent.

Solution:

$$\frac{1}{3} = \frac{N}{100}$$

$$3 \times N = 1 \times 100 = 100$$

$$N = \frac{100}{3} = 33\frac{1}{3}$$

Therefore, $\frac{1}{3} = 33\frac{1}{3}\%$.

Example 2: Change $\frac{5}{8}$ to a percent.

Solution:

$$\frac{5}{8} = \frac{N}{100}$$

$$8 \times N = 5 \times 100$$

$$N = \frac{500}{8} = 62.5$$

Therefore, $\frac{5}{8} = 62.5\%$.

Example 3: Change $2\frac{1}{3}$ to a percent.

Solution:

Since $2\frac{1}{3} = \frac{7}{3}$ we have

$$\frac{7}{3} = \frac{N}{100}$$

$$3 \times N = 7 \times 100$$

$$N = \frac{700}{3} = 233\frac{1}{3}$$

Therefore, $2\frac{1}{3} = 233\frac{1}{3}\%$.

Do exercises 25 through 30.

In exercises 25–30 write the given fraction as a percent.

25. $\frac{7}{8}$

26. $\frac{5}{16}$

27. $\frac{2}{3}$

28. $\frac{19}{500}$

29. $\frac{23}{200}$

30. $2\frac{5}{12}$

STOP

Exercises 5.3

Name ____________________

Class ____________________

In examples 1-30 complete the following chart. (The first line is completed as an example.)

Ratio in simplest form	Ratio with 100 as a denominator	Ratio as a decimal	Ratio as a percent	
$\frac{1}{2}$	$\frac{50}{100}$	0.50	50%	
$\frac{1}{4}$				1.
	$\frac{75}{100}$			2.
		0.20		3.
			40%	4.
	$\frac{60}{100}$			5.
$\frac{4}{5}$				6.
		$0.33\frac{1}{3}$		7.
	$\frac{70}{100}$			8.
		$0.66\frac{2}{3}$		9.
$\frac{3}{10}$				10.
		0.01		11.
			90%	12.
$\frac{1}{8}$				13.

14.		$\frac{300}{100}$		
15.			0.375	
16.				150%
17.		$\frac{625}{100}$		
18.			0.0001	
19.	$\frac{7}{8}$			
20.				100%
21.		$\frac{16\frac{2}{3}}{100}$		
22.	$\frac{5}{6}$			
23.	$\frac{1}{9}$			
24.		$\frac{62\frac{1}{2}}{100}$		
25.			0.015	
26.				$\frac{1}{2}$%
27.	$\frac{1}{300}$			
28.		$\frac{125}{100}$		
29.			0.0005	
30.				$\frac{1}{5}$%

For exercises 31–40 imagine a square divided into 100 smaller squares of the same area. For each given percent, state how many small squares would be needed to indicate this percent.

31. 25%

32. 20%

33. 1%

34. $37\frac{1}{2}\%$

35. 150%

36. 8%

37. 200%

38. $\frac{1}{2}\%$

39. $\frac{1}{4}\%$

40. 0.5%

31.

32.

33.

34.

35.

36.

37.

38.

39.

40.

In exercises 41–50 express the given percent as a decimal.

41. 71%

42. 47%

43. 93%

44. 1%

45. 100%

46. 0.5%

47. 0.01%

48. 165%

49. 975%

50. 1000%

41.

42.

43.

44.

45.

46.

47.

48.

49.

50.

In exercises 51–60 write, in lowest terms, the fraction equal to the given percent.

51. 32%

52. 90%

53. 120%

54. 175%

55. $6\frac{2}{3}\%$

56. $43\frac{3}{4}\%$

57. $\frac{5}{6}\%$

58. $\frac{16}{5}\%$

59. 0.96%

60. 0.625%

51.

52.

53.

54.

55.

56.

57.

58.

59.

60.

61. 62.

63. 64.

65. 66.

67. 68.

69. 70.

71. 72.

73. 74.

75. 76.

77. 78.

79. 80.

In exercises 61–70 write the given fraction as a percent.

61. $\frac{13}{25}$

62. $\frac{7}{20}$

63. $\frac{3}{10}$

64. $\frac{2}{5}$

65. $\frac{62}{200}$

66. $\frac{21}{50}$

67. $\frac{18}{300}$

68. $\frac{11}{15}$

69. $\frac{3}{40}$

70. $\frac{1}{125}$

In exercises 71–80 write the given number as a percent.

71. 0.0002

72. 0.001

73. 2.178

74. 3.3

75. 0.014

76. 0.7

77. 5.16

78. 0.005

79. 10

80. 1.1

5.4 Solving Percent Problems Using Proportions

We have seen that a percent is a comparison with 100 and can be expressed as a decimal or as a fraction. The statement

25% of 16 is 4

can be thought of as a statement of equality as shown by the proportion

$$\frac{25}{100} = \frac{4}{16}$$

This means that the ratio of 25 to 100 is the same as the ratio of 4 to 16; 100 and 16, the denominators, refer to the whole quantities, whereas 25 and 4, the numerators, represent parts of the wholes.

The statement "25% of 16 is 4" can lead to three types of questions:

(1) What percent of 16 is 4?

or

R% of 16 is 4. What is *R*?

(2) 25% of what number is 4?

or

25% of *B* is 4. What is *B*?

(3) 25% of 16 is what number?

or

25% of 16 is *P*. What is *P*?

Examine each of these questions closely and study the following relationships:

(1) *R*% of 16 is 4 means that

$$\frac{R}{100} = \frac{4}{16}$$

Solving for *R* we have

$$16 \times R = 4 \times 100$$

$$R = \frac{4 \times 100}{16} = 25$$

Thus 25% of 16 is 4.

OBJECTIVES

You should:

(a) Know how to translate a statement about percent into a proportion.

(b) Be able to solve problems involving percent by using proportions.

****

In exercises 1–7 solve each problem by using a proportion.

1. What is 15% of 240?

2. 9% of what number is 25.2?

3. 37% of what number is 233.1?

4. 114 is what percent of 950?

5. 3.87 is what percent of 129?

6. What is $\frac{3}{4}$% of 850?

7. What is 112% of 46?

STOP

(2) 25% of B is 4 means that

$$\frac{25}{100} = \frac{4}{B}$$

Solving for B, we have

$$25 \times B = 4 \times 100$$

$$B = \frac{4 \times 100}{25} = 16$$

Thus $B = 16$.

(3) 25% of 16 is P means that

$$\frac{25}{100} = \frac{P}{16}$$

Solving for P, we have

$$100 \times P = 25 \times 16$$

$$P = \frac{25 \times 16}{100} = 4$$

Thus $P = 4$.

Do exercises 1 through 7 in the adjoining column.

Exercises 5.4

Name ______________________

Class ______________________

In exercises 1–6 first estimate the answer by choosing what you think is the correct answer from the list of choices in parentheses. Then check your estimate by actual computation.

1. 1% of 394 is (0.394) (3.94) (39.4) (394)

   1. Estimate:
   Calculation:

2. 1% of 1265 is (1.265) (12.65) (126.5) (1265)

   2. Estimate:
   Calculation:

3. 4% of 285 is about (1.2) (12) (120) (1200)

   3. Estimate:
   Calculation:

4. 1.8% of 150 is about (0.27) (2.70) (27) (270)

   4. Estimate:
   Calculation:

5. 108% of 32 is about (0.35) (3.5) (35) (350)

   5. Estimate:
   Calculation:

6. 0.09% of 500 is about (0.05) (0.5) (5) (50)

   6. Estimate:
   Calculation:

In exercises 7–45 find the missing number in the given statement.

7. 20% of 350 = ____________. 7.
8. $37\frac{1}{2}$% of 320 = ____________. 8.
9. 100% of 256 = ____________. 9.
10. 1% of 450 = ____________. 10.
11. 67% of 603.25 = ____________. 11.
12. 1.7% of 88,916.7 = ____________. 12.
13. 21.28% of 524.76 = ____________. 13.
14. 0.008% of 591,400 = ____________. 14.
15. 183% of 268.57 = ____________. 15.
16. ____________% of 400 = 200. 16.
17. ____________% of 40 = 10. 17.
18. ____________% of 1000 = 200. 18.
19. ____________% of 93.2 is 14.9. 19.
20. ____________% of 6.88 is 3.829. 20.

21.	21. ____________% of 76,400 is 522.5.
22.	22. ____________% of 6230 is 9.84.
23.	23. ____________% of 0.888 is 0.037.
24.	24. 30 is 16% of ____________.
25.	25. 75% of ____________ is 24.
26.	26. 63 is 60% of ____________.
27.	27. 225% of 450 is ____________.
28.	28. 61.7% of ____________ is 488.96.
29.	29. 92.5 is 21.2% of ____________.
30.	30. 28.108 is 16.5% of ____________.
31.	31. 90.14% of ____________ is 0.0699.
32.	32. 0.094% of ____________ is 6.425.
33.	33. 281.08 is 165.4% of ____________.
34.	34. 36% of 1762 is ____________.
35.	35. 42 is 6% of ____________.
36.	36. 1.9% is 89.75 is ____________.
37.	37. 150 is ____________% of 750.
38.	38. 0.52% of 2.93 is ____________.
39.	39. 126 is $31\frac{1}{2}$% of ____________.
40.	40. 36% of ____________ is 198.
41.	41. 1.9% of 139.7 is ____________.
42.	42. 3573 is 32.8% of ____________.
43.	43. ____________% of 7320 is 83.102.
44.	44. ____________ is $\frac{1}{2}$% of 6.974.
45.	45. 617.8 is 21.2% of ____________.

5.5 Solving Percent Problems Using a Formula

We can re-phrase the statement

$$P \text{ is } R\% \text{ of } B$$

as the proportion

$$\frac{R}{100} = \frac{P}{B}$$

so that

$$P = \frac{R}{100} \times B$$

Often this last equality is written as

$$P = r \times B$$

where $r = \dfrac{R}{100}$.

Using the formula $P = r \times B$ to solve the three problems of section 5.4, we have

(1) What percent of 16 is 4?

Here $P = 4$, $B = 16$, and so

$$4 = r \times 16$$

$$r = \frac{4}{16} = 0.25$$

Thus 4 is 25% of 16.

(2) 25% of what number is 4?

Here $r = \dfrac{25}{100} = 0.25$, $P = 4$ and so

$$4 = 0.25 \times B$$

$$B = \frac{4}{0.25} = 16$$

Thus 25% of 16 is 4.

OBJECTIVES

You should:

(a) Know how to translate a statement about percent into an equality.

(b) Be able to solve problems involving percent by using a formula.

****

(3) 25% of 16 is what number?

Here $r = \dfrac{25}{100} = 0.25$, $B = 16$ and so

$$P = 0.25 \times 16 = 4$$

Thus 25% of 16 is 4.

Here are three additional examples:

Example 1: 2% of what number is 18?

This means that

$$0.02 \times B = 18$$

$$B = \frac{18}{0.02} = 900$$

Thus 18 is 2% of 900.

Example 2: What percent of 27 is 18?

This means that

$$r \times 27 = 18$$

$$r = \frac{18}{27} = \frac{2}{3} = 0.66\frac{2}{3}$$

Thus $66\frac{2}{3}$% of 27 is 18.

Example 3: Find 60% of 180.

This means that

$$P = 60\% \text{ of } 180.$$

$$P = 0.60 \times 180$$

$$P = 108$$

Do exercises 1 through 8 in the adjoining column.

In exercises 1-8 solve the given problem by using the formula $P = r \times B$.

1. 6% of 42.89 = ?

2. 3.8% of 96 = ?

3. 11.64 is 24% of what number?

4. 17% of what number is 416.5?

5. What percent of 105 is 7?

6. What percent of 78 is 39?

7. What is $\frac{1}{3}$% of 396?

8. What is 325% of 420?

STOP

Exercises 5.5

Name ______________________

Class ______________________

In exercises 1–6 first estimate the answer by choosing what you think is the correct answer from the list of choices in parentheses. Then check your estimate by actual computation.

1. 5% of 394 is (1.97) (19.7) (197) (1970)

2. 5% of 1265 is (6.325) (63.25) (632.5) (6325)

3. 19% of 20 is about (0.04) (0.4) (4) (40)

4. 12.5% of 72 is about (0.09) (0.9) (9) (90)

5. 196% of 44 is about (0.88) (8.8) (88) (880)

6. $\frac{1}{2}$% of 12.65 is about (0.006) (0.06) (0.6) (6)

1. Estimate:
   Calculation:
2. Estimate:
   Calculation:
3. Estimate:
   Calculation:
4. Estimate:
   Calculation:
5. Estimate:
   Calculation:
6. Estimate:
   Calculation:

In exercises 7–45 find the missing number in the given statement.

7. 25% of 640 = ____________.
8. 75% of 160 = ____________.
9. 150% of 200 = ____________.
10. $\frac{1}{2}$% of 140 = ____________.
11. 15.8% of 9.2261 = ____________.
12. 61.5% of 48.988 = ____________.
13. 0.52% of 73.76 = ____________.
14. 224% of 83.229 = ____________.
15. 505% of 60.098 = ____________.
16. ____________% of 500 = 100.
17. ____________% of 500 = 750.
18. ____________% of 500 = 1000.
19. ____________% of 762.16 is 70.883.
20. ____________% of 1744.6 is 725.13.

7.
8.
9.
10.
11.
12.
13.
14.
15.
16.
17.
18.
19.
20.

21.	21. __________% of 923 is 0.62.
22.	22. __________% of 297.4 is 3316.5.
23.	23. 50% of __________ is 32.
24.	24. 12 is 40% of __________.
25.	25. 150% of __________ is 250.
26.	26. 15.8% of __________ is 92.61.
27.	27. 172.9 is 88.1% of __________.
28.	28. 60.07% of __________ is 6.44.
29.	29. 26.87 is 0.026% of __________.
30.	30. 160.1% of __________ is 0.042.
31.	31. 103.3 is 656% of __________.
32.	32. __________% of 360 is 90.
33.	33. 113% of 1640 is __________.
34.	34. 112 is __________% of 280.
35.	35. 387% of 0.857 is __________.
36.	36. 44.5% of __________ is 507.8.
37.	37. 61.105 is __________% of 574.8.
38.	38. $\frac{3}{4}$% of 91.06 is __________.
39.	39. 0.782 is 0.56% of __________.
40.	40. __________% of 28.16 is 0.736.
41.	41. 0.0867 is __________% of 0.981.
42.	42. __________ is $7\frac{1}{4}$% of 200.84.
43.	43. 7036 is __________% of 418.
44.	44. 0.5027 is 350% of __________.
45.	45. __________% of 21,443 is 689.

5.6 Applications of Percent

There are many applications of the concept of percent, especially in the business world. Different kinds of businesses, however, frequently use a specialized language related to percent. Thus to understand how percent is used in the business world you must be familiar with certain special terms.

People who work as salesmen are often paid a *commission* instead of, or in addition to, a salary. For example, a real estate salesperson may receive a commission of 6% of the selling price of a house. Thus if a house were to sell for \$30,000 the commission would be 6% of 30,000 or \$1800. Frequently the percent of the selling price which gives the commission is called the *rate of commission.* Thus in the above example 6% is the rate of commission.

Merchants sometimes sell articles at a *discount.* During a sale an advertisement may state "all sweaters will be sold at a discount of 30%" or "all sweaters 30% off." A sweater marked \$20 then sells at a discount of \$6, which is 30% of \$20. The sale price (sometimes called the *net price*) is \$14 (\$20 – \$6 = \$14).

If you rent anything—a car, a house, or an apartment—you must pay a certain sum of money regularly to the owner for the use of his or her property. In the same way, if you borrow money you must pay for the use of it. The money that you borrow is called the *principal* and the money that you pay for using it is called the *interest.* The amount of interest is usually indicated as a certain percentage of the principal and is normally based on one year's time. This is called the *rate* of interest per year.

For example, if Ms. Clock borrowed \$800 from Mr. Mizner for one year and paid interest at the rate of 6% per year, Ms. Clock would pay Mr. Mizner \$(0.06 × 800) or \$48 for the use of the money for one year. At the end of the year she would pay back an amount equal to the principal plus the interest, or \$800 + \$48 which would amount to \$848. If Ms. Clock had borrowed the money for 6 months (or $\frac{1}{2}$ year), the interest would be $\frac{1}{2}$ of \$48, or \$24.

On the other hand, if you place your money in a savings account at a bank, the bank will use your money and pay you interest.

Here are four more examples of application of percent.

Example 1: What interest would you receive at the end of one year on a deposit of \$1200 if the interest rate is $5\frac{1}{4}\%$?

Solution: If I is the interest in dollars, $I = 5\frac{1}{4}\%$ of \$1200 = 63. The interest would be \$63.

Do exercises 1 and 2 in the adjoining column.

OBJECTIVE

You should be able to solve problems that involve applications of percent.

****

1. How much interest would you receive on a \$5000 savings certificate at $8\frac{1}{4}\%$ at the end of one year?

2. For the purchase of an automobile, a 30% down payment was needed. If the purchase price was \$2950, how much down payment was needed?

STOP

3. The price of hamburger at the supermarket increased from 80¢ to 95¢ a pound. What was the percent of increase in the price of hamburger?

4. When Ron went to summer camp, he weighed 120 pounds. At the end of summer camp, he weighed 126 pounds. By what percent did his weight increase at summer camp?

STOP

5. A rancher lost 72 sheep in one winter. If this was 18% of his flock, how many sheep did he have originally?

6. The rental of equipment on a construction project amounts to $5520. If this represents 3% of the cost of the project, what was the cost of the project?

STOP

7. The Podunk Community College has an enrollment of 8568 students. If this enrollment is an increase of 12% over last year, what was last year's enrollment?

8. Jack went on a diet and now weighs 186 pounds. If this is a 7% loss in weight, what was his weight before going on the diet?

STOP

Example 2: Joe Doakes' salary was increased from $12,000 to $15,000. By what percent was his salary increased?

Solution: The increase was $3000, so we want to find out what percent 3000 is of 12,000. We have

$$3000 = r \times 12{,}000$$

and so

$$r = 3000 \div 12{,}000 = 0.25$$

His salary was increased by 25%.

Do exercises 3 and 4.

Example 3: Felicia Arietta purchased some stock on January 1, 1976. A year later she sold the stock for 6% less than she paid. If her loss was $3000, what did she pay for the stock?

Solution: We have

$$3000 = 0.06 \times B$$

so that the purchase price of the stock, in dollars, was

$$B = 3000 \div 0.06 = 50{,}000$$

She paid $50,000 for the stock.

Do exercises 5 and 6.

Example 4: After receiving a 6% increase in salary, Janet Doakes' salary was $14,840. What was her salary before the increase?

Solution: If her salary before the increase was S dollars, we have

$$\begin{aligned} S + 0.06 \times S &= 14{,}840 \\ 1 \times S + 0.06 \times S &= 14{,}840 \\ (1 + 0.06) \times S &= 14{,}840 \quad \text{(Why?)} \\ 1.06\,S &= 14{,}840 \\ S &= 14{,}840 \div 1.06 \\ S &= 14{,}000 \end{aligned}$$

Her salary before the increase was $14,000.

Do exercises 7 and 8.

Exercises 5.6

Name ____________________

Class ____________________

1. Alice saved \$50. If she then bought a model road racer for \$7.50, what percent of her savings did she spend?

2. Gravity on the moon is such that the weight of an object is 16% of its weight on earth. If a man weighs 180 pounds on earth, how much would he weigh on the moon?

3. One hundred and twenty persons were invited to a wedding reception, but only 100 were able to come. What percent of those invited were able to come?

4. To qualify as stainless steel, steel must contain a minimum of $11\frac{1}{2}\%$ chromium. What is the minimum amount of chromium in 450 kilograms of stainless steel?

5. A used auto is sold at \$1225 less than it cost new. If it sold at 75% of the cost, what was the cost of the new auto?

6. What is the price of 12 baseball suits listed at \$45.40 each with a 20% discount?

7. A state has a sales tax of 4.2%. (a) What is the amount of tax on a purchase of \$125? (b) What did the person actually pay for the article?

8. In a mathematics test of 25 problems, Sue received a score of 80%. How many problems did she answer correctly?

1.

2.

3.

4.

5.

6.

7. (a)

(b)

8.

9.

10.

11.

12.

13.

14.

15.

16.

17.

18.

19.

20.

21.

9. One day, during a flu epidemic, 60% of Mr. Hyde's class was absent. If 18 persons were absent, how many people were enrolled in Mr. Hyde's class?

10. A quarterback completed 16 out of 24 passes in a football game. What percent of his passes were completed?

11. During a sale, Jody bought a dress for 20% less than the regular price. If she saved $16 by buying the dress on sale, what was the original price of the dress?

12. Ray told Alice that his 55 folk song records were only 20% of his record collection. How many records were in Ray's collection?

13. A basketball team played 22 games and won 82% of them. How many games did they win?

14. Mr. Gonzales' secretary was given 82 letters to type. By noon he had completed 27. What percent of the letters to be typed had he completed?

15. A man earns $675 a month and spends $150 a month for the rent on his apartment. What percent of his earnings go for rent?

16. Mr. Marconi bought a television set and paid $90 down. This represented 15% of the cost of the set. What was the cost of the set?

17. Ms. Anderson rents a house for $225 a month. This is 1% of the value of the house. What is the value of the house?

18. At a "one of a kind" sale, a $230 sofa was offered at a 45% reduction. How much could be saved by buying the sofa on sale?

19. Jerry was paid $52.50 interest per month for one year on a principal of $7500. If none of the principal has been paid back, what rate of interest is Jerry paying?

20. Ten of 43 pupils in a class were on the honor roll. What percent of the pupils were on the honor roll?

21. Janice had a score of 80% on a mathematics test. If she had 16 problems correct, how many problems did the test contain?

22. At sea level, sound travels at a speed of 742 miles per hour. If a supersonic airliner travels at a speed of Mach 2.2 (220% of the speed of sound), what is its speed in miles per hour? 22.

23. A certain kind of steel contains 2% carbon. How many kilograms of carbon are there in a bar of this kind of steel that weighs 75 kilograms? 23.

24. Boy's sweaters marked at $18.95 were on sale at a discount of 20%. Find the sale price. 24.

25. Joe Rankin needs $200 to complete the payments on his car. He borrows the money from a friend at 6% interest per year and repays this loan in 3 months. How much did he need to pay his friend at that time? 25.

26. What rate of yearly interest did Sally pay if she borrowed $3600 for 60 days and paid $36 interest? 26.

27. A car battery is guaranteed by a certain company for 36 months. If the battery were to wear out after 27 months, on what percent of the battery life would the company have to make good? 27.

28. A man's brain weighs between 2% and 3% of his body weight. If his brain weighs four pounds, (a) what is his least possible weight? (b) what is his greatest possible weight? 28. (a)

(b)

29. The Ajax Record Shop pays a commission of 25¢ on each record sold. If the commission rate is $12\frac{1}{2}$%, at what price do the records sell? 29.

30. Charlie said that his blood contains $4\frac{1}{2}$ quarts of water. If blood is 90% water, how many quarts of blood does Charlie think he has? 30.

31. Honey is 71.5% sugar. How many pounds of sugar are there in a 5-pound can of honey? 31.

32. Mrs. O'Hara borrowed $7540 at 8% interest to enlarge her hat shop. At the end of 3 years she paid off the debt. How much interest did she have to pay? 32.

33. Ms. Scott is a saleswoman whose rate of commission is 12%. How many dollars worth of goods did she have to sell to earn a commission of $720? 33.

34.

35. (a)

(b)

36.

37.

38.

39.

40.

41.

42.

43.

44.

34. Mr. Jacobs arranged a loan from his bank for \$2000. At the end of the year he paid the bank \$2140, which included the total amount of principal and interest. What was the rate of interest charged by the bank?

35. Robert sold Christmas cards to his friends and relatives. Each box cost him 79¢ and he sold them for \$2 each. (a) What was his percent of commission on the selling price? (b) What was the percent of increase in price?

36. Two successive census reports showed that the population of a certain city had risen from 320,000 to 400,000. What was its percent of increase?

37. A merchant insured a cargo on a ship worth \$12,560 for 75% of its value at a cost of 4% of the insured value. What did the insurance cost her?

38. A manufacturing firm predicted that its sales for the year would be \$265,000,000. At the end of the year the actual sales were 107% of the predicted sales. What was the actual amount of their sales?

39. A mechanic purchases an overhaul gasket set, at a 40% discount from the list price, for \$12.60. He charges his customer the list price for the set. What should he charge for this item?

40. The list price of the parts used to repair an engine is \$35.70. Of this, \$29.40 is subject to a $33\frac{1}{3}\%$ discount from the list price. What is the net cost of the parts?

41. By changing the casting pattern slightly, a savings of 8 pounds can be made in the weight of an aluminum casting that was originally 64 pounds. What percent of the original weight was saved?

42. The cutting speed in turning steel is increased 22.5% with the proper application of a cutting fluid. If the cutting speed of a piece of steel being turned is 40 feet per minute without cutting fluid, find the cutting speed with cutting fluid.

43. Brazing metal is composed of 85% copper and 15% zinc. How many pounds of brazing metal can be made from 34 pounds of copper?

44. Oregon produced 95 million seedling trees during the four years 1968 to 1972. Twenty-four percent of these were planted during 1970–72 at a density of 454 seedlings per acre. How many seedlings were planted during the years 1970–72?

45. In a timber sale in a national forest there were 18,500 marked trees. 1400 of those marked trees were Lodgepole Pine. What percent of the trees in the sale were Lodgepole Pine? | 45.

46. The growth in volume of trees can be expressed as a percent of the volume originally measured. The result of a timber cruise in 1975 indicated 4700 MBF (million board feet). If the growth rate is approximately 5% a year, what would be the amount of timber available in (a) 1976? (b) 1977? | 46. (a) (b)

47. If you bought a home last year for $32,900 and its value appreciated 15% this year, what is the value of the home now? | 47.

48. If hamburger costs $1.15 a pound and is rated at "not more than 28% fat," what are you paying per pound for the non-fat portion of the hamburger? | 48.

49. If a family has a before taxes income of $10,800 per year and spends $175 per month for food, what percent of their budget goes for food? | 49.

50. The average American family spends 22.3% of its after taxes income for food. If the family in exercise 49 pays 9.2% of its income in taxes, how does the amount they pay for food each month compare with the average American family? | 50.

51. If your Federal income tax amounts to $1380 plus 22% of the excess over $8000 of taxable income, how much tax would you pay on a taxable income of $9238? | 51.

*52. What percent above cost must a merchant mark an article, in order to be able to sell it at a discount of 15% from the list price, and still make a profit of 20%? | *52.

*53. Andy and Bert started working for different firms at the same salary. Last year Andy had a raise of 10%. This year Andy had a cut of 10%; Bert's salary remained the same. Are they now making the same amount? If not, who is making more money now? Prove that your answer is correct. | *53.

*54.

*54. Mr. Jones read the following statements in different parts of his evening paper:

- Have your engine tuned and save 33% of the gasoline you use in your car.
- Save 28% of your gasoline by driving at 55 mph on the highway.
- Save 20% of your gasoline consumption by not making "jack rabbit" starts and stops.
- Keeping your tires properly inflated saves you 15% of your gasoline consumption.
- Save 10% on your gasoline use by using radial tires.

Mr. Jones concluded that if he did all these things he could retire by selling the gasoline his car produced. What is wrong with his reasoning?

Review Exercises 5

Name ______________________

Class ______________________

In exercises 1–15 express each relationship as a ratio $\frac{a}{b}$ in lowest terms using the same units for a and b.

1. 21 to 24
2. 150 to 130
3. $\frac{1}{2}$ to $\frac{7}{10}$
4. $\frac{5}{6}$ to $\frac{10}{9}$
5. 3 yards to 2 feet
6. 45 centimeters to 105 centimeters
7. 6 pounds, 4 ounces to 8 pounds, 6 ounces
8. 35 centimeters to 3 meters (1 meter = 100 centimeters)
9. 16 gallons to 8 quarts
10. 120 grams to 0.5 kilograms (1 kilogram = 1000 grams)
11. 300 cubic centimeters to 2 liters (1 liter = 1000 cubic centimeters)
12. 2 feet, 8 inches to 1 foot, 10 inches
13. 80 hits to 240 times at bat
14. Interest of \$36 to loan of \$600
15. 35 students to 50 chairs

1.
2.
3.
4.
5.
6.
7.
8.
9.
10.
11.
12.
13.
14.
15.

16. 17.

18. 19.

20. 21.

22. 23.

24. 25.

26. 27.

28. 29.

30.

31.

32.

33.

34. (a)

(b)

In exercises 16–30 find the missing number in the given proportion.

16. $\frac{20}{120} = \frac{N}{720}$

17. $\frac{120}{80} = \frac{36}{N}$

18. $\frac{440}{64} = \frac{11}{N}$

19. $\frac{68}{N} = \frac{170}{90}$

20. $\frac{12}{N} = \frac{108}{172}$

21. $\frac{2000}{120} = \frac{N}{300}$

22. $\frac{2000}{N} = \frac{3000}{1725}$

23. $\frac{55}{11} = \frac{41}{N}$

24. $\frac{300}{120} = \frac{N}{14}$

25. $\frac{N}{0.5} = \frac{40}{0.25}$

26. $\frac{N}{550} = \frac{7800}{2750}$

27. $\frac{51}{N} = \frac{4590}{6300}$

28. $\frac{N}{20} = \frac{5400}{180}$

29. $\frac{2500}{300} = \frac{25}{N}$

30. $\frac{N}{450} = \frac{1800}{90{,}000}$

In exercises 31–50 solve the problem by using a proportion.

31. Two numbers are to each other as 15:24. If the smaller number is 45, what is the larger?

32. An airplane pilot found that he had flown a distance of 60 miles in 15 minutes. At the same rate, how long would it take him to reach a check point 150 miles away?

33. In a school election, the votes cast for the two candidates were in a ratio of 5 to 6. If the winning candidate received 822 votes, how many votes were cast for his opponent?

34. If apples are 4 pounds for 97¢, find (a) how much 10 pounds of apples would cost; (b) how many pounds should I have if the clerk charged me $1.70.

35. The scale on a drawing is $\frac{1}{8}$ inch for 1 foot.
   (a) How long a line must be made for a dimension of 35 feet?
   (b) If a line of $3\frac{1}{16}$ inches is made on the drawing, what is the dimension of the object?

35. (a)

(b)

36. If 18 square yards of carpeting cost $342, how much will 24 square yards of the same carpeting cost?

36.

37. If oranges sell at 3 for 25¢,
   (a) how much will two dozen oranges cost?
   (b) how many oranges could you buy and not spend more than $1.35?

37. (a)

(b)

38. A manufacturer finds that the ratio between the selling price and the manufacturing cost of an item is 13:2.
   (a) If it costs 25¢ to manufacture the item, how much will it sell for?
   (b) If the item were to sell for $4.85, what would be the maximum cost of manufacturing it?

28. (a)

(b)

39. An automobile has a 20-quart cooling system. If the ratio of anti-freeze solution to water is 3:1, how much of the solution does the radiator contain?

39.

40. With a tail wind, Canadian geese sometimes travel as far as 130 miles in 2 hours.
   (a) At this rate, how far would they travel in 7 hours?
   (b) How long would it take them to fly 650 miles?

40. (a)

(b)

41. A thrust of 7,700,000 pounds is produced in a Saturn V launch vehicle when 4,735,000 pounds of propellant are burned in 161 seconds.
   (a) At this rate, how much propellant would be burned in one minute?
   (b) At this rate, how long would a ton of propellant burn?

41. (a)

(b)

42. A steel I-beam 8 feet long weighs 312 pounds.
   (a) What is the weight of a beam of the same kind that is 13.5 feet long?
   (b) How long is a beam that weighs 390 pounds?

42. (a)

(b)

43. (a)

(b)

43. Four hundred eighty square feet of wall can be painted with 3 quarts of paint.
(a) How many gallons of the same kind of paint should be purchased if 2500 square feet of wall are to be painted?
(b) How much area will a 5-gallon can of paint cover?

44. (a)

(b)

44. The pitch of a roof is the ratio between the rise and the run of the rafters.
(a) If the pitch is an 8-inch rise for each foot of run, how long is the run if the rise of the roof is 5 feet?
(b) Would you be able to stand in an attic if the roof had the pitch in (a) and a run of 10 feet?

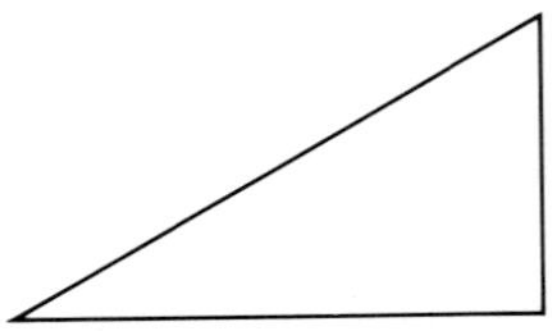

45. (a)

(b)

45. In a certain pictorial graph a picture of one horse represents 450 horses.
(a) How many pictures of horses would be needed to represent 90,000 horses?
(b) At this same rate how many horses are represented by $7\frac{3}{4}$ pictures of horses?

46.

46. If a 5-grain aspirin tablet is about the right strength for a 120-pound woman, how much aspirin should be given to a 60-pound child?

47. (a)

(b)

47. Water is 2 parts hydrogen to 16 parts oxygen (by weight).
(a) How much hydrogen is in 50 grams of water?
(b) How much oxygen is in 75 grams of water?

48. (a)

(b)

48. An investor earns $7.50 on every $100 he invests.
(a) What will he have to invest to earn $500?
(b) If he invests $5250, how much would he earn from the investment?

49. (a)

(b)

(c)

49. When tranquilizers are administered by a veterinarian intramuscularly, a dosage of 1 to 3 milligrams per kilogram of body weight is required. Calculate, to the nearest milligram, the maximum and minimum dosages of the drug for each of the following animals:
(a) a beagle weighing 10 kilograms (1 kilogram = 2.2 pounds)
(b) a tom cat weighing 9 pounds
(c) a 1300-pound steer

50. In a Morse no. 1 taper the diameter of a steel shaft is reduced 0.6 inches per foot of length. How much should a machinist reduce the diameter if it is tapered for a length of 2 feet, 4 inches? 50.

In exercises 51–71 find the missing number in each statement.

51. 15% of 240 = ____________. 51.

52. 3.8% of 96 = ____________. 52.

53. $\frac{3}{4}$% of 850 = ____________. 53.

54. $162\frac{1}{2}$% of 800 = ____________. 54.

55. 780 is 16% of ____________. 55.

56. 286 is 134% of ____________. 56.

57. $\frac{2}{5}$ is 45% of ____________. 57.

58. 35 is $1\frac{1}{4}$% of ____________. 58.

59. 7.3 is ____________ % of 21.9. 59.

60. 6 is ____________ % of 1000. 60.

61. 53 is ____________ % of 620. 61.

62. $2\frac{1}{2}$ is ____________ % of 15. 62.

63. 120% of 960 is ____________ . 63.

64. 5 is $\frac{1}{2}$ % of ____________. 64.

65. 21 is ____________ % of 57. 65.

66. 45 is 120% of ____________. 66.

67. 67. $\frac{1}{10}$% of 1000 is ______________.

68. 68. 275 is 5% of ______________.

69. 69. ______________ is $37\frac{1}{2}$% of 160.

70. 70. 8.8 is $5\frac{1}{2}$% of ______________.

71. 71. 72 is ______________ % of 24.

72. 72. From an income of \$650 a month, a family spends \$175 for food. What percent of their monthly income is spent on food?

73. 73. Mr. Rasmussen planted 15 new rose bushes in his garden. Two of them died. What percent of the bushes survived?

74. 74. The Ajax Record Company says that $\frac{1}{2}$% of the discs which a certain machine manufactured were defective. How many defective records were produced if the machine put out 300,000 records?

75. 75. Joan bought 50 yards of washable drapery material. If the material shrinks 1% when washed, how long will the material be after it is washed?

76. 76. The milk from the Bossy Dairy tests 5.26% butterfat. How many pounds of this milk would be needed for 200 pounds of butterfat?

77. 77. In a certain company, 156 workers, 12% of the work force, were less than 25 years of age. How many workers were employed by the company?

78. 78. A certain store sells merchandise at 10% above cost. What percent discount did Mrs. Harvey receive when she purchased a washing machine that cost the store \$250 and is sold elsewhere for \$350?

79. 79. Ken paid \$.99 for a book that was marked down from \$1.79. What was the percent of discount?

80. 80. Mr. Wright sells clothing in a department store. His daily salary is \$30 a day plus a commission of 5% on all merchandise he sells after the first \$200. If his salary for one day was \$60, what was the total cost of clothing he sold?

81. By selling an article for $40, Mr. Perez made a profit of 20% of the selling price. What was the cost of the article to Mr. Perez? | 81.

82. Mrs. Nelson borrowed $800 from a friend and agreed to pay back the sum after $1\frac{1}{2}$ years with interest at the rate of 7% per year. What amount did she have to pay at the end of the year and a half? | 82.

83. In a toy factory 250 dolls were inspected. If the usual rate of defective dolls is 1.2%, how many defective dolls would be expected? | 83.

84. A wholesaler gives a discount of $1\frac{1}{2}$% to any buyer who pays for an order in 30 days. Mr. Jenkins paid his bill and received a discount of $50. How many dollars worth of merchandise did Mr. Jenkins purchase? | 84.

85. An instrument package of a satellite weighs 250 grams and represents 0.015% of the weight of the satellite. What was the weight of the satellite in kilograms? (1 kilogram = 1000 grams) | 85.

86. A stick is 28 inches long.
(a) What percent of a yard is this?
(b) What percent of a foot is this?

86. (a)

(b)

87. There are two pints in a quart and four quarts in a gallon.
(a) What percent of a gallon is a pint?
(b) What percent of a pint is a gallon?

87. (a)

(b)

88. The planet Venus has a diameter which is 91% of that of the earth. If the earth has a diameter of 7927 miles, what is the diameter of Venus? | 88.

In exercises 89–93 refer to the information in the table on the following page.

89. (a) What percent of the total money spent for automobile transportation in 1955 was spent for repairs, parking, and rentals?
(b) Compare the percentage in 1955 with that in 1970 for the same items. Was there an increase or a decrease in the percentage?

89. (a)

(b)

90. Compare the percentages spent for automobile purchase in 1955 and 1970. Was there an increase or a decrease? | 90.

91. In the five-year period, 1966–70, what percent of the total auto transportation cost was due to gas and oil? Was this more or less than the percent from 1956–60? | 91.

CONSUMER SPENDING FOR AUTOMOBILE TRANSPORTATION (Millions of dollars)

Year	Total	Automobile Purchase[1]	Tires, Tubes, Accessories, Parts	Auto Repairs, Parking, Rental, etc.	Gasoline and Oil	Tolls	Auto Insurance[2]
1955	32,589	16,840	1,581	3,619	9,000	171	1,378
1956	31,725	14,686	1,733	3,918	9,783	196	1,409
1957	34,755	16,281	1,972	4,225	10,642	225	1,410
1958	32,572	13,258	2,120	4,387	10,951	250	1,606
1959	37,980	17,154	2,362	4,805	11,571	292	1,796
1960	39,825	17,748	2,353	5,198	12,252	308	1,966
1961	38,135	15,991	2,430	4,973	12,386	315	2,040
1962	42,516	19,486	2,557	5,200	12,908	327	2,038
1963	45,695	21,549	2,707	5,591	13,457	344	2,047
1964	47,842	22,822	2,946	5,590	14,023	381	2,080
1965	54,356	27,153	3,155	6,122	15,261	431	2,234
1966	56,446	26,805	3,520	6,419	16,562	447	2,693
1967	58,163	26,646	3,848	6,637	17,619	464	2,949
1968	67,265	32,979	4,479	7,292	18,992	504	3,019
1969	72,843	35,270	5,084	7,864	21,095	517	3,013
1970	72,354	31,535	5,581	8,672	22,862	537	3,167

[1]New cars and net purchases of used cars [2]Auto insurance premiums less claims paid Source: Department of Commerce

92.

92. In 1970 what percent of all consumer spending for automobile transportation except automobile purchases was spent on gas and oil?

93.

93. What was the percent of increase or decrease in the amount of money spent for gas and oil when comparing the years 1955 and 1970 and leaving out automobile purchases?

94.

94. Three chairs are listed at $48 apiece. The price of the poorest selling chair is reduced to $32. The price of the other two chairs is not reduced. If all three chairs are purchased together, a reduction of $16\frac{2}{3}\%$ of the list price can be obtained. If you want to purchase all three chairs, which is the best way to purchase them—separately, or together? How much can you save by purchasing them in the most economical fashion?

*95.

*95. A certain store gives a 10% discount for cash and an additional 5% discount for purchases made on Monday. That is, if a customer purchased an article priced at $100 and paid cash on any other day than Monday, it would cost him $(100 – 10) = $90. If, however, he purchased the article on Monday, he would get a further 5% discount. Thus the article would have cost him $85.50 (5% of 90 = 4.50 and 90 – 4.50 = 85.50). Suppose the 5% discount had been computed first and the 10% discount second. Would the final prices have been the same? If not, which one would have been less and by how much? Would these two ways of computing the final cost give the same conclusion for an article priced at $250?

Just for Fun

A. PATTERNS AND SEQUENCES

Write the next five terms in each sequence.

1. $\frac{1}{3}, \frac{1}{15}, \frac{1}{35}, \frac{1}{63},$ ______, ______, ______, ______, ______

2. $\frac{4}{5}, \frac{5}{6}, \frac{6}{7}, \frac{7}{8},$ ______, ______, ______, ______, ______

3. 1, 2, 6, 24, 120, ______, ______, ______, ______, ______

4. $\frac{3}{4}, \frac{1}{2}, \frac{1}{3}, \frac{2}{9}, \frac{3}{16},$ ______, ______, ______, ______, ______

B. MORE PATTERNS

For those items indicated by an asterisk (*) you are to examine the pattern indicated by the previous computations and then write what you think is the result.

5.

$7 \times 7 =$

$67 \times 67 =$

$667 \times 667 =$

$*6667 \times 6667 =$

$*66667 \times 66667 =$

6.

$\frac{22 \times 22}{1+2+1} =$

$\frac{333 \times 333}{1+2+3+2+1} =$

$* \frac{4444 \times 4444}{1+2+3+4+3+2+1} =$

$* \frac{55555 \times 55555}{1+2+3+4+5+4+3+2+1} =$

C. FIND THE MISSING DIGITS

The following addition problems are correct as written. They are also correct if a different digit is substituted for each letter. Find a proper substitution.

7.
```
   O N E
 + O N E
 -------
   T W O
```

8.
```
     O N E
 + F O U R
 ---------
   F I V E
```

D. TRY THESE

9. If you spent \$5 per minute, how much would you spend in four years?

10. Three men rented a hotel room for \$30, each paying \$10. Later the manager found that they had been overcharged \$5 and sent the bellhop to the room with \$5. The men each kept one dollar and tipped the bellhop the other two. The room cost each man \$9 and a \$2 tip, but $(3 \times 9) + 2 = \$29$. What happened to the other dollar?

11. Using only the keys 0, 1, 2, 3 on your HHC, find the product 2997×48 in the fewest number of operations ($+, -, \times, \div$). Would you believe six?

E. MAGIC SQUARE

12. Complete the following magic square.

129		135	142
		130	
	131		133
143		137	132

F. TRIANGLE PUZZLE

13. Place the numbers 1 through 15 (using each one only once) in the circles so that the sum of the numbers on the three sides of the triangle is 34 and the sum of the three numbers in the center of the triangle also equals 34.

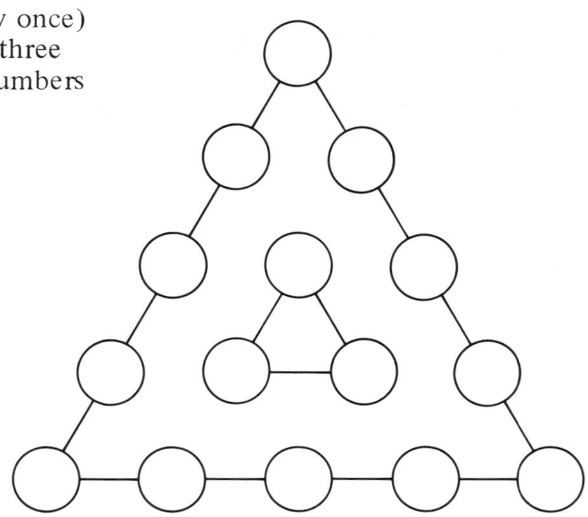

G. CALCULATOR GAME—"SWITCH"

14. Play this game on your calculator.

	Sample Game:	*Display:*
a. Each player enters a 6-digit number with the first number divided by the second.	A enters 256,493.	256,493
	B divides A's number by 753,468.	0.3404165
b. The last three digits on the right of the quotient is the key number.	Key number is 165.	
c. Each player now has 30 seconds to obtain the number named by reversing the digits of the key number or the number 777. If A chooses reversing the digits, B must do 777. Players alternate first choice each round.	A chooses 777;	
	enters 165;	165
	adds 612.	777
	A scores 1 point.	
	B must now try to obtain 561;	
	enters 165;	165
	adds 396.	561
	B scores 1 point.	
d. Each player who does it correctly scores 1 point. If a player does it incorrectly or takes more than 30 seconds, his opponent scores 1 point. First player to score 11 points wins.		

H. THE TALKING CALCULATOR

15. Perform the following calculations to find what a vagrant or tramp who rides the rails is called.

$$[0.9864 + (396 + 106) \times 0.003] \div 3.1$$

Mathematics and the Automobile

1. Choose three cars from those listed in the table on page 324 and compare them by completing the following table (round off any decimal answers to the nearest thousandth in line 2 and otherwise to the nearest tenth):

	Make:	Make:	Make:
Cost of accessories			
HP per pound			
Pounds per HP			
Cruising range:* (195 mi trip mileage)			
Open road mileage			
City mileage			
% reduction or increase of EPA estimates Open road			
City			

**Distance car can travel on a tank of gas.*

2. If you drove an average of 12,000 miles per year with 40% of your mileage on highways (open road) and the rest in the city and paid 75 cents for gasoline, what would it cost you each year for gasoline for each of the three cars you chose in exercise 1?

3. Choose one of the cars listed and calculate the cost per mile for each year for ten years by completing the following table. At what point does it seem to be most economical to trade in your car? Assume that the car will depreciate (lose value) 30% the first year, 20% of the value after the first year in the second year, 15% of the value after the preceding year in the third through fifth years, and 10% of the value after the preceding year in each year after that.

MAKE	1st year	2nd year	3rd year	4th year	5th year	6th year	7th year	8th year	9th year	10th year
Miles driven	12,000	24,000	36,000	48,000	60,000	72,000	84,000	96,000	108,000	120,000
a. Value (this year)*										
b. Depreciation (this year)										
c. Cost of gasoline (this year)**										
d. Cost of repairs (this year)										
e. Total cost (this year)†										
f. Total cost (previous years)	0									
g. Total cost††										
h. Cost per mile										

*For the first year this is the cost of the car; each year after that, the value is that of the previous year minus depreciation during the year.
**See exercise 2.
†This is the sum of (b), (c), and (d).
††This is the cost for this year plus depreciation plus previous years' costs.

4. Choose a tire size listed in all three categories (50,000, 35,000, and 30,000) in the table on page 325 and compute the cost of tires to drive a car 100,000 miles using the tires in each category. Remember that a car needs four tires!

SPECIFICATIONS OF SELECTED 1977 CARS

	Chevrolet Chevette	Dodge Colt	Subaru	AMC Gremlin	Volks-wagen Rabbit	Toyota Corolla	Ford Mustang II
Base price	\$3225	\$3341	\$3299	\$3248	\$4079	\$3778	\$3877
Price with acc.	\$3522*	\$3410*	\$3368*	\$3542*	\$4474†	\$4163†	\$4354†
Wheel base (in.)	94	92	97	96	94	93	96
Weight (lb)	2051	2075	1985	2660	1940	2265	2755
Horsepower	63	83	67	80	78	75	89
Fuel capacity (gal)	13.0	13.2	13.2	15.0	10.9	13.2	13.0
Mileage (mpg): 195 mile trip	30.5	33.5	30.0	26.5	29.0	26.0	25.5
Open road	33.0	37.5	34.5	29.5	31.0	27.0	26.5
City	21.0	19.0	18.5	17.5	20.0	17.5	15.0
EPA est.: Highway	43	45	41	33	33	31	29
City	31	29	28	21	24	25	21

Source: Everything except the EPA data is from *Consumer Reports* magazine. Copyright 1977 by Consumers Union of United States, Inc., Mount Vernon, New York 10550. Excerpted by permission from *Consumer Reports*, June, July, 1977.

*Accessories consist of AM radio, rear window defroster, and radial tires.

†Accessories consist of automatic transmission, AM radio, rear window defroster, and radial tires.

78 SERIES STEEL RADIALS

50,000-MILE TREADWEAR WARRANTY

Whitewalls. Made from 2 strong belts of steel and 2 polyester cord plies for superb handling, great traction, long mileage and fuel economy. Lifetime road hazard, workmanship and materials warranties.

Size	Cash & Carry Group Price	Fed. Excise Tax
BR78-13	35.60	2.11
DR78-14	37.77	2.42
ER78-14	40.94	2.49
FR78-14	41.92	2.69
GR78-14	42.97	2.89
HR78-14	45.99	3.07
GR78-15	43.86	2.97
HR78-15	46.97	3.15
JR78-15	47.97	3.31
LR78-15	49.90	3.47

78 SERIES, 4+2, WHITEWALLS

35,000-MILE TREADWEAR WARRANTY

Four polyester body plies plus two fiberglass belts, six-ply tread for extra strength. Lifetime road hazard, workmanship and materials warranties.

Size	Cash & Carry Group Price	Fed. Excise Tax
A78-13	24.46	1.74
C78-14	26.64	2.04
E78-14	29.77	2.25
F78-14	30.94	2.39
G78-14	31.87	2.55
H78-14	33.74	2.75
J78-14	34.84	2.96
G78-15	32.82	2.58
H78-15	33.96	2.80
J78-15	35.60	3.00
L78-15	36.86	3.08

4-PLY POLYESTER

30,000-MILE TREADWEAR WARRANTY

Premium quality . . . full 78 series whitewalls Lifetime road hazard, workmanship and materials warranties.

Size	Cash & Carry Group Price	Fed. Fed. Tax
A78-13	19.91	1.74
B78-13	20.60	1.84
C78-13	21.62	1.98
B78-14*	21.94	1.88
C78-14	22.74	2.04
E78-14	23.76	2.25
F78-14	24.72	2.39
G78-14	25.69	2.55
H78-14	26.72	2.75
C78-15	23.54	2.06
F78-15	25.51	2.43
G78-15	26.47	2.58
H78-15	27.49	2.80
J78-15	28.62	3.00
L78-15	29.57	3.08

*2 ply, 4 ply rated

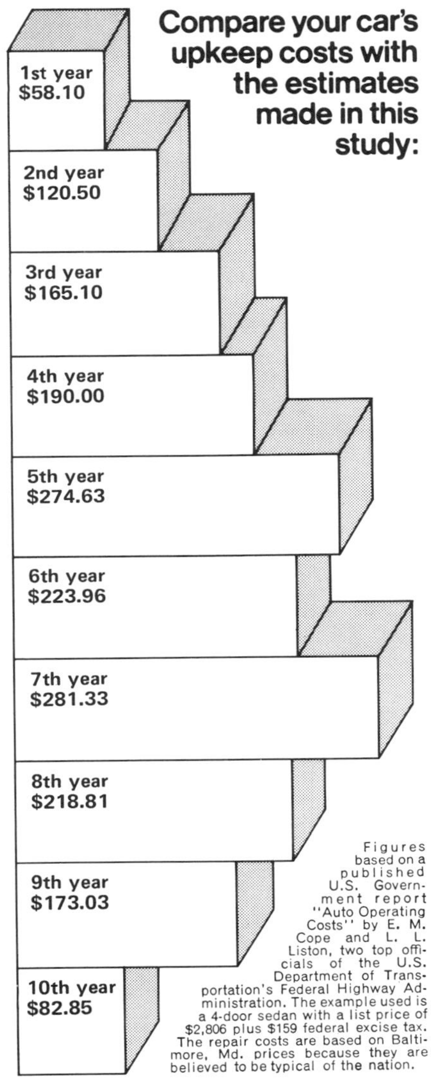

Figures based on a published U.S. Government report "Auto Operating Costs" by E. M. Cope and L. L. Liston, two top officials of the U.S. Department of Transportation's Federal Highway Administration. The example used is a 4-door sedan with a list price of $2,806 plus $159 federal excise tax. The repair costs are based on Baltimore, Md. prices because they are believed to be typical of the nation.

6

Using Formulas

6.1 Some Examples of the Use of Formulas

In section 5.5, the statement "P is $R\%$ of B" was expressed as the equality

(1) $$P = r \times B$$

where $r = \frac{R}{100}$. This equality is an example of a *formula*—a statement of equality that relates two or more quantities.

Some other examples of formulas are

(2) $$F = 1.8C + 32$$

(F = temperature in degrees Fahrenheit and C = temperature in degrees Celsius),

(3) $$A = l \times w$$

(A = area in square units of a rectangle where l = the number of units in the length and w = the number of units in the width), and

(4) $$d = rt$$

(d = distance an object travels in, for example, feet, r = speed of the object in feet per second, and t = time of travel in seconds). Note here that we have written $d = rt$ rather than $d = r \times t$, following the usual practice in algebra of indicating multiplication simply by placing the letters next to each other.

Do exercises 1 through 4 in the adjoining column.

Exercises 1–4 involve a *direct* use of formulas (1) to (4). Section 5.5 also showed how formula (1), $P = r \times B$, can be used *indirectly* when either P and B are given and we want r, or when r and P are given and we want B.

OBJECTIVES

You should be able to:

(a) Use simple formulas.

(b) Round off answers to problems in a reasonable fashion.

****

1. Use formula (1) to find P if $R = 7.2\%$ and $B = \$775$.

2. Use formula (2) to find the temperature in degrees Fahrenheit if the temperature in degrees Celsius is 45.

3. Use formula (3) to find the area in square centimeters of a rectangle whose length is 5.2 cm (centimeters) and whose width is 1.2 cm.

4. Use formula (4) to find the distance in feet that an object will travel in 13 seconds at a speed of 7.4 feet per second.

STOP

Do exercises 5 and 6.

5. Given $P = 30$ and $B = 150$, find r by using formula (1).

6. Given $R = 5.5\%$ and $P = 71.50$, find B by using formula (1).

STOP

Likewise, we can use formulas (2) through (4) indirectly. For example, if $F = 212$ in formula (2), we have

$$212 = 1.8C + 32$$

$$212 - 32 = (1.8C + 32) - 32$$

$$180 = 1.8C$$

$$C = \frac{180}{1.8} = 100$$

Do exercises 7 through 9.

7. If the temperature is 77° Fahrenheit, use formula (2) to find the temperature in degrees Celsius.

8. Use formula (3) to find the width of a rectangle whose area is 375 square centimeters and whose width is 12.5 cm.

9. Use formula (4) to find the speed of an object in feet per second given that it travels 125 feet in 0.2 seconds.

STOP

Let us return to exercise 3, where you computed the area of a rectangle 5.2 cm long and 1.2 cm wide as

$$5.2 \times 1.2 = 6.24$$

square centimeters (abbreviated as cm^2). How meaningful is this answer? Should we do any rounding off, as discussed in section 2.1? We will consider the answers to these and similar questions in this section. (It might be a good idea for you to review section 2.1 at this time.)

Now one very valuable aspect of the metric system of measurement is the ease with which the degree of accuracy of a measurement can be reported. This is because the subdivisions of units of measurements in the metric system are always tenths, hundredths, thousandths, etc. in sharp contrast to, for example, 12 inches in a foot, four quarts in a gallon, 16 ounces in a pound, etc.

Thus in the metric system we can have measurements reported as

5.2 cm, 5.20 cm, and 5.200 cm

and each of these means something different. The 5.2 cm means that the length is 5.2 cm to the nearest tenth of a centimeter and hence the length may actually be as much as 5.24 cm (since 5.24 rounded off is 5.2) and as little as 5.15 cm (since 5.15 rounded off is also 5.2). This is illustrated on the number line that follows.

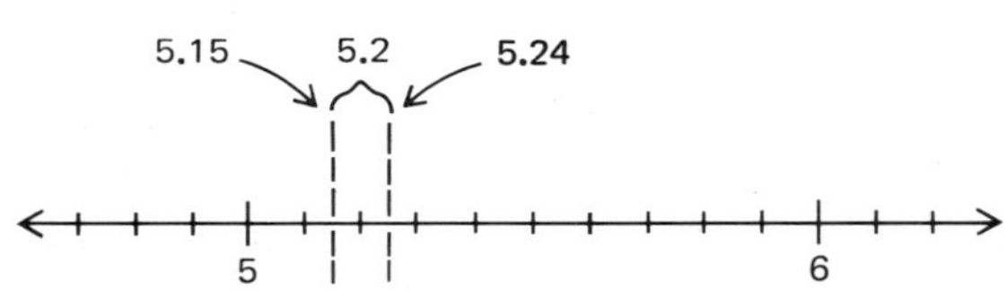

Similarly, the 5.20 cm means that the length is 5.2 centimeters to the nearest hundredth of a centimeter and hence, in this case, the length may actually be as much as 5.204 cm (why?) and as little as 5.195 cm (why?). A number line interpretation of this situation is shown in the following figure, where a portion of the previous number line has been "magnified."

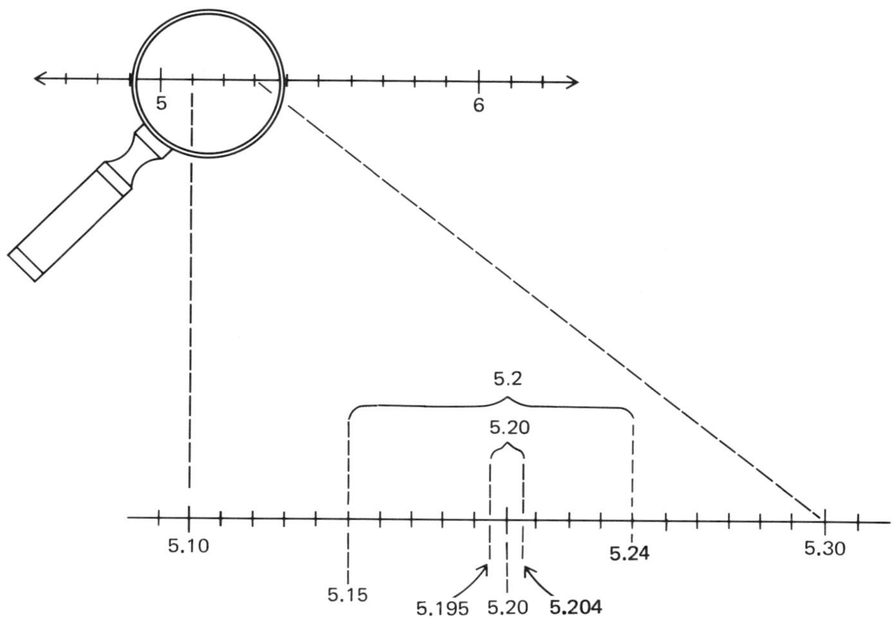

Do exercises 10 and 11.

So, although we computed the area of the rectangle in exercise 3 as

$$5.2 \times 1.2 = 6.24 \text{ (cm}^2\text{)}$$

we now see that its actual area could be as much as

$$5.24 \times 1.24 = 6.4976 \text{ (cm}^2\text{)}$$

or as little as

$$5.15 \times 1.15 = 5.9225 \text{ (cm}^2\text{)}$$

In exercises 10 and 11 fill in the blanks to make correct statements.

10. If a measurement is reported as 5.200 cm, this means that the length is 5.2 cm to the nearest ____________ of a centimeter. Hence the length may actually be as much as 5.200 ___ cm and as little as 5.199 ___ cm.

11. Since the width of the rectangle in exercise 3 is given as 1.2 cm, we know that the actual width could be as much as ______ cm and as little as ______ cm.

STOP

12. Fill in the following blanks so as to make correct statements: In exercise 4, the calculations show that the distance d in feet is

$$7.4 \times 13 = 96.2$$

However, the actual speed might be as much as 7.44 feet per second and as little as 7.3 ___ feet per second. Similarly, the actual time might be as much as 13. ____ seconds and as little as 12.5 seconds. Hence the actual distance in feet might be as much as

_________ × _________

= ___________

or as little as

_________ × _________

= ___________

STOP

In exercises 13–17 round off the given number as indicated.

13. 17.482 to the nearest hundredth and also to the nearest tenth.

14. 125.37 to the nearest tenth and also to the nearest unit (ones).

15. 3047.05 to the nearest ten and also to the nearest unit.

16. 0.003486 to the nearest ten thousandth and also to the nearest hundred thousandth.

17. 10.050 to the nearest tenth and also to the nearest unit.

STOP

These two possible extremes of area are shown in the following figure. We see that our answer of 6.24 (which is supposed to certify that the area is 6.24 cm², correct to the nearest hundredth of a square centimeter) is clearly not appropriate; it simply grossly exaggerates our knowledge of what the area is. A common practice would have us report the area as 6.2 cm² by rounding 6.24 to 6.2, even though our calculations show that we really can't know the answer correct to the nearest tenth of a square centimeter (as the answer of 6.2 implies we do).

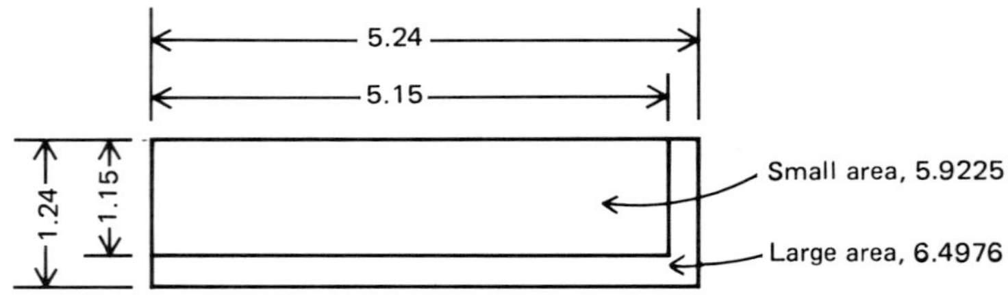

Do exercise 12.

The results of exercise 12 again show that direct computation exaggerates the correctness of the answer. We do not know that the distance is 96.2 feet, correct to the nearest tenth of a foot, since the actual distance may be as much as

$$7.44 \times 13.4 = 99.696 \text{ (feet)}$$

and as little as

$$7.35 \times 12.5 = 91.875 \text{ (feet)}$$

Again, common practice would have us report the distance as 96 feet by rounding off 96.2 to 96, even though our calculations show that we can't really know the answer correct to the nearest foot (as our answer of 96 implies we do).

Do exercises 13 through 17.

Clearly, computations on an HHC involving measurements need some rounding off because of the approximate nature of all measurement. The commonly used rules for rounding off the answer to such computations are, however, rather complicated and vary somewhat in different branches of science and industry; they are discussed in Appendix B. In the exercises of this chapter, we will indicate the rounding off that should be done to conform with the rules given in Appendix B.

Exercises 6.1

Name ______________________

Class ______________________

In these exercises use the given formula and the given information to find the missing number and, if necessary, round off your answer as directed.

Mechanical advantage formula: $M = \dfrac{W}{E}$

1. $W = 4.5$ grams, $E = 16$ grams (nearest hundredth) — 1.

2. $W = 4.25$ pounds, $M = 1.6$ (nearest tenth) — 2.

3. $E = 2.25$ kilograms, $M = 3200$ (nearest hundred) — 3.

Cost formula: $C = np$

4. $n = 27$, $p = \$0.127$ (nearest tenth) — 4.

5. $n = 63$, $C = \$42.13$ (nearest hundredth) — 5.

6. $C = \$127.18$, $p = \$0.0125$ (nearest unit) — 6.

Postage formula for the cost of mailing catalogs: $C = 0.05p + 0.07$

7. $p = 24$ pounds (nearest hundredth) — 7.

8. $C = \$3.62$ (nearest unit) — 8.

9. $C = \$2.77$ (nearest unit) — 9.

Distance formula: $d = rt$

10. $r = 682$ feet per second, $t = 3.5$ seconds (nearest ten) — 10.

11.

11. $r = 54.7$ miles per hour, $d = 568.4$ miles (nearest tenth)

12.

12. $r = 27.06$ centimeters per second, $d = 762.01$ centimeters (nearest hundredth)

Perimeter formula: $P = 2l + 2w$

13.

13. $l = 96.5$ centimeters, $w = 4.05$ centimeters (nearest unit)

14.

14. $l = 7.774$ inches, $P = 67.92$ inches (nearest hundredth)

15.

15. $w = 97.81$ meters, $P = 2281$ meters (nearest unit)

Formula for converting from temperature in Celsius to temperature in Fahrenheit:

$$F = \frac{9}{5}C + 32$$

16.

16. $C = 18$ (nearest unit)

17.

17. $C = 30.1$ (nearest tenth)

18.

18. $C = 54$ (nearest unit)

Formula for converting from temperature in Fahrenheit to temperature in Celsius:

$$C = \frac{5}{9}(F - 32)$$

19.

19. $F = 100$ (nearest unit)

20.

20. $F = 2980$ (nearest ten)

21.

21. $F = 50.1$ (nearest tenth)

Motion formula: $s = \frac{1}{2} gt^2$

22. $g = 32.6$ feet per second per second, $t = 3.10$ seconds (nearest unit) 22.

23. $g = 980$ centimeters per second per second, $t = 10.5$ seconds (nearest thousand) 23.

24. $t = 7.75$ seconds, $s = 3000$ kilometers (nearest tenth) 24.

Formula for resistance in an electrical circuit: $R = \frac{E}{I}$

25. $E = 32$ volts, $I = 0.75$ amperes (nearest unit) 25.

26. $E = 22.5$ volts, $R = 2.3$ ohms (nearest tenth) 26.

27. $R = 4.25$ ohms, $I = 1.41$ amperes (nearest hundredth) 27.

Formula for the area of a trapezoid: $A = \frac{h(b_1 + b_2)}{2}$

28. $h = 23.1$ cm, $b_1 = 5.02$ cm, $b_2 = 7.52$ cm (nearest unit) 28.

29. $h = 5.06$ inches, $b_1 = 78.7$ inches, $b_2 = 122.1$ inches (nearest unit) 29.

30. $A = 167.7$ square meters, $b_1 = 79.91$ meters, $b_2 = 6.33$ meters (nearest hundredth) 30.

Formula for Archimedes' Principle: $F = DgV$

31. $D = 0.81$ grams per cubic centimeter, $g = 980.6$, $V = 103.6$ cubic centimeters (nearest thousand) 31.

32. $F = 1.44$ poundals, $g = 32.2$, $V = 37$ cubic feet (nearest ten thousandth) 32.

33.

33. $F = 42.68$ newtons, $g = 980.6$, $D = 5.043$ grams per cubic centimeter (nearest millionth)

Formula for energy: $K = \dfrac{Wv^2}{2g}$

34.

34. $W = 30.4$ grams, $v = 56.5$ centimeters per second, $g = 980.6$ (nearest tenth)

35.

35. $K = 91.97$ ergs, $v = 43.9$ centimeters per second, $g = 980.6$ (nearest tenth)

36.

36. $K = 3.92$ ergs, $v = 13.4$, $W = 42.8$ grams (nearest unit)

Pendulum formula: $L = d + \dfrac{2r^3}{5d}$

37.

37. $d = 19.05$ cm, $r = 1.35$ cm (nearest tenth)

38.

38. $d = 215.07$ cm, $r = 10.16$ cm (nearest tenth)

39.

39. $d = 43.85$ inches, $r = 7.43$ inches (nearest tenth)

Formula for electrical power: $W = \dfrac{V^2}{R}$

40.

40. $V = 11.5$ volts, $R = 5.6$ ohms (nearest unit)

41.

41. $V = 12.3$ volts, $W = 1.75$ watts (nearest tenth)

42.

42. $V = 1.5$ volts, $W = 0.2$ watts (nearest ten)

Formula for electrical current: $I = \dfrac{2E}{R + 2r}$

43.

43. $E = 9.5$ volts, $R = 0.375$ ohms, $r = 1.1$ ohms (nearest tenth)

44. $i = 8.8$ amperes, $R = 3.4$ ohms, $r = 0.75$ ohms (nearest unit) 44.

45. $i = 2.78$ amperes, $E = 117$ volts, $r = 7.89$ ohms (nearest tenth) 45.

46. If $A = \frac{2\,PR}{R - r}$, find R if $A = 1400$, $r = 18$, $P = 600$ (nearest ten). 46.

47. If $A = \frac{h}{2}(a + b)$, find a if $A = 60$, $h = 6$, $b = 8$ (nearest unit). 47.

48. If $\frac{PV}{T} = \frac{pv}{t}$, find v if $P = 400$, $p = 500$, $V = 500$, $T = 32$, $t = 32$. 48.

49. If $c = \alpha \ell (T_2 - T_1)$, find ℓ if $T_2 = 92$, $T_1 = 62$, $\alpha = 6.0 \times 10^{-6}$, $c = 1.8$ (nearest thousand). 49.

50. If $\frac{1}{R} = \frac{1}{R_1} + \frac{1}{R_2} + \frac{1}{R_3}$, find R if $R_1 = 9$, $R_2 = 12$, $R_3 = 15$ (nearest tenth). 50.

51. If HP (horsepower) $= \frac{\text{weight (in pounds)} \times \text{height (in feet)}}{550 \times \text{time in seconds}}$ 51.

find the horsepower required to hoist 100 tons of coal per hour from a mine 200 feet deep.

52. If each of the citizens of a city of 400,000 population uses 5 gallons (40 pounds) of water each day, what horsepower engines must be used if the height to which the water must be pumped is 100 feet? (Use the formula in exercise 51.) 52.

53.

53. In calculating centrifugal force (C.F.) in pounds the following formula is used: $\text{C.F.} = \frac{mv^2}{32r}$ where m is the mass (weight) of the object in pounds, v is the velocity in feet per second, and r is the radius in feet of the curve.

A car weighing 3600 pounds rounds a curve of 80-foot radius at 20 miles per hour. What is the centrifugal force in pounds tending to throw the car off the road?

54. (a)

(b)

54. Suppose the car in exercise 53 is traveling 60 miles per hour around the curve, (a) what is the centrifugal force in pounds? (b) how much would you need to lengthen the radius to reduce the centrifugal force to the weight of the car at this speed?

55.

55. The number of foot candles (measure of the illumination of light) may be found by dividing the candle power (c.p.) of the light by the square of the distance (in feet) the object is from the light:

$$\text{Foot candles} = \frac{\text{c.p.}}{d^2}$$

If a 25-watt light bulb produces 20 c.p., how many foot candles of light does a man seated 2 feet from the light receive?

56.

56. If a man is seated 3 feet from a 40-watt light bulb (32 c.p.), how many foot candles of light does he receive? (Use the formula in exercise 55.)

6.2 Measurement of Length

THE METRIC SYSTEM

The various formulas introduced in the last section all involved measures of some kind: length, speed, weight, and so on. The one system of measurement now employed throughout the world is the *metric* system. Having already used some metric units of measurement (as the centimeter) in earlier parts of the text, we now turn to a systematic study of the metric system.

The basic unit of length in the metric system is the *meter* (slightly longer than a yard). Various prefixes, *milli-*, *centi-*, *kilo-*, etc., are used to indicate parts or multiples of a meter (and these same prefixes are used in connection with area, volume, and weight). This is shown in the two tables below.

TABLE 8.1 MEASURE OF LENGTH

1 *milli*meter	(mm)	=	0.001 meter
1 *centi*meter	(cm)	=	0.01 meter
1 *deci*meter	(dm)	=	0.1 meter
1 meter	(m)	=	1.0 meter
1 *deka*meter	(dam)	=	10 meters
1 *hecto*meter	(hm)	=	100 meters
1 *kilo*meter	(km)	=	1000 meters

TABLE 8.2 METRIC PREFIXES AND THEIR VALUE

Prefix	Value	
milli	0.001	thousandths
centi	0.01	hundredths
deci	0.1	tenths
deka	10	tens
hecto	100	hundreds
kilo	1000	thousands

Do exercise 1 in the adjoining column.

OBJECTIVES

You should:

(a) Understand the meaning of the various units of measurement of length in the metric system and the relationships between them.

(b) Be able to estimate with some degree of accuracy lengths in centimeters and meters.

(c) Be able to find the perimeters of various geometric figures.

****

For exercise 1 examine a centimeter ruler. Observe that the centimeter divisions are numbered and that between each centimeter mark there are nine shorter marks that indicate millimeters. First *estimate* the length of the following line segment in centimeters. Then *measure* it to the nearest millimeter with your ruler.

1. ______________________

Estimated length: ____________

Measured length: ____________

STOP

In exercises 2–20 fill in the blanks.

2. 2.146 km = ________ hm

3. 6.2 cm = ________ mm

4. 8.27 hm = ________ dam

STOP

5. 29.8 mm = ________ cm

6. 0.23 dm = ________ m

7. 0.034 cm = ________ dm

8. 1940.13 mm = ________ cm

STOP

9. 1940.13 mm = ________ dm

10. 1940.13 mm = ________ m

11. 1940.13 mm = ________ dam

12. 1940.13 mm = ________ hm

13. 1940.13 mm = ________ km

14. 8.203 km = ________ mm

15. 8.203 km = ________ cm

16. 8.203 km = ________ dm

17. 8.203 km = ________ m

18. 8.203 km = ________ dam

19. 8.203 km = ________ hm

20. 1.21 km = ________ hm

STOP

Note that to change from a measurement in a given unit to a measurement in the next smaller unit we multiply by 10. For example,

$$6.3 \text{ km} = 6.3 \times 10 \text{ hm} = 63 \text{ hm}$$

$$5.74 \text{ m} = 5.74 \times 10 \text{ dm} = 57.4 \text{ dm}$$

$$16 \text{ cm} = 16 \times 10 \text{ mm} = 160 \text{ mm}$$

Do exercises 2 through 4.

Going the other way, to change from a measurement in a given unit to a measurement in the next larger unit, we divide by 10. For example,

$$24.3 \text{ cm} = 24.3 \div 10 \text{ dm} = 2.43 \text{ dm}$$

$$5.6 \text{ hm} = 5.6 \div 10 \text{ km} = 0.56 \text{ km}$$

Do exercises 5 through 8.

In these examples, we have gone only one stage at a time. To go two stages involves multiplying or dividing by 100; to go three stages, multiplying or dividing by 1000, etc. For example,

Two stages

$$0.23 \text{ m} = 0.23 \times 100 \text{ cm} = 23 \text{ cm}$$

$$43 \text{ mm} = 43 \div 100 \text{ dm} = 4.3 \text{ dm}$$

Three stages

$$0.23 \text{ m} = 0.23 \times 1000 \text{ mm} = 230 \text{ mm}$$

$$5324 \text{ m} = 5324 \div 1000 \text{ km} = 5.324 \text{ km}$$

Do exercises 9 through 20.

Although the progression *milli, centi, . . . , hecto, kilo* is a systematic "by tens" one, the fact is that the prefixes *deka* and *hecto* are practically never used and *deci* is used but rarely. Hence from now on we will use only the prefixes *milli, centi,* and *kilo* in the exercises.

Do exercises 21 through 30.

PERIMETERS

One use of linear measure is in determining the *perimeter* of geometric figures, the distance around the figure. If the figure is a polygon, then the perimeter is the sum of the lengths of its sides.

Example: A rectangular swimming pool is to be fenced as shown in the following figure. How many meters of fencing will be needed?

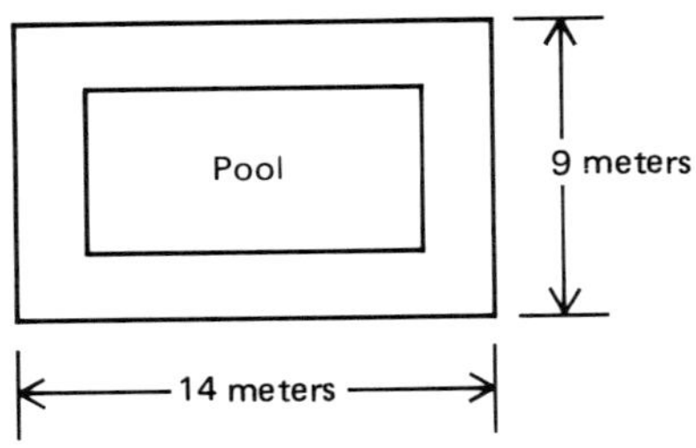

Solution: The amount of fencing needed will be equal to the perimeter of a 14-meter by 9-meter rectangle:

$$2 \times 14 \text{ meters} = 28 \text{ meters}$$
$$2 \times 9 \text{ meters} = 18 \text{ meters}$$
$$\text{Perimeter} = 46 \text{ meters}$$

The amount of fencing needed is 46 meters.

Formulas for the perimeters of some common geometric figures are as follows.

Perimeter of a rectangle:

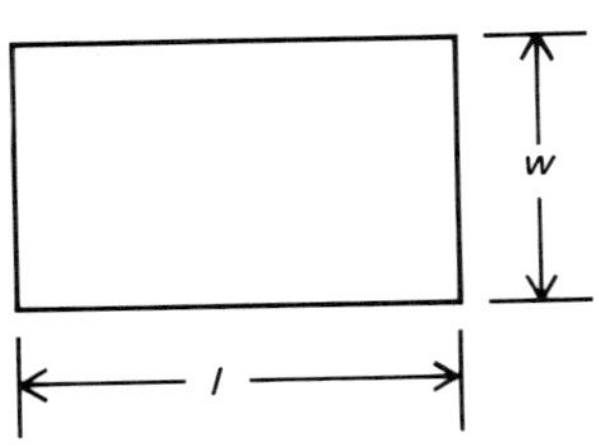

$$P = l + w + l + w$$
$$P = 2l + 2w$$
$$P = 2(l + w)$$

Do exercise 31.

Perimeter of a square:

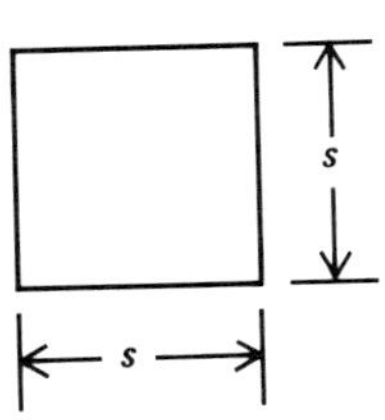

$$P = s + s + s + s$$
$$P = 4s$$

Do exercise 32.

In exercises 21–30 fill in the blanks.

21. 3122 m = ________ km
22. 6743 mm = ________ m
23. 0.524 m = ________ mm
24. 0.968 km = ________ m
25. 0.78 km = ________ m
26. 3.41 m = ________ cm
27. 853 cm = ________ m
28. 20,095 cm = ________ km
29. 5.23 km = ________ cm
30. 0.000024 km = ________ mm

STOP

31. Find the perimeter, to the nearest tenth of a meter, of a rectangle whose length is 137.2 m and whose width is 87.6 m

STOP

32. How long, to the nearest tenth of a centimeter, is the side of a square whose perimeter is 156.9 cm?

STOP

Perimeter of a triangle: $P = a + b + c$

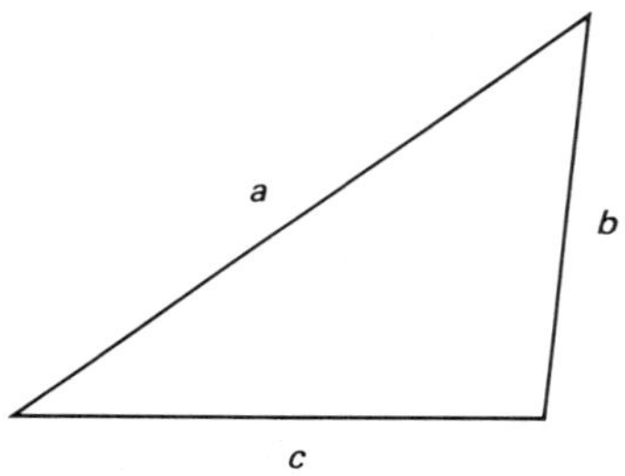

An *isosceles* triangle has two sides of the same length, so its perimeter can be found by the formula $P = a + 2b$.

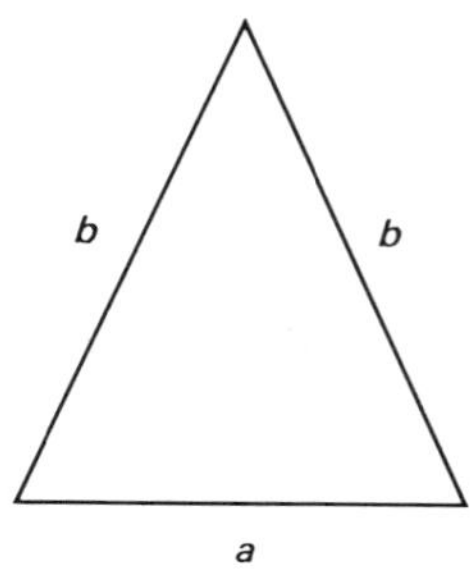

An *equilateral* triangle has all three sides of the same length, so the formula for its perimeter is $P = 3s$.

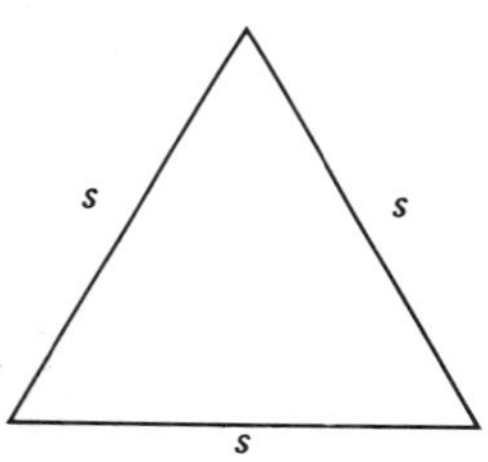

Do exercises 33 through 35.

33. Find the perimeter, to the nearest hundredth of a centimeter, of a triangle whose three sides measure 21.46 cm, 15.25 cm, and 18.37 cm.

34. Find the perimeter, to the nearest tenth of a centimeter, of an isosceles triangle whose *base* is of length 57.6 cm and whose two equal sides are each of length 33.9 cm.

35. How long, to the nearest hundredth of a meter, are the sides of an equilateral triangle whose perimeter is 1492.87 meters?

Sometimes it is necessary to find the perimeter of an irregularly shaped figure. One must then add each of the lengths involved.

Example: The house whose outline and dimensions are shown in the accompanying illustration is to have the gutters on its outside perimeter replaced. How many meters of gutter will be required?

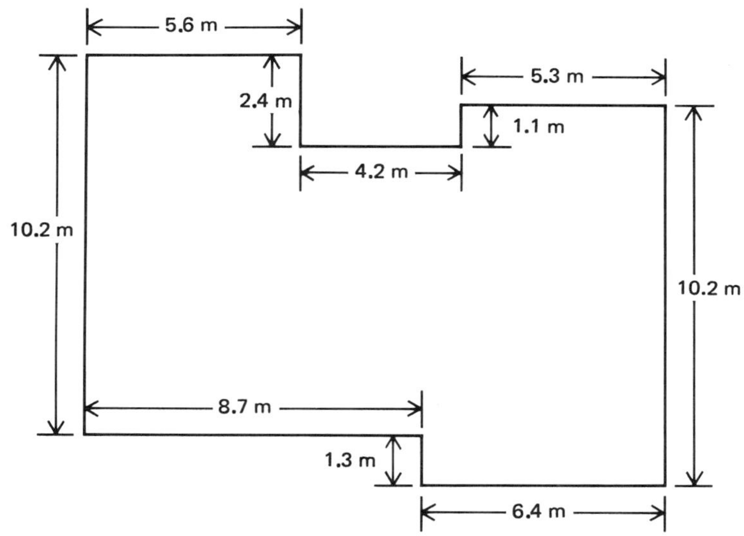

Solution:

$$P = (5.6 + 2.4 + 4.2 + 1.1 + 5.3 + 10.2 + 6.4 + 1.3 + 8.7 + 10.2) \text{ meters}$$

$$= 55.4 \text{ meters}$$

Thus 55.4 meters of gutter will be needed.

Do exercise 36.

36. Find the perimeter, to the nearest tenth of a centimeter, of the following figure.

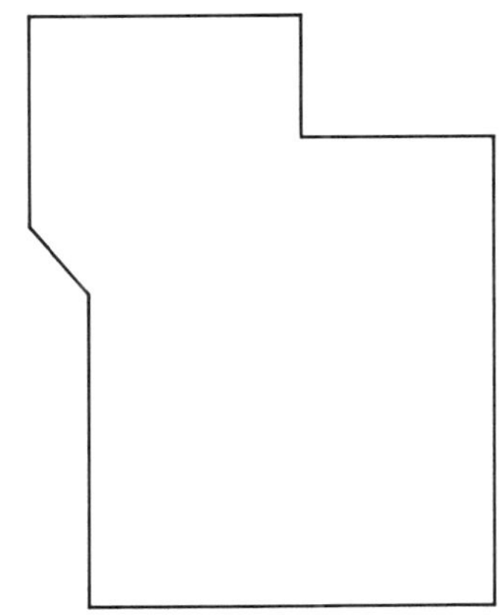

Exercises 6.2

Name ______________________

Class ______________________

In exercises 1–5 (a) estimate the length in centimeters of the given line segment and (b) use a ruler to determine the length to the nearest millimeter.

1. 

2. 

3. 

4. 

5. 

In exercises 6–10 (a) estimate the perimeter of the given figure in centimeters, (b) find the perimeter by measurement to the nearest millimeter, (c) write the measure of the perimeter in centimeters, and (d) write the measure of the perimeter in meters.

6. 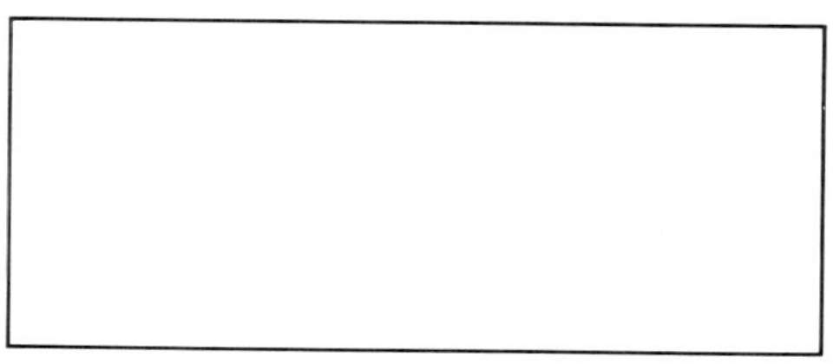

7. 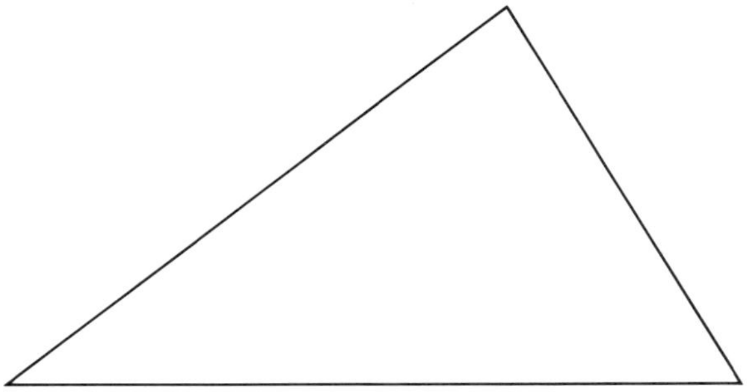

8. 

1. (a)
   (b)
2. (a)
   (b)
3. (a)
   (b)
4. (a)
   (b)
5. (a)
   (b)
6. (a)
   (b)
   (c)
   (d)
7. (a)
   (b)
   (c)
   (d)
8. (a)
   (b)
   (c)
   (d)

9. (a)
(b)
(c)
(d)

9.

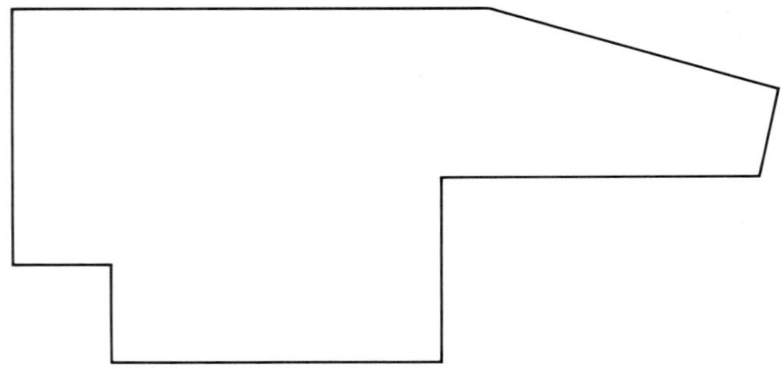

10. (a)
(b)
(c)
(d)

10.

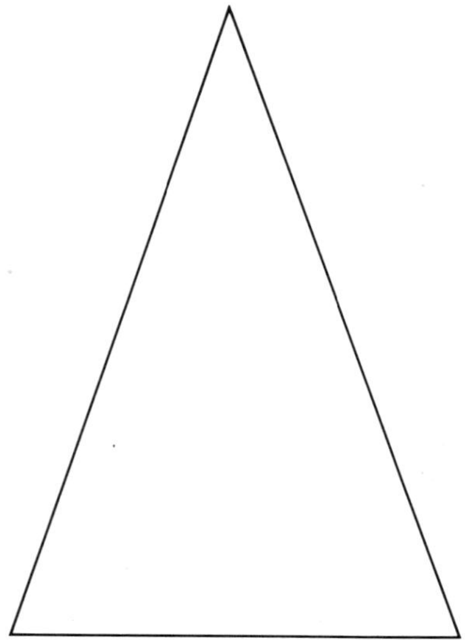

In exercises 11–20 (a) estimate the length or height of the object in centimeters and (b) measure the object and record its length in centimeters, millimeters, and meters.

	Object	Estimate of length in centimeters	Actual length in centimeters	Actual length in millimeters	Actual length in meters
11.	Paper clip				
12.	Ball point pen				
13.	Classroom door				
14.	Chalkboard eraser				
15.	Your right foot				
16.	Penny				
17.	Length of the classroom				
18.	Two-liter coke bottle				
19.	This book				
20.	Height of the classroom				

In exercises 21–35 complete the following table. The first line is filled in as an example.

Kilometers	Meters	Centimeters	Millimeters
0.001	1	100	1000
21.	5	22.	23.
24.	2.4	25.	26.
27.	28.	29.	500
30.	31.	10	32.
0.5	33.	34.	35.

In exercises 35–45 find the perimeter of the given figure. (Be sure to express all dimensions in the same units.)

36. 36.

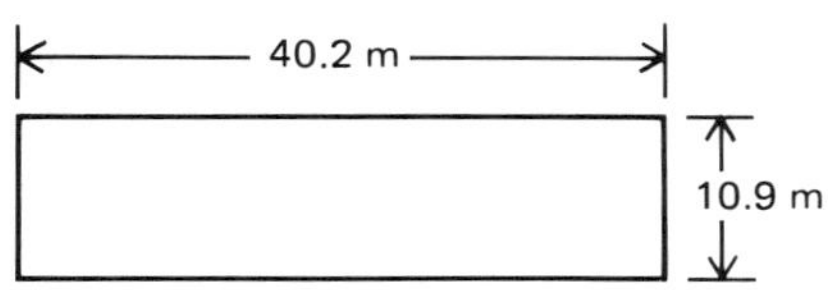

37. 37.

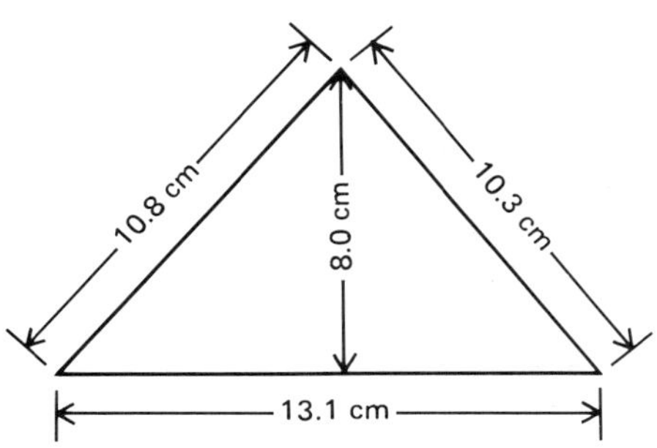

38. 38.

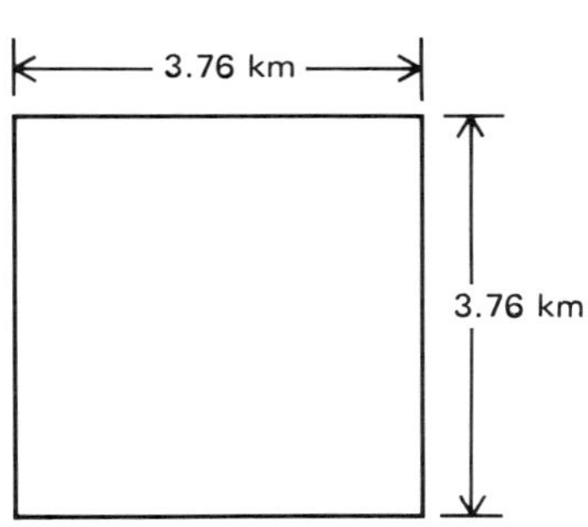

39.

39.

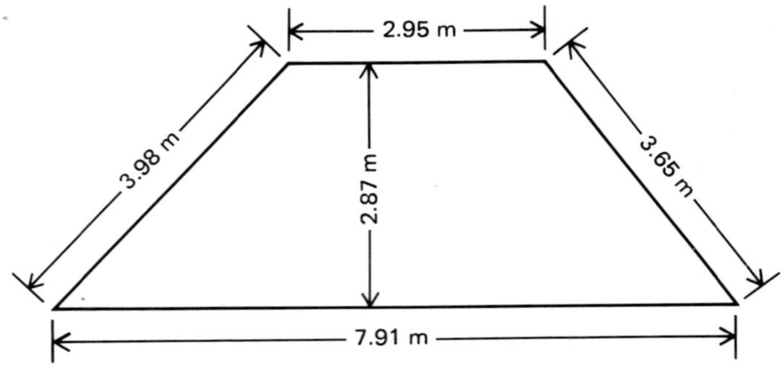

40.

40.

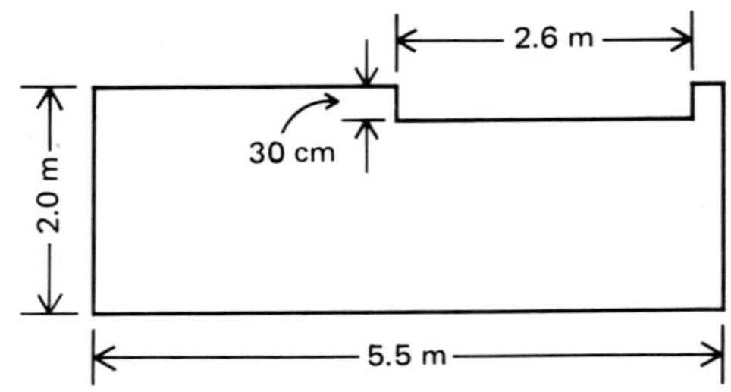

41.

41.

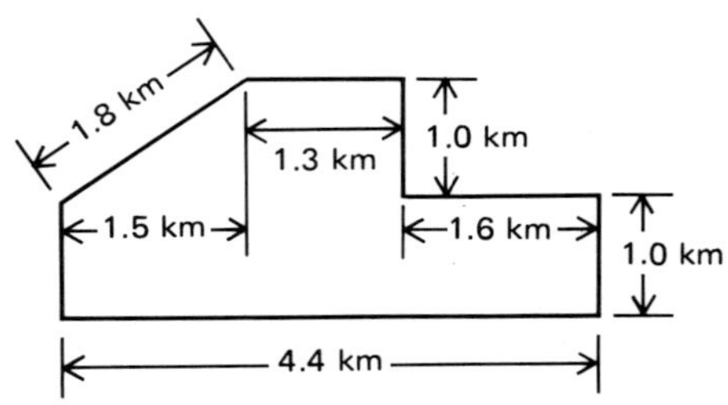

42.

42.

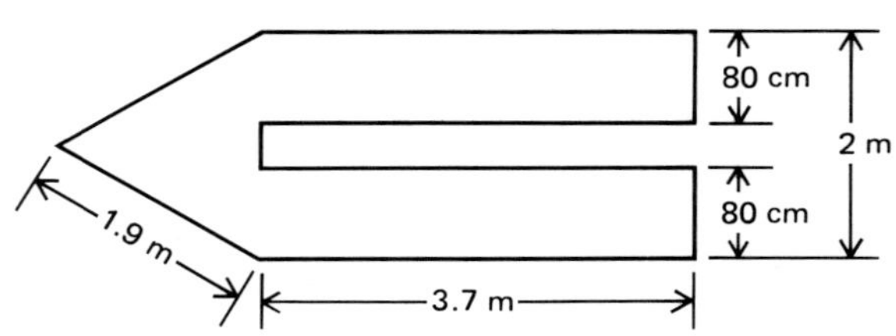

43.

43.

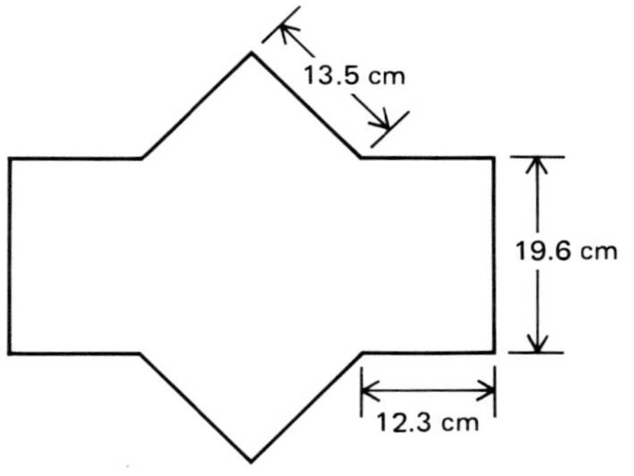

44. 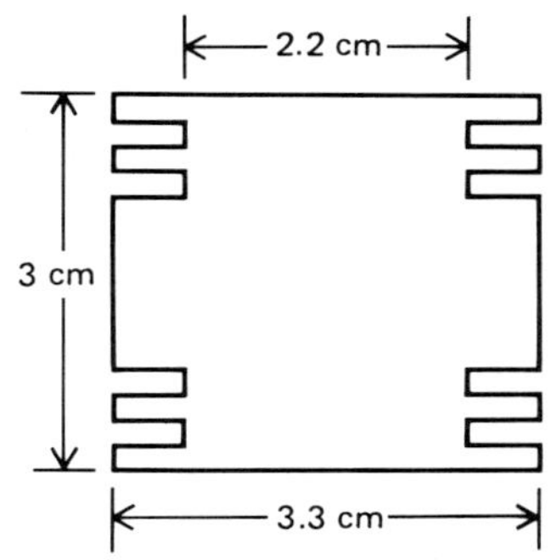

44.

45. 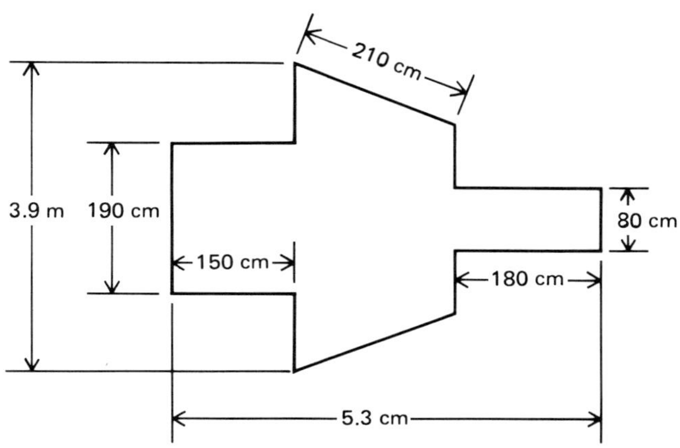

45.

46. If the standard width of a piston ring is 4.76 mm and a certain ring measures 4.68 mm, how much has it worn?

46.

47. A piston wore a groove in a cylinder wall 0.34 mm deep in one spot and 0.28 mm deep in another. (a) To what size should the cylinder be rebored in order to clean up the grooves if the original size was 98.425 mm? (b) What size piston should be used, allowing 0.05 mm for clearance?

47. (a)

(b)

48. If the diameter of a new piston is 92 mm and the average wear per 10,000 km is 0.09 mm at all points of the piston, what will be the total wear after 100,000 km?

48.

49. A metric blueprint calls for a dimension of 3.5 cm ±0.2 mm. What are the large and small limits of this dimension?

49.

50. Three equally spaced holes are to be drilled in a metal strap 22 cm long, with 15 mm from the center of the end holes to each end of the strap. How far apart should the centers of the holes be?

50.

51.

52.

53.

54.

55. (a)

(b)

56.

57.

58.

51. Jack's height at the beginning of summer vacation was 1.39 meters. During the summer he grew 3.2 centimeters. How tall was he at the end of summer?

52. In arranging a 21.5-centimeter by 27.8-centimeter paper for a drawing, a line is to be drawn 12 millimeters from each edge. What are the inside dimensions between the lines?

53. The outside diameter of a piece of copper tubing is 3.75 centimeters and its wall thickness is 3 millimeters. What is its inside diameter?

54. Al walked all the way around the boundary of a rectangular plot of ground 350 meters long and 160 meters wide. How many kilometers did he walk?

55. A ranch is 36 kilometers long and 23 kilometers wide and is rectangular in shape. (a) How many meters of fencing would be required to fence in the entire ranch? (b) How much would it cost to fence the ranch with two strands of barbed wire if a 400-meter reel of $12\frac{1}{2}$-gauge, 113-cm spacing, barbed wire costs $15.50?

56. A rectangular painting is 104 cm by 74 cm. How much will the material cost to frame the picture using molding that is 8 cm wide if the price of this molding is $9.00 per meter? (Remember that the corners must be mitered).

57. A room is 5.3 meters wide and 6.1 meters long and has two doors that are each 72 cm wide. If baseboard molding costs 48¢ a meter, what will it cost to buy material for baseboards for the room?

58. At a party, each guest received a gift box. Each box was tied with ribbon as shown in the accompanying figure. How many rolls of ribbon (each roll containing 12 meters of ribbon) will be needed if 15 guests are expected and each bow requires an additional 45 cm of ribbon?

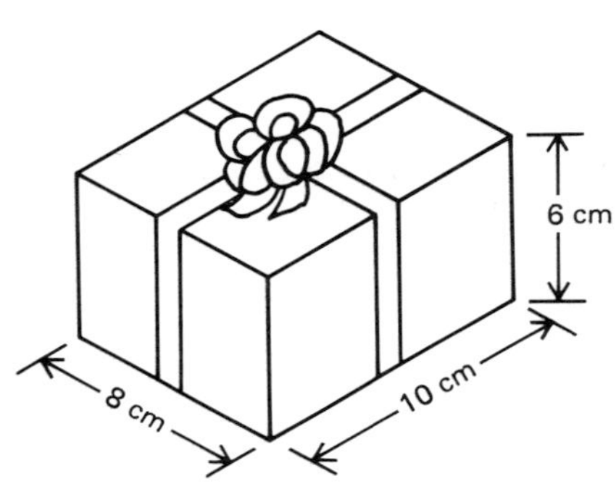

59. For shipping, the box in the illustration below was bound with three bands of strap steel. How many meters of strap steel would be needed if 150 of these boxes were shipped and an additional 3 cm was needed for fastening each band? 59.

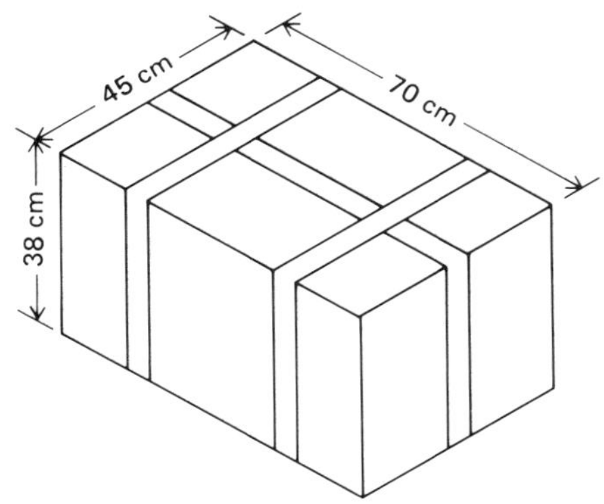

*60. An equilateral triangle with sides of 1 unit has a perimeter of 3 units. *60.

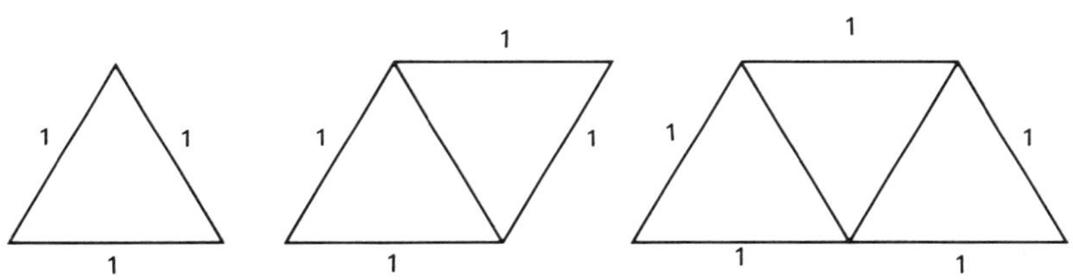

$P = 3$ units $P = 4$ units $P = 5$ units

When 2 triangles are used, $P = 4$. When 3 triangles are used, $P = 5$.

(a) Find the perimeter for 5 triangles placed in the same manner. (a)

(b) Find the perimeter for 12 triangles. (b)

(c) Find the number of triangles it would take for a perimeter of 18 units. (c)

(d) Find the number of triangles it would take for a perimeter of 33 units. (d)

*61. A *pentominoe* is an arrangement of five squares in such a way that a full side of one square must also be a full side of another square that touches it. One example of a pentominoe is shown here. There are 12 possible such pentominoes. *61.

(a) Show the 12 possible pentominoes on the next page. (a)

(b) If each square used is 2 cm on a side, which of the 12 pentominoes has the smallest perimeter? What is it? (b)

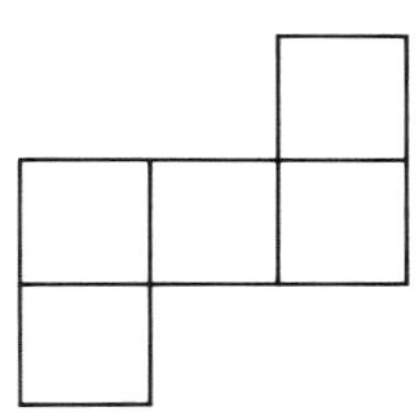

(c) Which of the 12 pentominoes has the largest perimeter. What is it? (c)

6.3 Measurement of Area

The following table shows the commonly used units of measures of area, all based on the area of a square.

Length of side of square	Area of square	Abbreviation
1 cm	1 square centimeter	1 cm^2
1 m	1 square meter	1 m^2
1 hm	1 hectare	1 ha
1 km	1 square kilometer	1 km^2

Because the hectometer (100 m) is not commonly used as a unit of length, it is customary to describe a hectare as equal to the area of a square 100 m by 100 m. Typically, where we now talk about acres under cultivation, metric terminology will refer to hectares. (A hectare is slightly less than 2½ acres.)

There is rarely a need to convert either from or to square centimeters. Converting from hectares to square meters, we know that a hectare is the area of a square 100 m by 100 m. Hence

$$1 \text{ ha} = (100 \times 100) \text{ m}^2 = 10{,}000 \text{ m}^2$$

Do exercises 1 through 4 in the adjoining column.

Similarly, since 1 km = 10 hm, we have

$$1 \text{ km}^2 = (10 \times 10) \text{ ha} = 100 \text{ ha}$$

Do exercises 5 through 12.

By the area of any figure we mean the number of unit squares or parts of unit squares that make up the figure.

OBJECTIVES

You should:

(a) Understand the meaning of the various units of measurement of area in the metric system and the relationships between them.

(b) Know and be able to use formulas for the area of various common geometric shapes.

****

In exercises 1–4 fill in the blanks.

1. 1.21 ha = ________ m^2
2. 0.03 ha = ________ m^2
3. 123 m^2 = ________ ha
4. 4527 m^2 = ________ ha

STOP

In exercises 5–12 fill in the blanks.

5. 0.85 km^2 = ________ ha
6. 5.24 km^2 = ________ ha
7. 1 ha = ________ km^2
8. 0.75 ha = ________ km^2
9. 1 km^2 = ________ m^2
10. 500,000 m^2 = ________ km^2
11. 1.35 m^2 = ________ km^2
12. 0.025 km^2 = ________ m^2

STOP

In exercises 13–14 the side of each square is 1 cm long. Find the area of the given figure by counting squares or parts of squares.

13. 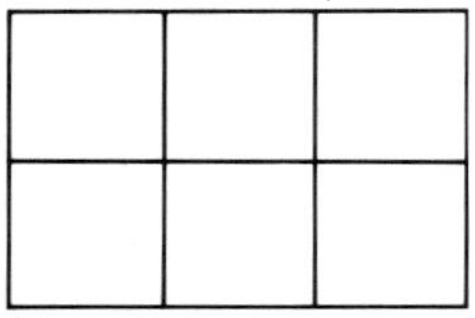

14. 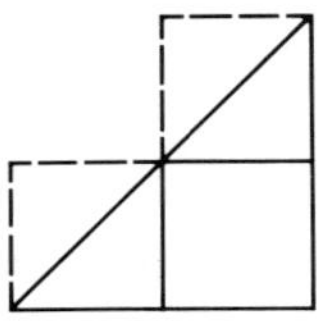

STOP

15. Find, to the nearest unit, the area of a rectangle whose width is 12.6 m and whose length is 6.8 m.

16. Find, to the nearest tenth, the area of a triangle whose base is 15.8 cm and whose altitude is 7.6 cm.

STOP

17. Find, to the nearest hundred, the area of a parallelogram whose base is 45.3 m and whose altitude is 33.7 m.

18. Find, to the nearest unit, the area of a trapezoid whose bases are 137.52 cm and 85.37 cm and whose altitude is 47.20 cm.

STOP

Do exercises 13 and 14.

Of course formulas exist for areas of various geometric figures. Some of these are as follows:

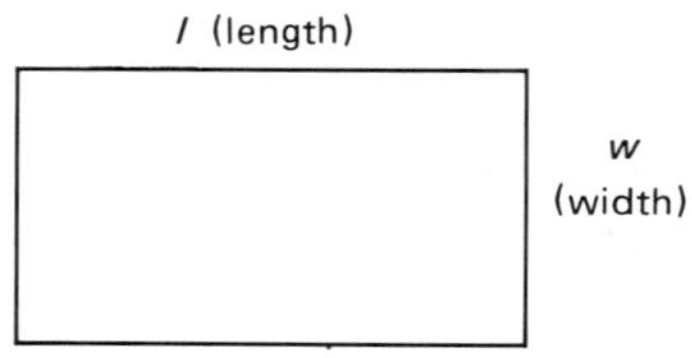

Rectangle

$A = l \times w$

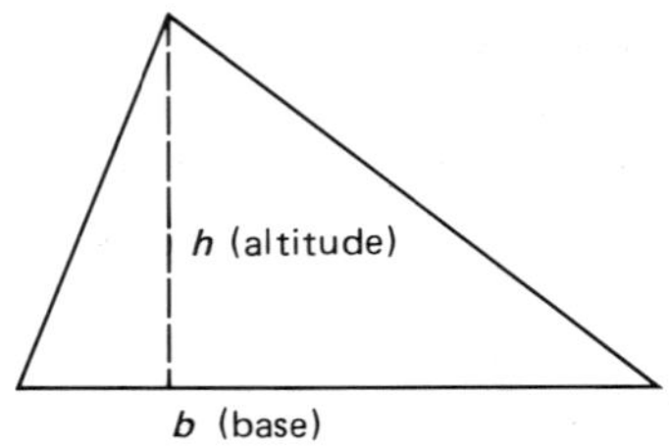

Triangle

$A = \frac{1}{2} b \times h$

Do exercises 15 and 16.

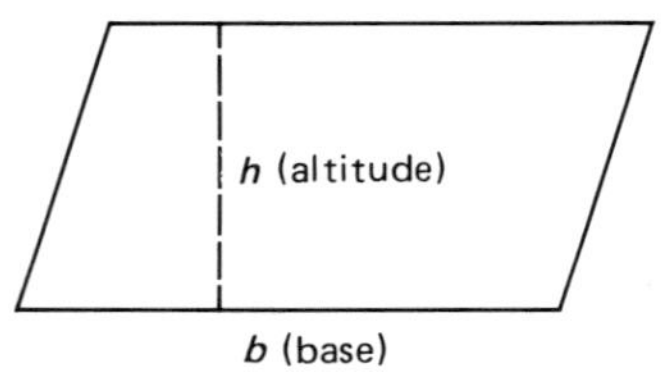

Parallelogram

$A = b \times h$

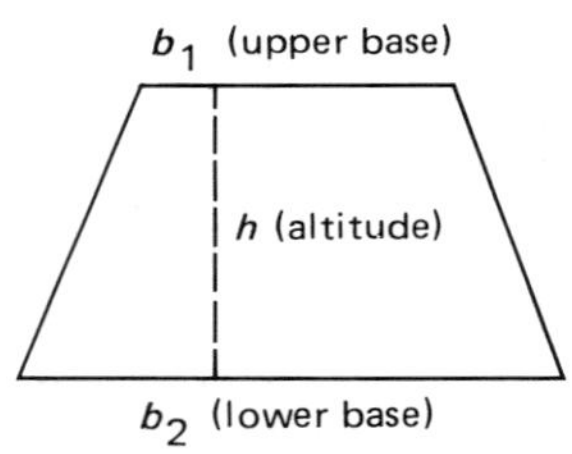

Trapezoid

$A = \frac{1}{2} h \times (b_1 + b_2)$

Do exercises 17 and 18.

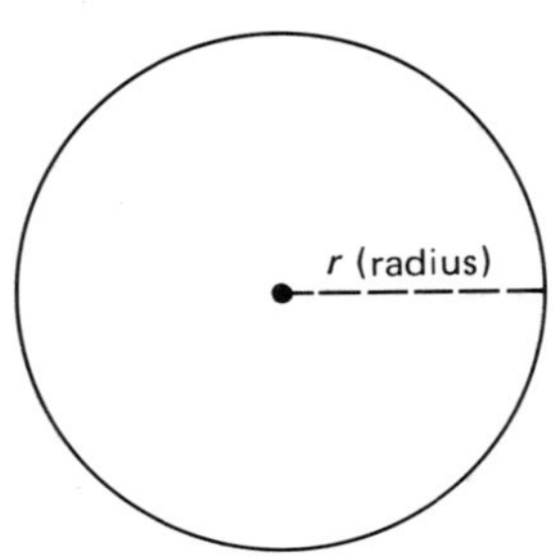

Circle

$A = \pi r^2$

In the formula for the area of a circle, we have used a number represented by the Greek letter π (pi). It is a number, like $\sqrt{2}$, $\sqrt{3}$, etc., in that it can only be approximated by a fraction or decimal. To four decimal places we have

$$\pi \approx 3.1416$$

Sometimes the fraction 22/7 is used as an approximation to π. Since

$$\frac{22}{7} = 3.1429 \text{ to four decimal places}$$

we see that 22/7 is an approximation to π correct to two decimal places.

Do exercise 19.

The number π also occurs in the formula for the *circumference* (perimeter) of a circle. We have for the circumference C of a circle of radius r,

$$C = 2\pi r$$

or, using the diameter d (= $2r$),

$$C = \pi d$$

Do exercises 20 and 21.

19. Using 3.142 for π find, to the nearest tenth, the area of a circle whose radius is 4.67 cm.

STOP

20. Find, to the nearest unit, the circumference of a circle whose radius is 27.8 cm. Use 3.142 for π.

21. Find, to the nearest unit, the circumference of a circle whose diameter is 35 mm. Use 3.14 for π.

STOP

Exercises 6.3

Name ______________________

Class ______________________

In exercises 1-20 complete the following table dealing with the relationships between units of area. The first line has been filled in as an example. (Note: Answers such as 5,000,000,000 can be recorded in exponential notation as 5×10^9.)

mm^2	cm^2	m^2	ha	km^2
1,000,000	10,000	1	0.0001	0.000001
1.	1000	2.	3.	4.
10,000	5.	6.	7.	8.
9.	10.	5000	11.	12.
13.	14.	15.	16.	1,000,000
17.	18.	19.	10,000	20.

In exercises 21-32 find the area of the given figure. Round off your answer as indicated.

21. (nearest ten)

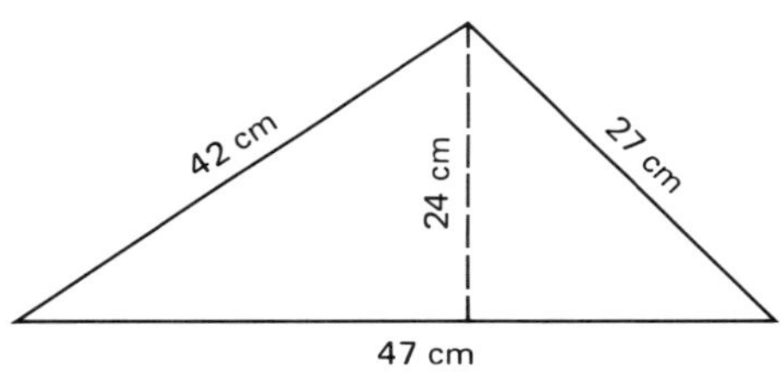

21.

22. (nearest hundred)

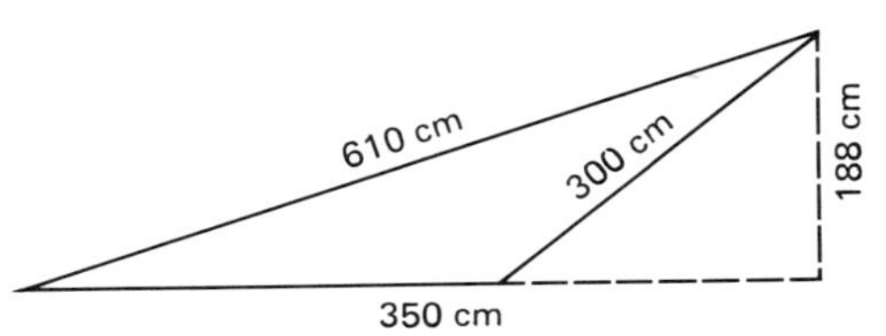

22.

23. (nearest ten)

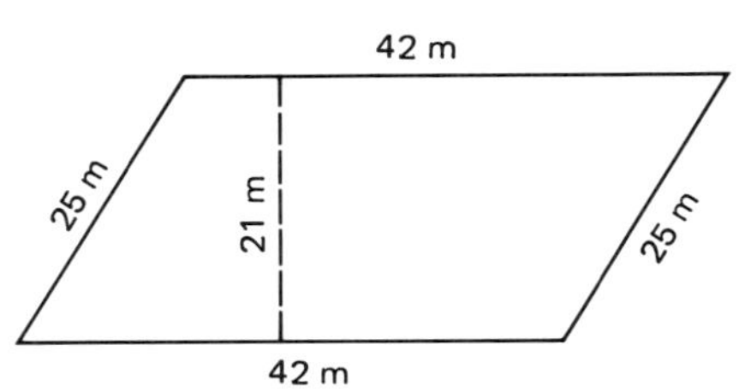

23.

24.

24. (nearest tenth)

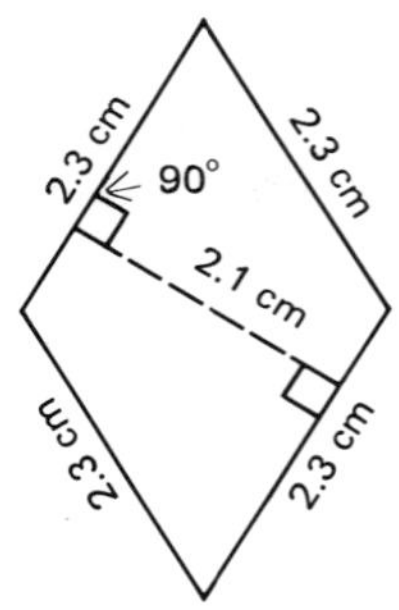

25.

25. (nearest unit)

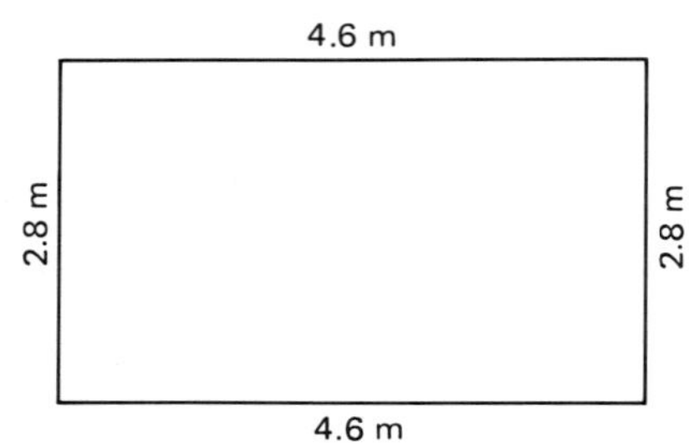

26.

26. (nearest hundred)

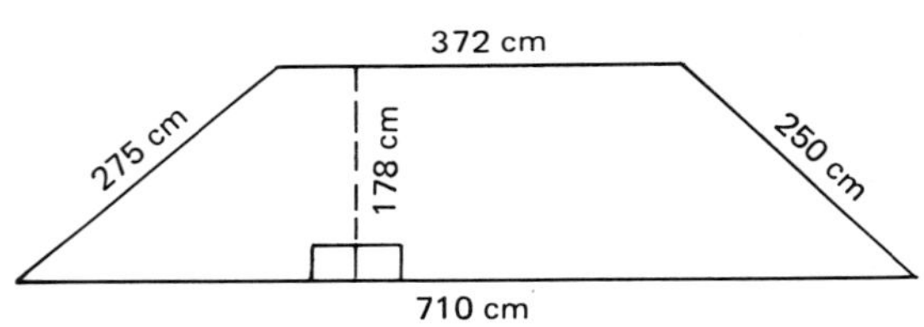

27.

27. (nearest tenth)

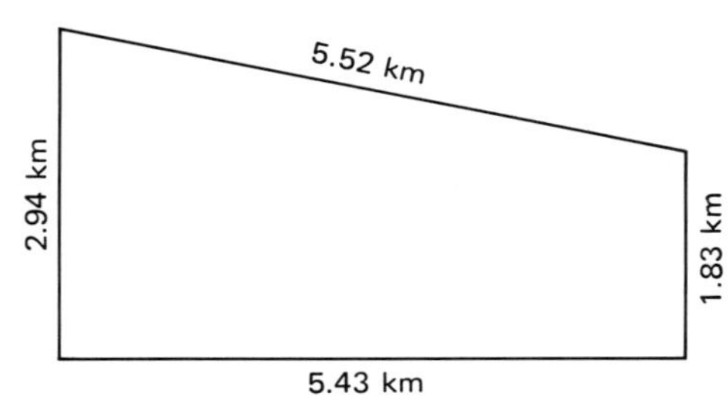

28.

28. (nearest thousandth)

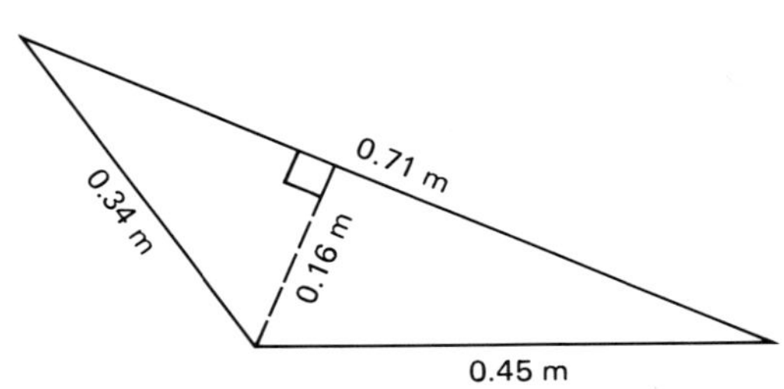

29. (nearest unit in cm^2)

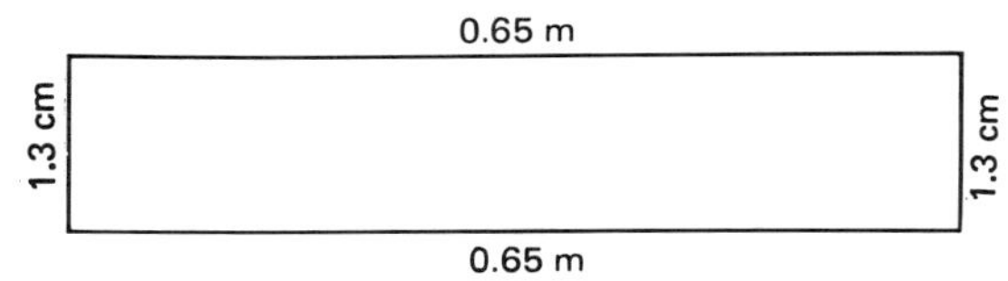

30. (nearest ten thousandth; use $\pi = 3.142$)

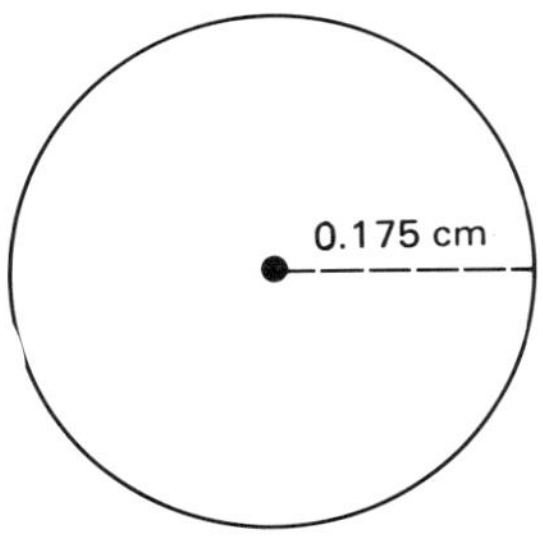

31. (nearest ten; use $\pi = 3.142$)

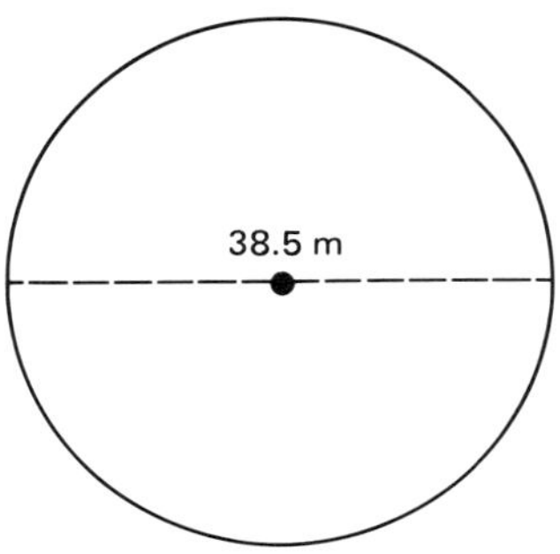

32. (nearest tenth)

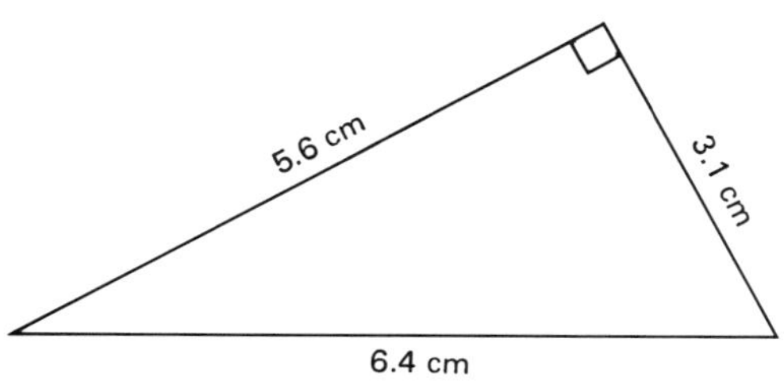

29.

30.

31.

32.

In exercises 33–42 find the area of the given irregular figure. Do not round off your answers. Note that in exercises such as these you will need to break up each figure into a series of smaller figures whose areas you can find by the formulas we have discussed and then find the total of these areas. Study the example.

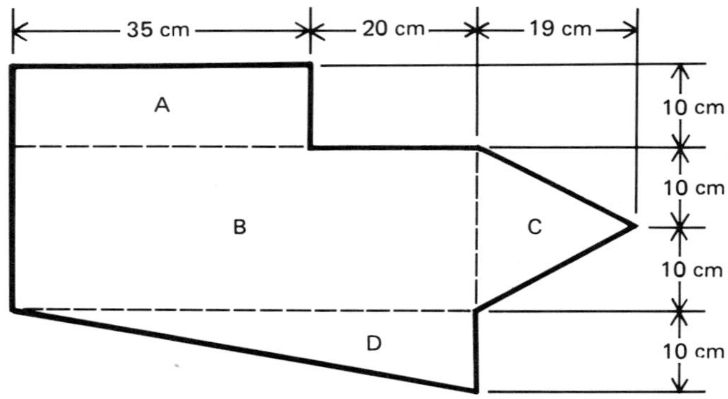

Dividing the figure into the four regions A, B, C, and D, we find the sum of the areas of these regions for a total area of 1915 cm^2.

rectangle A:	Area $= 35 \times 10 = 350$ cm^2
rectangle B:	Area $= 55 \times 20 = 1100$ cm^2
triangle C:	Area $= \dfrac{20 \times 19}{2} = 190$ cm^2
triangle D:	Area $= \dfrac{55 \times 10}{2} = 275$ cm^2
	Total area $= 1915$ cm^2

33.

33.

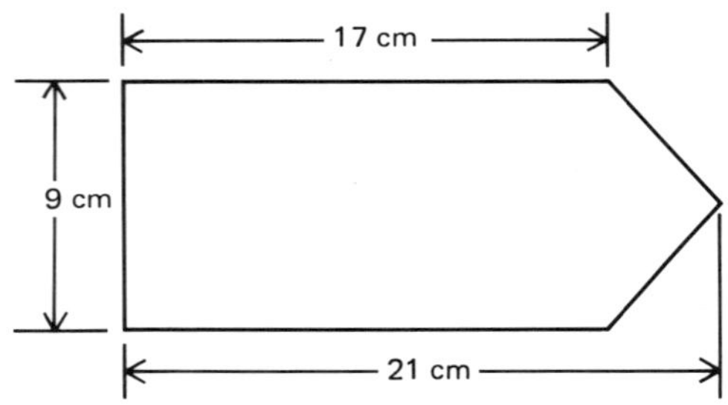

34.

34.

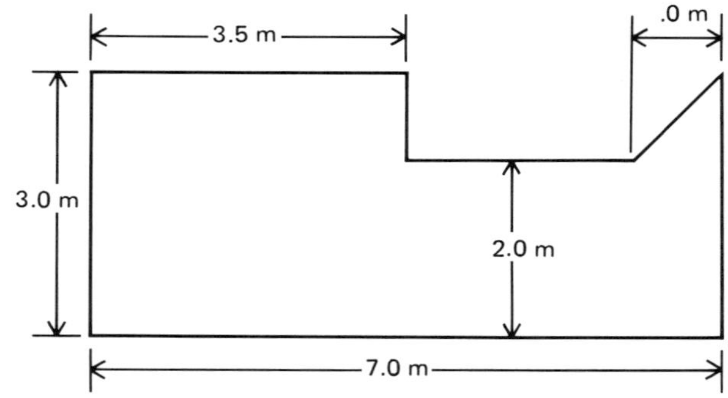

35.

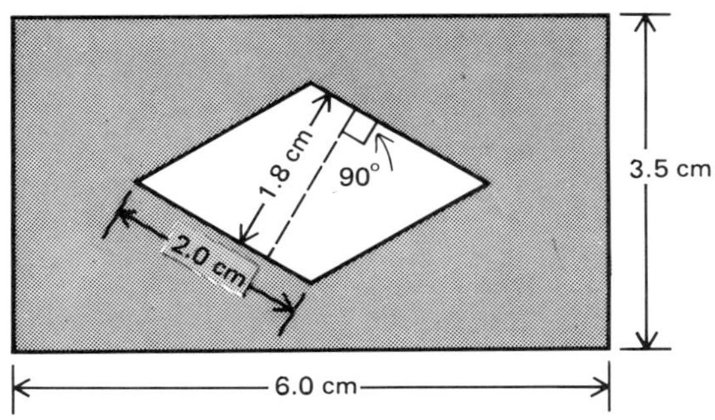

35.

36.

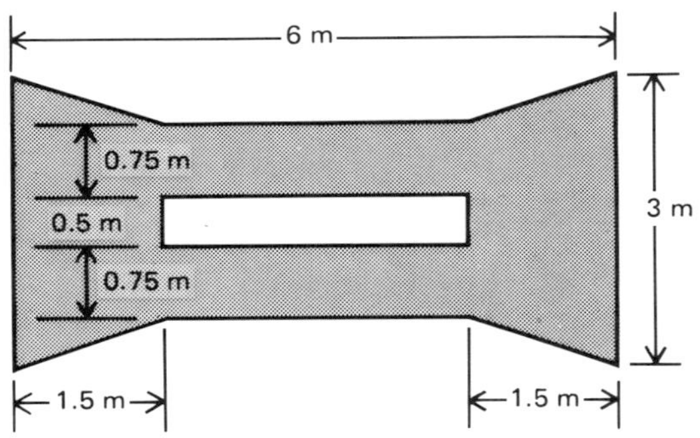

36.

37.

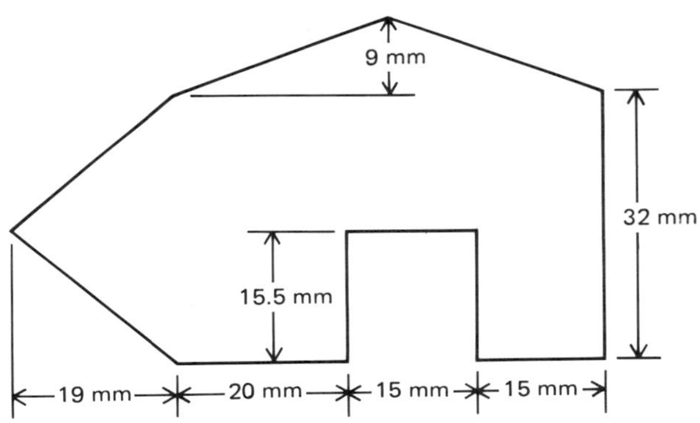

37.

38.

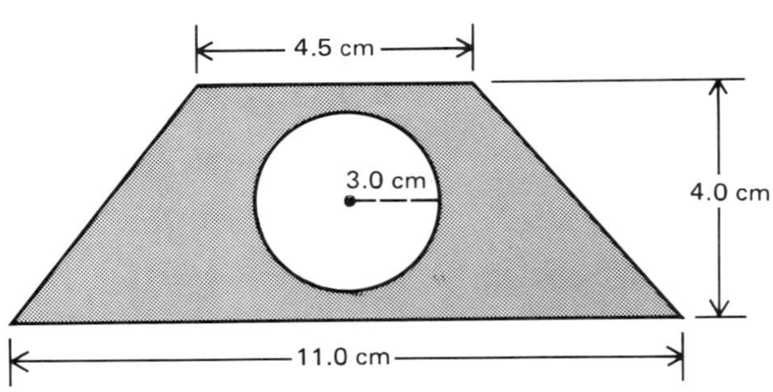

(Use 3.14 for π.)

38.

39.

40.

41.

42.

39.

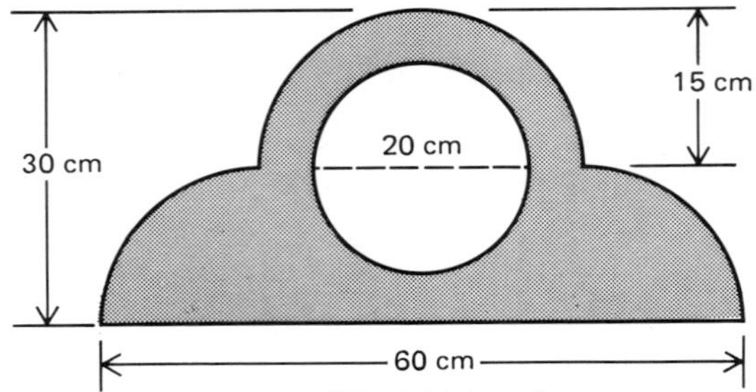

(Use 3.14 for π.)

40.

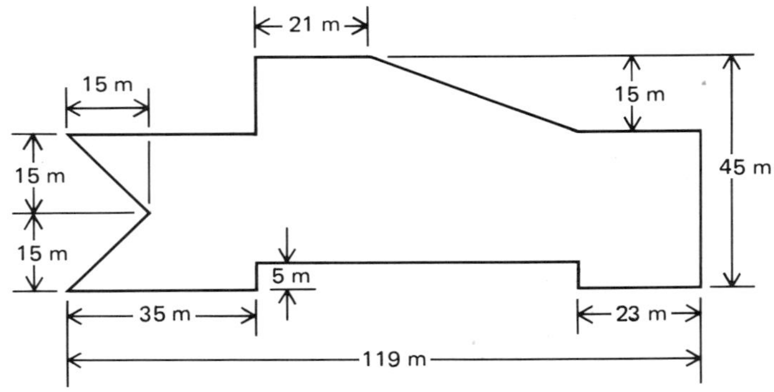

41.

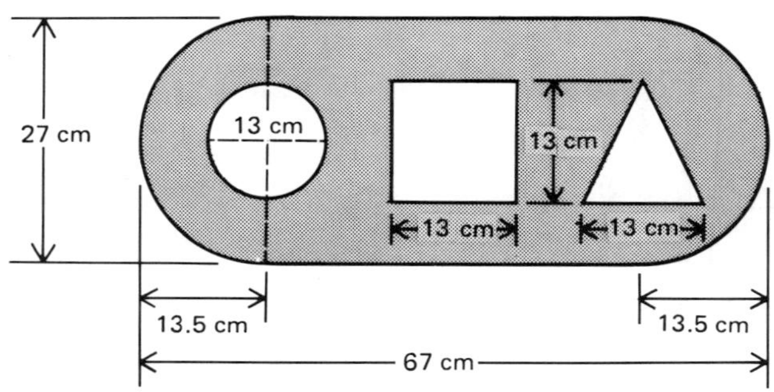

(Use 3.14 for π.)

42.

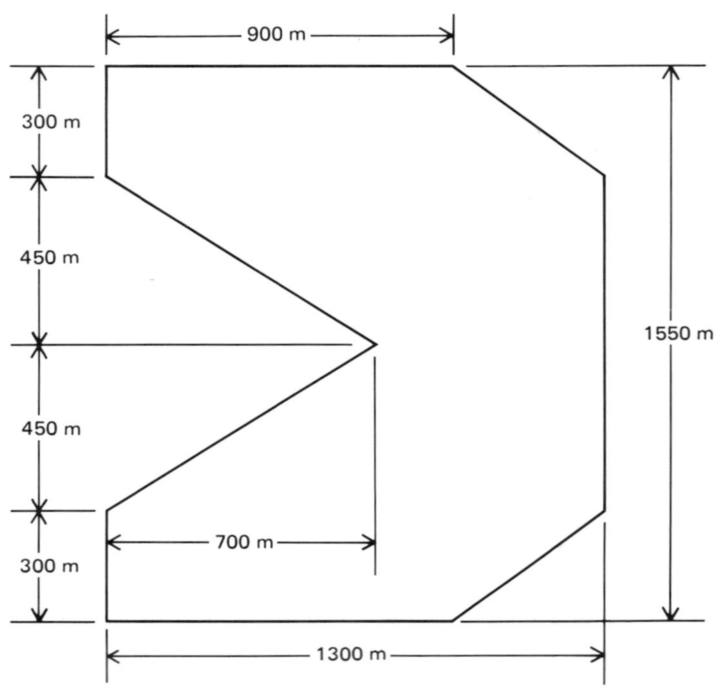

In exercises 43–71 round off your answer in a reasonable fashion.

43. A sidewalk is 12 meters long and 0.90 meters wide. What would be the cost of concrete for the sidewalk if concrete cost $4.50 a square meter?

43.

44. A kitchen floor is 4.2 meters long and 3.1 meters wide. If new plastic floor covering costs $5.98 a square meter, what would it cost to cover the floor with it?

44.

45. Victor has a garden plot 7.65 meters long and 4.43 meters wide in which he wishes to grow tomatoes. If each tomato plant should have 1800 square centimeters of space, how many tomato plants should he purchase?

45.

46. A field is 241 meters long and 186 meters wide. (a) How many hectares does the field contain? (b) At a cost of $2785 a hectare, how much would it cost to purchase the land?

46. (a)

(b)

47. A lawn adjacent to a new home is 12.2 meters long and 10.7 meters wide. (a) If a kilogram of lawn seed is required for every 22 square meters of lawn, how much grass seed will be required to seed the entire lawn? (b) If the cost of the best grass seed is $8.75 a kilogram and the cost of the cheapest grass seed is $2.22 a kilogram, what is the difference in the cost of seeding the lawn with the best grass seed as compared with the cheapest?

47. (a)

(b)

48. The length of an official U.S. flag is 1.9 times its width. What is the cost of a flag 1 meter wide if the material costs $4.95 a square meter?

48.

49. How many square meters of tarpaulin are needed to cover the infield of a baseball diamond if the distance between the bases is 27.43 meters?

49.

50.

51.

52.

53.

54.

55.

56.

57. (a)

(b)

58.

50. A triangular sail has a base of 4.3 meters and an altitude of 5.1 meters. How many square meters of surface does one side of the sail expose?

51. A porch roof consists of two sections, each shaped like a trapezoid. The bases of one section are 3.4 m and 3.9 m and of the other 5.2 m and 5.8 m. The distance between the bases in both sections is 2.4 m. At a spreading rate of 23 square meters per gallon, will a gallon of paint be sufficient to cover the entire roof? How much too much or too little paint would there be?

52. How many bricks 20 cm by 11 cm will be required for a brick walk that is to be 67 m long and 176 cm wide?

53. Find the width of a rectangular field whose area is 15.0 hectares and whose length is 375 meters.

54. Find the altitude of a trapezoid whose area is 139 cm^2 and whose bases are 19.4 cm and 12.3 cm respectively.

55. A triangular piece of property contains 3 hectares. If its base is 500 meters across the front, how deep is the property perpendicular to the base?

56. Over how many square kilometers can a program be received from a TV station if the station can televise programs for a distance of 85 kilometers?

57. A circular fountain 9 meters in diameter has a circular walk around it that is 4 meters wide. (a) What is the area of the walk? (b) At a cost of \$5.25 per square meter, what would it cost to pave the walk? (Use 3.14 for π.)

58. How many times as large is the area of the 200-inch mirror of the world's largest telescope compared to the previous record 100-inch telescope?

59. How many double rolls of wallpaper are needed to cover the walls of a room 4.8 meters long, 3.7 meters wide, and 2.4 meters high if each double roll covers 6.5 square meters? (Deduct 4.75 square meters for openings and allow $\frac{1}{2}$ roll extra for paper matching.) 59.

60. A painting contractor is hired to paint the walls of three rooms. Two rooms measure 3.35 meters wide by 4.25 meters long. The third room is 3.65 meters by 4.6 meters long. The walls are 2.5 meters high. If a gallon of paint covers 50 square meters, how much paint will be needed to do the job if doors and windows are disregarded? 60.

61. Figure the cost to carpet a 4-meter by 6-meter office if the pad costs \$5.25 per square meter, the carpet \$19.95 per square meter, and installation is \$2.55 per square meter. 61.

62. A tracing for a drawing is 45 cm wide by 60 cm long. How many blueprints can be made from this tracing from a roll of blueprint paper 1 m wide and 50 m long? 62.

63. Two pulleys which have the same diameter have a distance of 2.4 meters between their centers. If the pulley diameters are 32.5 cm each, what is the length of a belt for the system? 63.

64. The center distance of the fan pulley and the crankshaft pulley is 33 cm. The two pulleys are both 10 cm in diameter. Find the length of fan belt needed. 64.

65. The rear wheels of a car are turning 650 rpm (revolutions per minute). If the wheels on the car have a diameter of 57 cm, how many kilometers per hour is the car traveling? 65.

66. How long a clamp strap is necessary to go around a hose connection if the hose is 5 cm in diameter? 66.

*67.

*68.

*69. (a)

(b)

*70.

*67. Imagine a wire stretched tightly around the earth at the equator. If the earth has a diameter of 13,000 km (assume it is a perfect sphere), how much more wire would be needed to place the wire exactly 1 m above the equator at all points?

*68. A carpenter cut seven boards; each is exactly 1 foot wide. Four boards are each 4 feet long, and three of them are 3 feet long. How can he arrange the pieces to cover a square area without cutting or overlapping the boards?

*69. An advertisement once offered to sell one square inch of land in the Klondike for \$0.25. For your quarter you received a deed which described the property, and for a small additional charge they would send you a small bag of dirt from the general vicinity. (a) Compute the gross income per acre. (b) What is the gross profit per acre to the sellers if the land cost them \$500?

*70. Consider a square piece of paper 1 unit on a side:

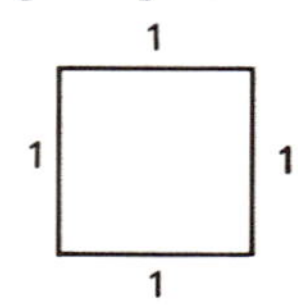

The area is 1 square unit.

The perimeter is 4 units.

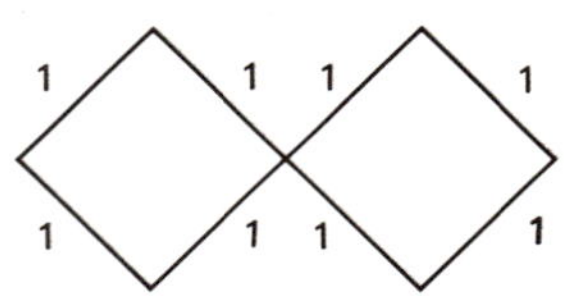

With 2 square pieces the area is 2 square units.

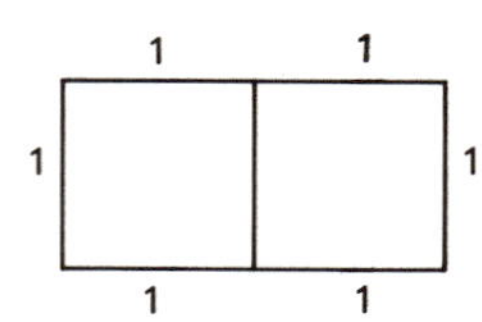

Maximum perimeter is 8 units.

Minimum perimeter is 6 units.

(a) Complete the following table:

Number of pieces	1	2	3	4	5	6	7	8	9
Area	1	2							
Maximum perimeter	4	8							
Minimum perimeter	4	6							

(b) Look for a pattern to the maximum perimeters.

(c) Look for a pattern to the minimum perimeters.

(d) Find the maximum perimeter for 100 pieces and for 1000 pieces.

(e) Find the minimum perimeter for 100 pieces and for 1000 pieces.

(f) Give a formula to obtain the maximum perimeter given the number of pieces.

(g) Give a formula to obtain the minimum perimeter given the number of pieces.

(b)

(c)

(d)

(e)

(f)

(g)

6.4 Measurement of Volume

Length is a measure of distance, area is a measure of surface, and volume is a measure of space. Thus the space occupied by a cube 1 centimeter on each side is 1 *cubic centimeter* (1 cm^3) as shown in the following figure. The other commonly used unit of volume is the *cubic meter* (m^3) which is, of course, the volume of a cube 1 meter on each side.

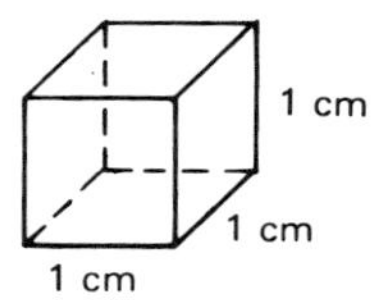

However, when we are thinking of volume in terms of *capacity* (the ability of a container to hold, for example, a certain amount of water), it is customary to use as units the *liter* (l) or the *milliliter* (ml). We have

$$1 \text{ l} = 1000 \text{ cm}^3 \text{ (slightly more than a quart)}$$

$$1 \text{ ml} = 0.0001 \text{ l} = 1 \text{ cm}^3$$

$$1 \text{ m}^3 = 1000 \text{ l}$$

Just as formulas exist for finding areas, so formulas exist for finding volumes. Some of these, together with examples of their use, are as follow.

Rectangular solid: $V = lwh$

Example: Find the volume of a rectangular solid of length 15 cm, width 12 cm, and height 19 cm.

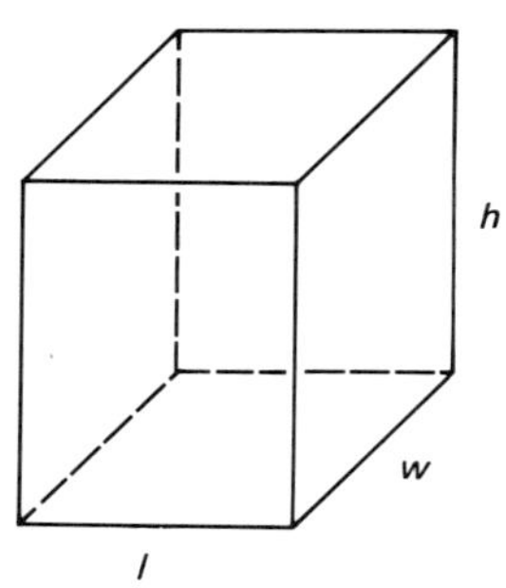

Solution: $V = 15 \times 12 \times 19 = 3420$

The volume is 3400 cm^3 to the nearest hundred. The volume in liters is

$$3400 \div 1000 = 3.4 \text{ (to the nearest tenth)}$$

Do exercises 1 and 2 in the adjoining column.

OBJECTIVES

You should:

(a) Understand the meaning of the various units of volume in the metric system and the relationships between them.

(b) Know and be able to use formulas for the volume of various common geometric solids.

****

1. Find the volume of a rectangular solid whose width is 1.43 m, whose height is 4.56 m, and whose length is 2.71 m. Express your answer in cubic meters rounded off to the nearest tenth.

2. A container is in the shape of the rectangular solid of exercise 1. What is its capacity in liters? (Use your rounded off answer in exercise 1.)

STOP

3. Find the volume of a rectangular pyramid whose base is 13.5 cm by 17.8 cm and whose height is 21.3 cm. Express your answer in both cubic centimeters and in milliliters, rounded off to the nearest ten.

4. What should be the height, to the nearest tenth of a meter, of a rectangular pyramid whose base is 2.1 m by 3.7 m if the pyramid is to have a volume of 13 m^3?

STOP

5. A yogurt container is in the shape of a right circular cone of height 12.3 cm and a base of radius 7.5 cm. How many milliliters of yogurt will it hold? Use 3.14 for π and round off your answer to the nearest unit.

STOP

Rectangular pyramid: $V = \frac{1}{3} lwh$

Example: Find the volume of a pyramid whose base is a rectangle of length 2.1 m and width 1.6 m and whose height is 3.7 m.

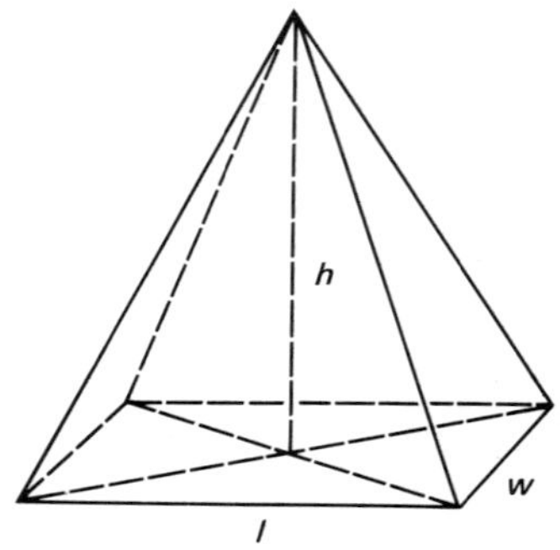

Solution: $V = \frac{2.1 \times 1.6 \times 3.7}{3} = 4.144$

The volume is 4.1 m^3 to the nearest tenth. The volume in liters is

$$4.1 \times 1000 = 4100 \text{ (to the nearest hundred)}$$

Do exercises 3 and 4.

Right circular cone: $V = \frac{1}{3} \pi r^2 h$

Example: Find the volume of a right circular cone of height 7.69 cm and a base of radius 5.24 cm.

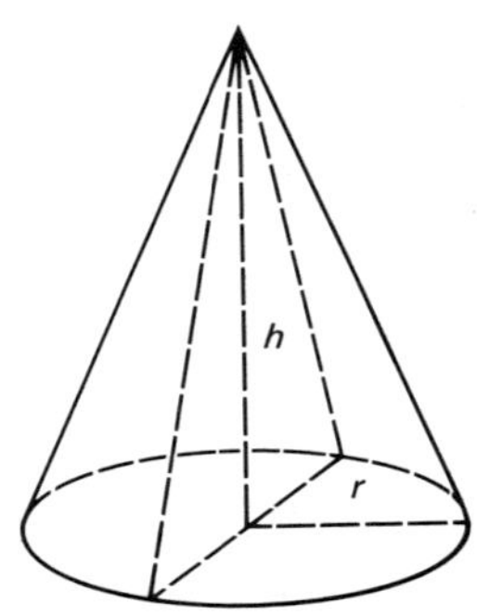

Solution: If we take π as 3.142, we have

$$V = \frac{1}{3} \times 3.142 \times (5.24)^2 \times 7.69 = 221.14332$$

In previous examples, although rounding off was needed, the need wasn't completely obvious. Here it certainly is and, to the nearest cubic centimeter, we have a volume of 221 cm^3.

Do exercise 5.

Sphere: $V = \frac{4}{3}\pi r^3$

Example: How many liters of hydrogen would be needed to fill a balloon in the shape of a sphere with a diameter of 20.2 cm?

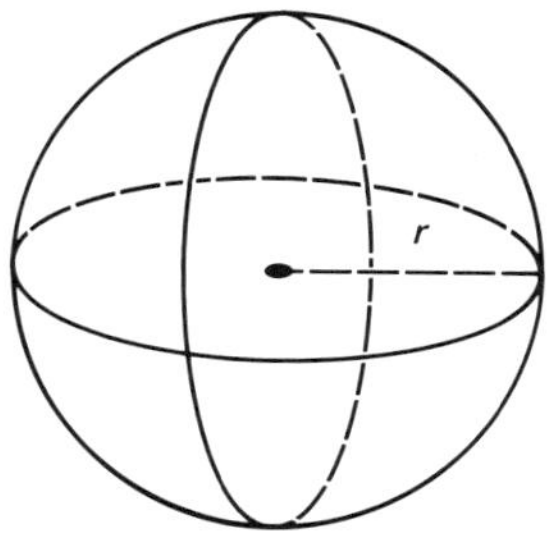

Solution: The radius of the balloon is 10.1 cm and so, taking 3.142 for π,

$$V = \frac{4}{3} \times 3.142 \times (10.1)^3 = 4316.274$$

Thus, to the nearest ten, the volume is 4320 cm³, and so

$$4320 \div 1000 = 4.32$$

liters of hydrogen would be needed (to the nearest hundredth of a liter). Here, however, we are dealing with a substance whose volume varies with changes in pressure and temperature, so that it would be much better to say "about 4 liters" or "between 4 and 5 liters."

Do exercise 6.

6. A golf ball must be not less than 1.68 inches in diameter. What, in cubic inches, is its minimum volume? Use 3.142 for π and round off your answer to the nearest hundredth of a cubic inch.

STOP

Right circular cylinder: $V = \pi r^2 h$

Example: What should be the height, to the nearest tenth of a centimeter, of a tin can that is to hold one liter of fruit juice if the diameter of the can is to be 10.0 cm?

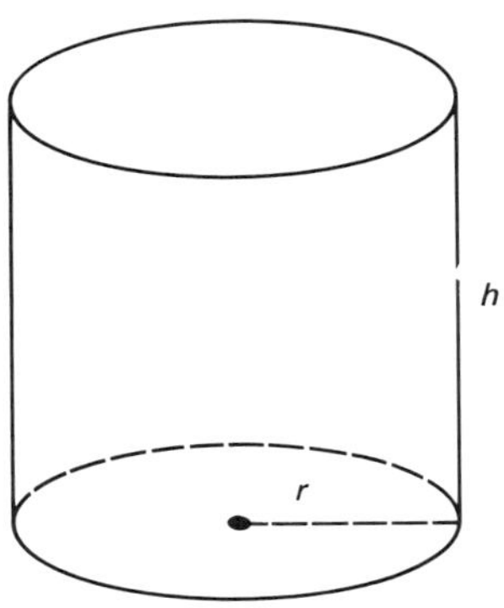

7. A hole in a golf course is $4\frac{1}{4}$ inches in diameter and at least 4 inches deep. Take $\frac{22}{7}$ for π and find, in fractional form, the minimum number of cubic inches in a hole.

STOP

Solution: The radius of the can is 5.0 cm and its volume is to be 1000 cm^3. Hence, using 3.142 for π, we must have

$$1000 = 3.142 \times 5^2 \times h$$

and so

$$h = \frac{1000}{3.142 \times 25} = 12.730744$$

Thus the height should be 12.7 cm to the nearest tenth of a centimeter.

Do exercise 7.

Exercises 6.4

Name ____________

Class ____________

In exercises 1–10 complete the following table dealing with the relationships between units of volume. The first line has been completed as an example.

ml	cm^3	l	m^3
1000	1000	1	0.001
500	500	1.	2.
3.	60	4.	5.
6.	7.	8.	0.05
7	7	9.	10.

11. Find the volume of this rectangular solid.

11.

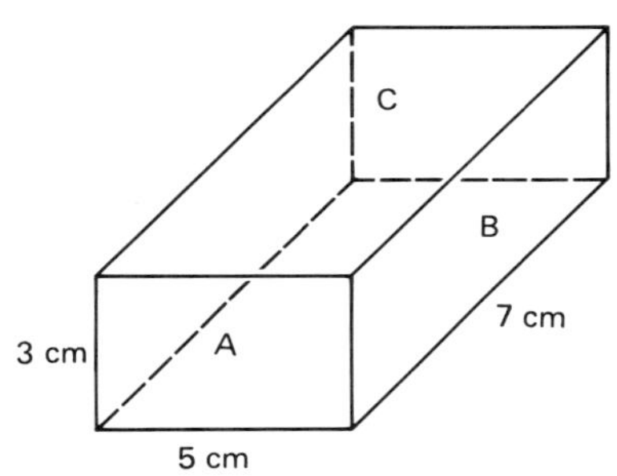

12. The *lateral area* of a solid is the area of its outer surface. Find the lateral area of the rectangular solid of exercise 11 as follows:

12.

(a) two surfaces (faces) the size of A: $2 \times (3 \times 5)$ = ____________ cm^2 (a)

(b) two surfaces (faces) the size of B: $2 \times (3 \times 7)$ = ____________ cm^2 (b)

(c) two surfaces (faces) the size of C: $2 \times (7 \times 5)$ = ____________ cm^2 (c)

(d) lateral area = total area of 6 faces = ____________ cm^2 (d)

In exercises 13–16 find (a) the volume and (b) the lateral area of the given solid. Round off your answer as indicated.

13. (a)

(b)

13. (nearest tenth)

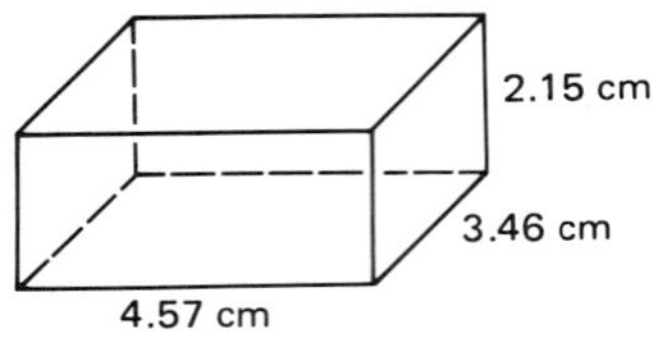

14. (a)

(b)

14. (nearest tenth)

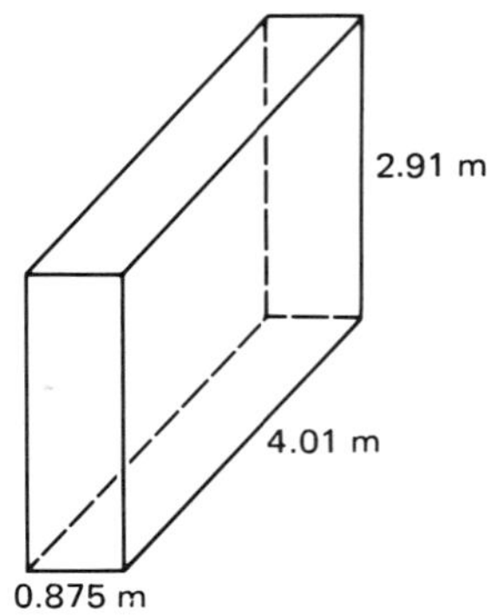

15. (a)

(b)

15. [nearest unit for (a); nearest tenth for (b)]

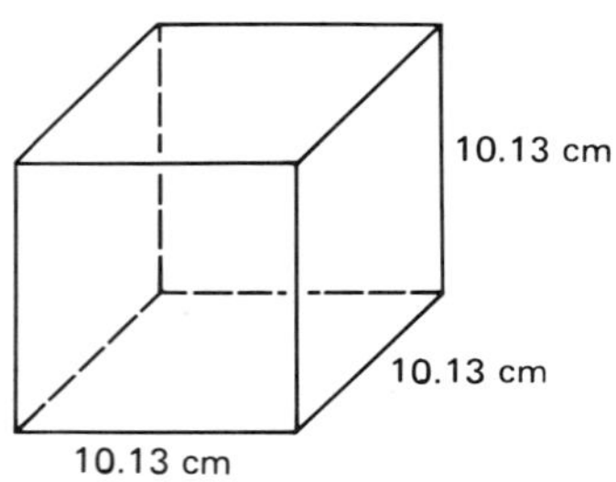

16. (a)

(b)

16. [nearest tenth for (a); nearest unit for (b)]

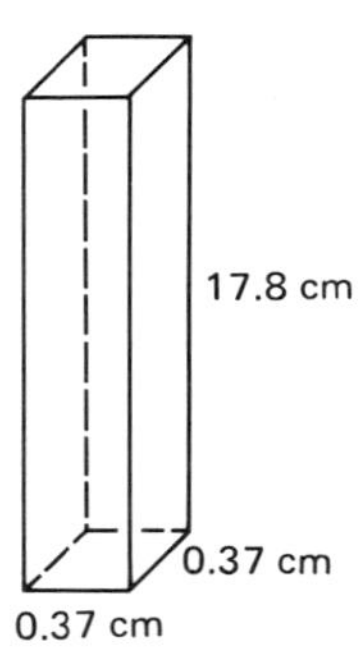

In exercises 17–20 find the volume of the given solid. Round off your answer as indicated.

17. (nearest tenth)

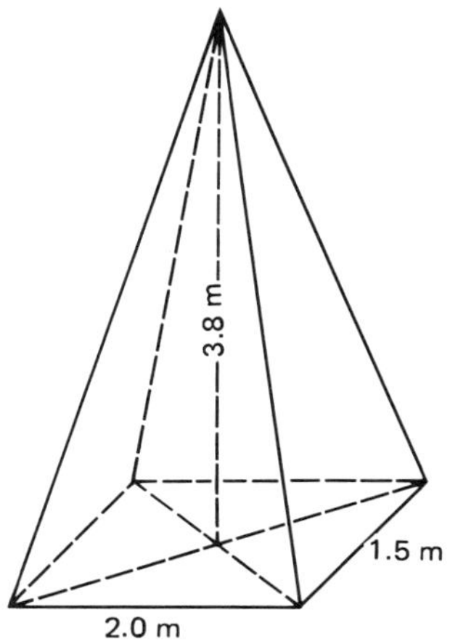

17.

18. (nearest ten)

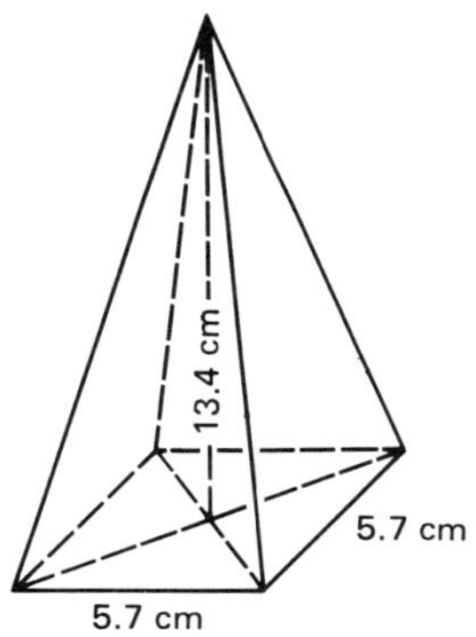

18.

19. (nearest ten)

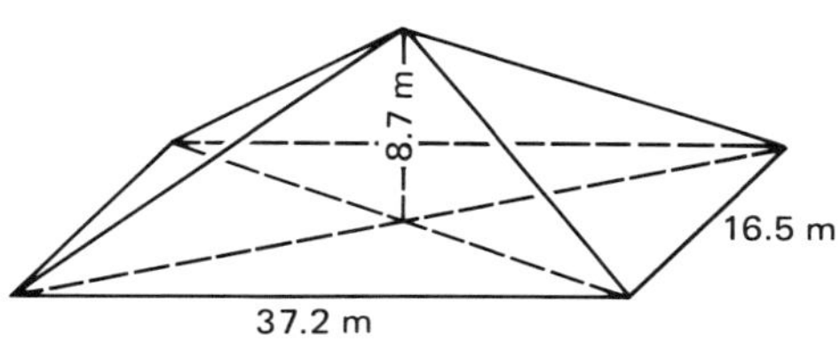

19.

20. (nearest ten)

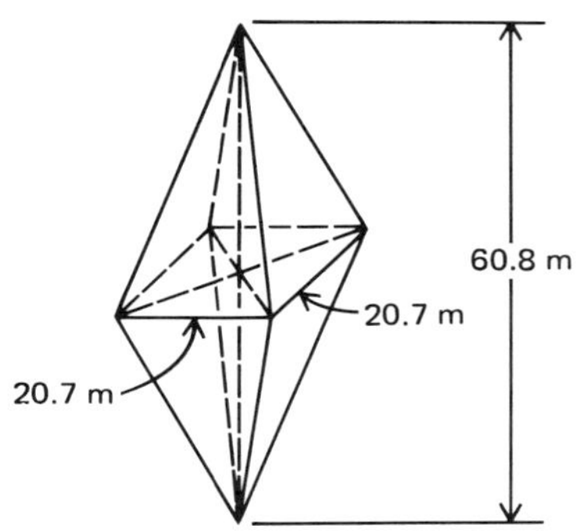

20.

In exercises 21–28 find the volume of the given solid in liters. If π is needed, use 3.142. Round off your answer as indicated.

21.

21. (nearest hundredth)

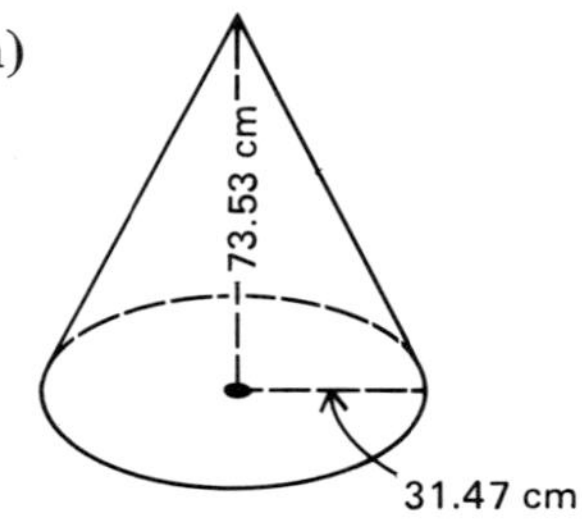

22.

22. (nearest unit)

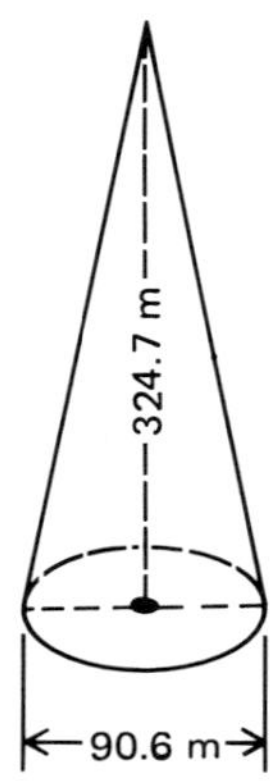

23.

23. (nearest thousand)

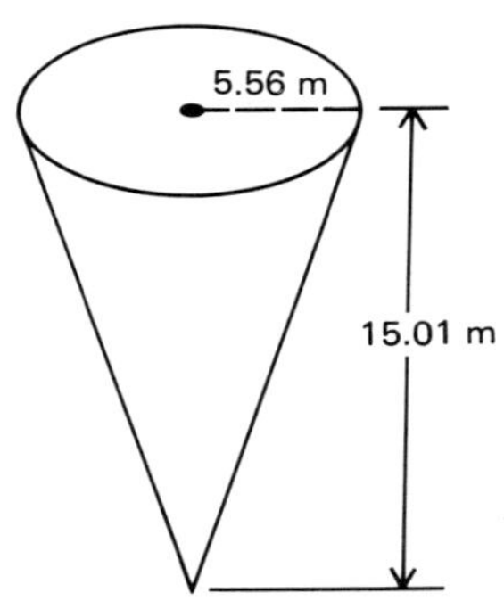

24.

24. (nearest thousandth)

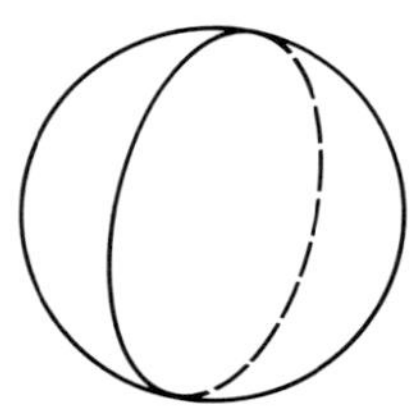

radius = 2.91 cm

25.

25. (nearest ten thousand)

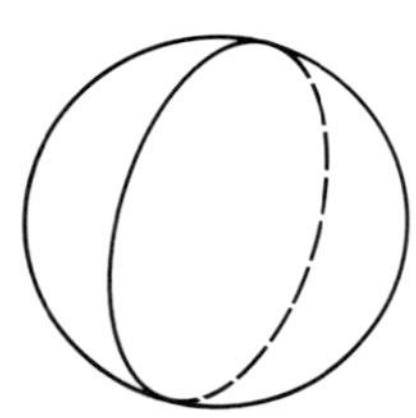

diameter = 48.47 m

26. (nearest hundredth) 26.

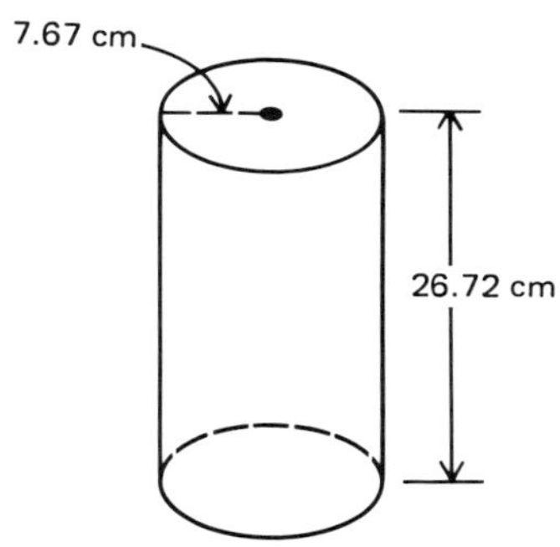

27. (nearest thousandth) 27.

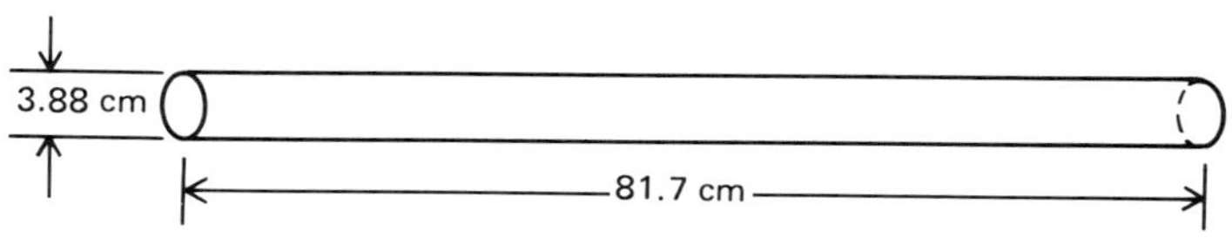

28. (nearest ten thousand) 28.

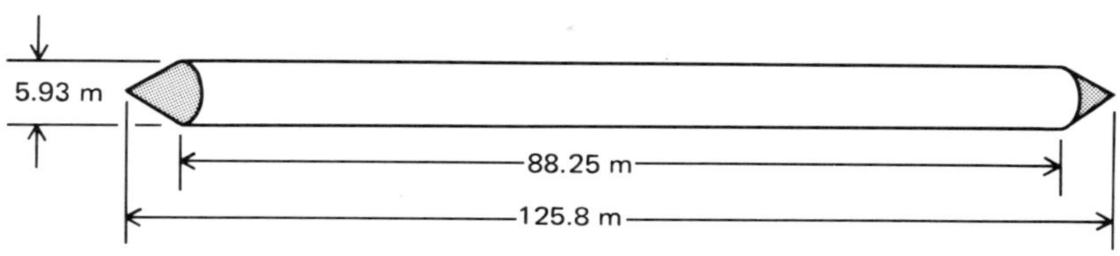

In exercises 29–42 round off your answer in a reasonable fashion.

29. A truck body used for hauling coal is 3 meters long, 1.9 meters wide, and 0.85 meters deep. How many metric tons (1000 kilograms) can it haul in one load if coal weighs 177 kilograms per cubic meter? 29.

30. A freight car is 13 meters long and 2.6 meters wide. If it is filled to a depth of 1.4 meters with wheat and one bushel occupies 0.035 cubic meters, how many bushels (to nearest ten) of wheat are in the freight car? 30.

31. A concrete retaining wall is to be built along one side of a building lot. The wall is to be 25.0 meters long, 762 centimeters high, and 381 centimeters wide. How many cubic meters of concrete will be required? 31.

32. A hot water tank is 300 cm in diameter and 1.5 m long. How many liters of water will the tank hold? 32.

33.

33. A gasoline truck has a cylindrical tank whose inside measurements are 1.8 m in diameter and 4.26 m in length. How many liters of gasoline will the truck hold?

34.

34. A silo is a cylindrical structure used on farms to store finely cut corn or other grain fodder for cattle. Each of two silos is 4.5 m in diameter and 12 m high. When both silos are filled to a height of 9 m and a cow is fed 0.09 m^3 of silage a day, how many days could this silage feed a herd of 36 cows?

35.

35. To what depth will 40 liters of water fill an aquarium 76.0 cm long and 35.5 cm wide?

36.

36. A pipeline in Texas is 2158 km long and 610 cm in diameter. Oil moves through the pipeline at the rate of 4.8 km per hour. How many liters of oil can this pipeline deliver in an hour?

37.

37. What would be the weight in grams of a hailstone 5 centimeters in diameter if ice weighs 439 kilograms per cubic meter?

38.

38. A granite monument shaped as a square pyramid is 3 meters high and 2 meters square at the base. If granite weighs 890 kilograms per cubic meter, what is the weight of the monument?

39.

39. A marble sphere used on a monument in a park is 760 centimeters in diameter. How much does the sphere weigh if marble weighs 850 kilograms per cubic meter?

40.

40. In an orchard there is a conical pile of apples about 4.5 m in diameter and 1.5 m high. How many bushels of apples are in the pile if one bushel occupies 0.035 m^3?

41. A truck hauls 2.5 m^3 of sand in each load. How many loads are contained in a conical shaped pile of sand 2.3 m high and 3.7 m across at the base? 41.

42. When camping in a forest service campground, Grace pumped a cylindrical bucket full of water in 3 minutes. The bucket was 35.5 cm in diameter and 30.4 cm deep. If she could pump continuously at the same rate for an hour, how many liters of water would she pump? 42.

43. A rectangular dipping vat is 10 m long, 4 m deep, and 5 m across the top. Dip flows into the vat at the rate of 200 liters per minute. Find the time in hours and minutes that is required to fill the vat to a depth of 3 m. 43.

44. How many liters of water per second will be discharged by a pipe 20 cm in diameter if the water has a velocity of 1.3 m per second? 44.

45. Find the total displacement in liters of a 6-cylinder engine with a bore of 86 mm and a stroke of 8 cm. 45.

46. A steel pipe is 7 meters long with an inside diameter of 6 centimeters and a thickness of 50 millimeters. What will the pipe weigh if steel weighs 7.8 grams per cubic centimeter? 46.

47. A newspaper article is 1977 indicated that Navy weathermen, via satellite, were watching an Antarctic iceberg big enough to provide California enough fresh water to last 1100 years–if it could be towed to the then drought-plagued state. If the iceberg is 72.4 km long, 42.3 km wide, and 305 m thick, and the population of California remained at 20,000,000, how many liters of water would it supply each Californian per day during the 1100 years? (Assume 365 days in each year.) 47.

48. An open tray is constructed by starting with a rectangular sheet of metal 20 centimeters by 28 centimeters. Pieces are cut from each corner so that when the sides are folded up the tray will be 3 centimeters deep. What is the volume of the tray?

49. A box open at the top is 5.7 centimeters wide, 20.2 centimeters long, and 8.9 centimeters high. How would you cut down the box so it would have a volume of 1000 cubic centimeters?

50. How many cubic meters of rock are required to surface 5 kilometers of road to a depth of 25 centimeters if the overall width of the surface is 9 meters and 2.5% is added for loss of rock at the edges?

*51. Determine the measurements of a box that has a surface area to volume ratio of 1.

*52. Consider a cube that has an edge of 3 meters. Suppose it is painted red and then cut into 1-meter cubes.

(a) How many cuts are necessary to separate the cube into 1-meter cubes?

(b) How many 1-meter cubes are there?

(c) How many cubes have paint on no faces?

(d) How many cubes have paint on exactly two faces?

(e) How many cubes have paint on exactly three faces?

(f) How many cubes have paint on exactly four faces?

(g) Is it possible to arrange these 1-meter cubes into a 3-meter by 3-meter cube so that no painted face is showing?

6.5 Mass, Weight, and Density

There is a technical difference between the *mass* of an object and its *weight.* For example, an object has the same mass on the moon as on the earth but will weigh less on the moon because the attraction of gravity is much less on the moon than it is on the earth. But for practical purposes we can, if we stay on earth, regard mass and weight as the same, and we will do so here.

The basic unit of mass is the *gram* (g). (A nickel weighs about 5 grams.) The other commonly used units of mass are the *kilogram* (kg) which is, of course, 1000 grams, and the *metric ton* which is 1000 kg. (A kilogram is about 2.2 pounds.)

The *density* of an object is defined as its mass per unit of volume. For example, the density of aluminum is 2.7 grams per cubic centimeter (g/cm^3) meaning that every cubic centimeter of aluminum weighs 2.7 grams. To find the weight of 124 cm^3 of aluminum, we calculate

$$2.7 \times 124 = 334.8$$

and round off to get 335 g.

Do exercises 1 and 2 in the adjoining column.

OBJECTIVES

You should:

(a) Understand the meaning of the various units of mass in the metric system and the relationships between them.

(b) Know what is meant by density and how to use it in calculating the mass of an object from its volume and density.

****

1. The density of gold is 19.3 g/cm^3. What would be the weight of 0.56 cm^3 of gold? Round off your answer to the nearest tenth of a gram.

2. The density of a certain kind of wood is 45 lb/ft^3. What would be the weight of a block of this wood that is 12 inches long, 4 inches wide, and 4 inches thick?

STOP

Exercises 6.5

Name ______________________

Class ______________________

In exercises 1–16 complete the following table dealing with the relationships between units of mass. The first line has been completed as an example.

g	kg	Metric tons
500	0.5	0.0005
1000	1.	2.
3.	4.	3
5.	5000	6.
150	7.	8.
120	9.	10.
11.	12.	0.16
13.	7	14.
18,000	15.	16.

In exercises 17–43 refer to the following table. Round off your answers in a reasonable fashion.

DENSITIES OF SOME COMMON SUBSTANCES

Substance	Density (g/cm^3)	Substance	Density (g/cm^3)
Aluminum	2.7	Lead	11.34
Brass	8.4	Mercury	13.6
Copper	8.8	Milk	1.029
Gasoline	0.69	Steel	7.79
Gold	19.3	Silver	10.5
Iron (cast)	7.5	Water	1.0

17. A copper rod is 5 cm in diameter and 200 cm long. What will be the mass of the rod in grams? In kilograms?

17.

18. A tank is 2 m long, 1.5 m wide, and 1.8 m deep. How many kilograms of water will it hold?

18.

19.

19. What is the mass of a cast-iron cube with sides of 3.4 cm?

20.

20. A sphere of aluminum is 13 cm in diameter. What is its mass?

21.

21. A pipe has an inside diameter of 15 cm and contains a column of milk 10 m high. What is the weight of the milk in the pipe?

22.

22. Suppose that a car has its gas tank filled and that it takes 19.5 liters to fill it. How much has the weight of the car increased by having its tank filled?

23.

23. A bar of silver is 6 cm long, 2 cm wide, and 1.5 cm thick. What is the mass of the bar?

24.

24. What is the weight of a cast-iron collar 10.5 cm in diameter and 5 cm thick if it has a 2.5-cm hole in the center?

25.

25. How much will the steel cost for 18 bushings 35 cm long and 9.5 cm in diameter with a 6.5-cm diameter hole if steel can be purchased for $1.49 per kilogram?

26.

26. An order was given for 500 motor shafts, each to be 23 cm long and 2.3 cm in diameter. If steel for these shafts costs $1.49 per kilogram, how much will the material for the shafts cost if the steel comes only in 2-, 3-, or 5-cm diameter stock?

27.

27. A brick, supposedly made of gold, is 15 cm long, 10 cm wide, and 5 cm thick. If the brick has a mass of 19.375 kg, would you suspect that it is not made of gold? Why?

28. A manufacturer wishes to make a 1000-gram brass weight. How many cubic centimeters of brass will be needed? 28.

29. You pour 35 ml of water into a graduated cylinder and then lower into it an irregularly shaped piece of metal with a mass of 125 g. If the water level rises to the 60-ml mark, what is the density of the material? 29.

30. The external diameter of a cast iron pipe is 40 cm and its internal diameter is 36 cm. Calculate its weight if it is 2 m long. 30.

31. A cast-iron ball, 19 cm in diameter, is covered with a coating of lead 2.5 cm thick. What is the mass of the coated ball? 31.

32. What is the mass in kilograms of a solid lead ball used by commercial fisherman if the ball is 15 cm in diameter? 32.

33. A solid metal ball 10 cm in diameter weighs 5.93 kg. What kind of metal would you suspect that the ball is made of? 33.

34. Five hollow copper balls are to be used as a decoration. If the copper balls are to be 20 cm in diameter and the copper is to be 1 mm in thickness, what will the copper cost at $1.43 per kilogram? 34.

35. A steel washer 19.5 cm in diameter with a 4-cm diameter hole is 0.5 cm thick. Find the number of these washers in a box of them weighing 51 kg if the box itself weighs 1 kg. 35.

36.

36. How many square meters would one ounce (28 g) of gold cover if it is beaten into foil 0.0000254 mm thick?

37.

37. Jack went to buy a canteen and asked the store clerk how much water the canteen would hold. The clerk weighed the empty canteen and found it to weigh 135 g. She then filled the canteen with water and found that the filled canteen weighed 1315 g. How much water will the canteen hold?

38.

38. What would be the weight of a cord of fir wood if it were a solid mass? (A cord of wood is a pile 8 feet long, 4 feet wide, and 4 feet high. The density of fir is 36.9 pounds per cubic foot.) If in actuality a cord of fir wood weighs 2500 pounds, what percent of the space in a cord of wood is occupied by the wood?

39.

39. What would be the weight of copper sheeting 1.6 mm thick, used to line the sides and bottom of a rectangular storage tank 3 m long, 1 m wide, and 1.2 m deep if an allowance of 5% of the area of the liner is needed for lapping purposes?

40.

40. When harvesting ice it is customary to calculate the metric tons of ice that can be harvested rather than to weigh each load. If the ice has a uniform thickness of 40 centimeters, how many metric tons of ice can be removed from a pond whose area is $\frac{1}{4}$ hectare? (Ice has a density of 0.917 gram per cubic centimeter.)

41.

41. A brass tube whose inner diameter is 5 cm and outer diameter is 7 cm is to be made out of 23 kg of brass. How long will it be?

42.

42. A certain brass spherical shell 2 cm thick and 25 cm in outside diameter is found to weigh 28 kg. What is the density of the brass used in its construction?

Review Exercises 6

Name ______________________

Class ______________________

In exercises 1–22 use the given formula and information to find the missing number and round off your answer as directed.

Velocity formula: $V = \frac{d}{t}$

1. $d = 135,\ t = 23$ (nearest tenth) 1.

2. $V = 425,\ t = 7.2$ (nearest hundred) 2.

3. $V = 35.67,\ d = 176.8$ (nearest thousandth) 3.

Acceleration formula: $a = \frac{V_2 - V_1}{t}$

4. $V_2 = 15.7,\ V_1 = 7.01,\ t = 6.95$ (nearest hundredth) 4.

5. $a = 40.8,\ V_1 = 11.6,\ V_2 = 20.0$ (nearest thousandth) 5.

6. $a = 84.2,\ V_1 = 0,\ t = 28.2$ (nearest ten) 6.

Mechanical equivalent of heat: $W = J \times Q$

7. $J = 6.46,\ Q = 3.63$ (nearest tenth) 7.

8. $W = 66.17,\ J = 17.37$ (nearest thousandth) 8.

9. $W = 854,\ Q = 51.4$ (nearest tenth) 9.

	Centripetal force: $F = \dfrac{mv^2}{r}$
10.	10. $m = 15.33,\ v = 94.1,\ r = 657.6$ (nearest unit)
11.	11. $F = 5.76,\ v = 2.94,\ r = 40.9$ (nearest tenth)
12.	12. $F = 57.2,\ m = 22.7,\ v = 7.14$ (nearest tenth)
	Kinetic energy formula: $\text{K.E.} = \frac{1}{2}mv^2$
13.	13. $m = 77.9,\ v = 9.68$ (nearest ten)
14.	14. $\text{K.E.} = 1211,\ v = 4.85$ (nearest unit)
15.	15. $\text{K.E.} = 693.5,\ v = 90.2$ (nearest thousandth)
	Capacitance formula: $C = \dfrac{q}{v}$
16.	16. $q = 16.7,\ v = 45.8$ (nearest thousandth)
17.	17. $C = 6.29,\ v = 1.67$ (nearest tenth)
18.	18. $C = 48.02,\ q = 2005$ (nearest hundredth)
	Number of board feet formula: $\text{B.F.} = \dfrac{twl}{12}$
19.	19. $t = 2,\ w = 6,\ l = 12$ (no round-off needed)

20. $t = 6$, $w = 12$, $l = 20$ (no round-off needed)

20.

21. B.F. = 600, $w = 6$, $l = 16$ (no round-off needed)

21.

In exercises 22–35 find (a) the perimeter and (b) the area of the given figure. Use 3.14 for π and round off your answer to the (b) part as indicated.

22. (nearest unit)

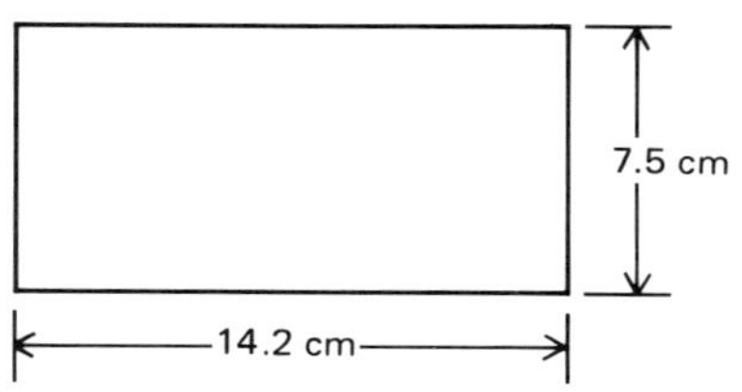

22. (a)

(b)

23. (nearest ten)

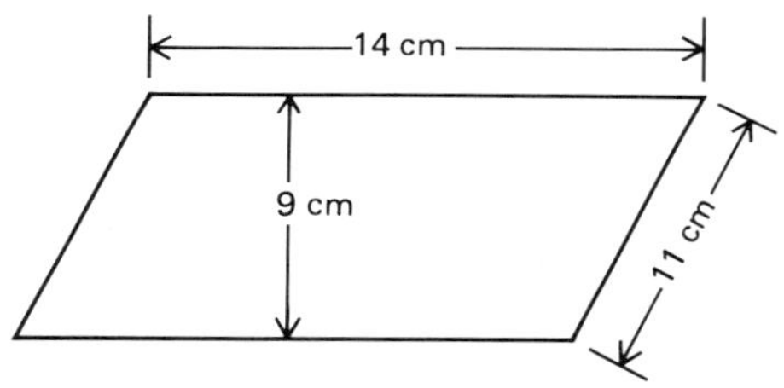

23. (a)

(b)

24. (nearest tenth)

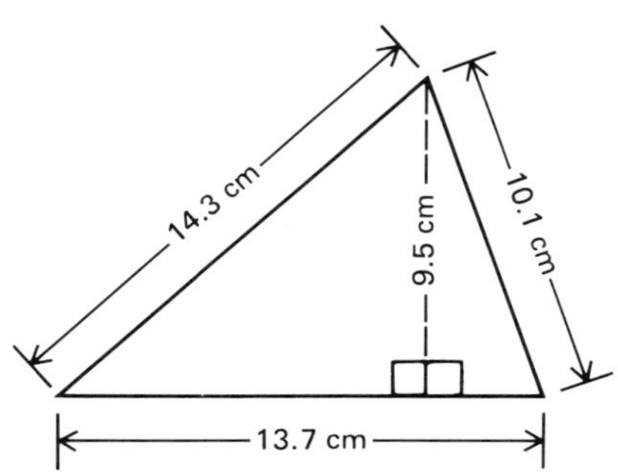

24. (a)

(b)

25. (nearest tenth)

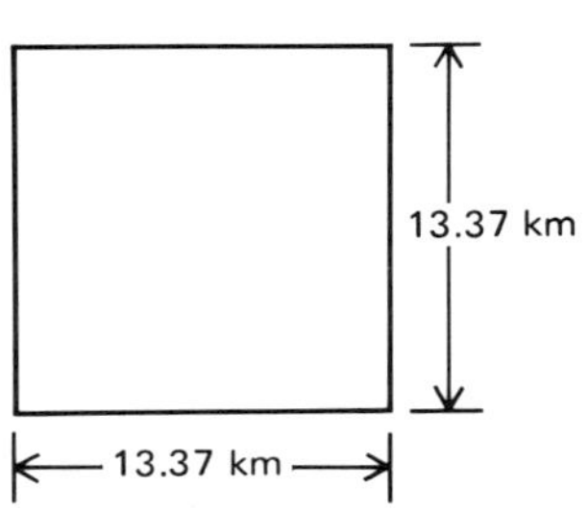

25. (a)

(b)

26. (a)

(b)

26. (nearest unit)

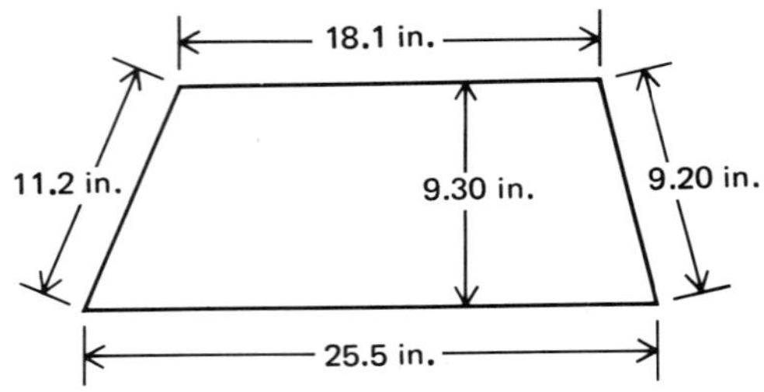

27. (a)

(b)

27. (nearest unit)

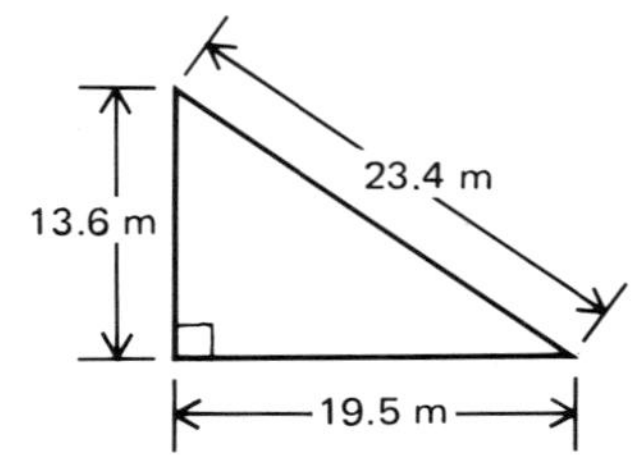

28. (a)

(b)

28. (nearest ten for [a]; nearest tenth for [b])

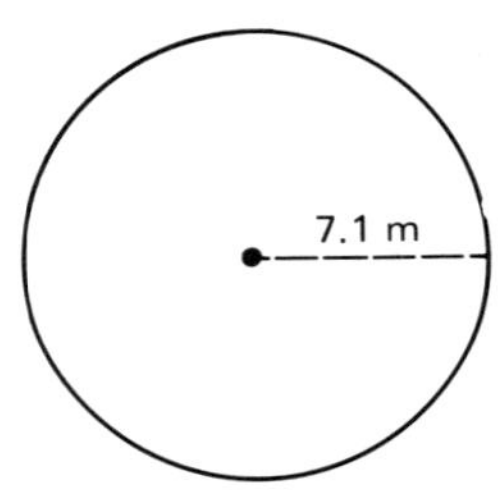

29. (a)

(b)

29. (nearest ten)

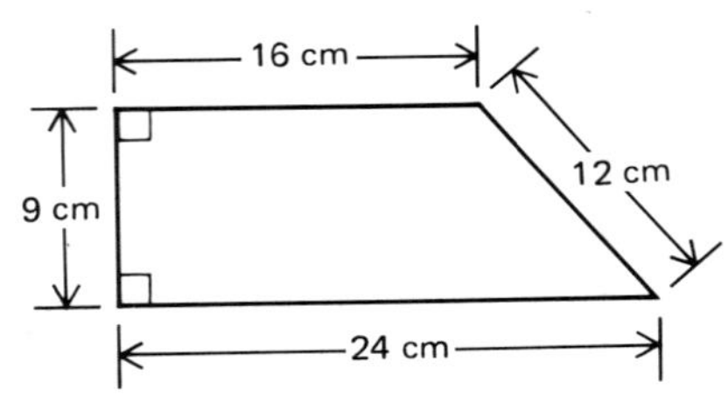

30. (a)

(b)

30. (nearest unit)

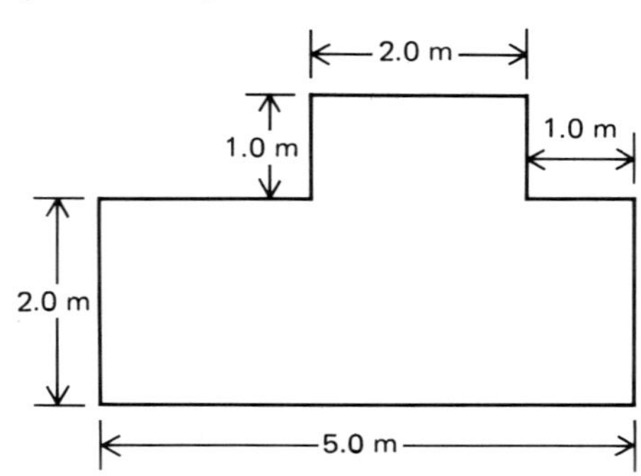

31. (nearest unit)

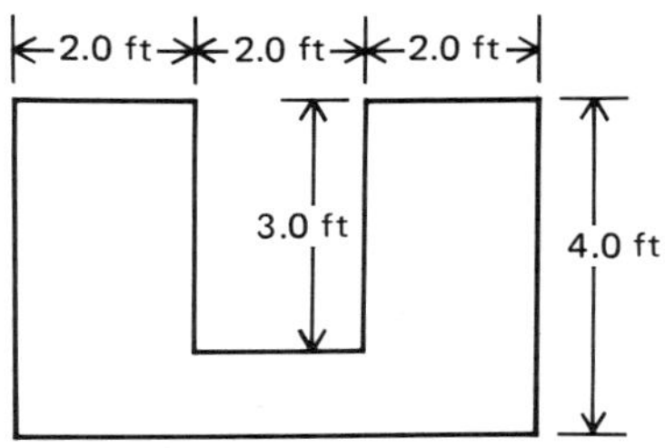

32. (nearest unit)

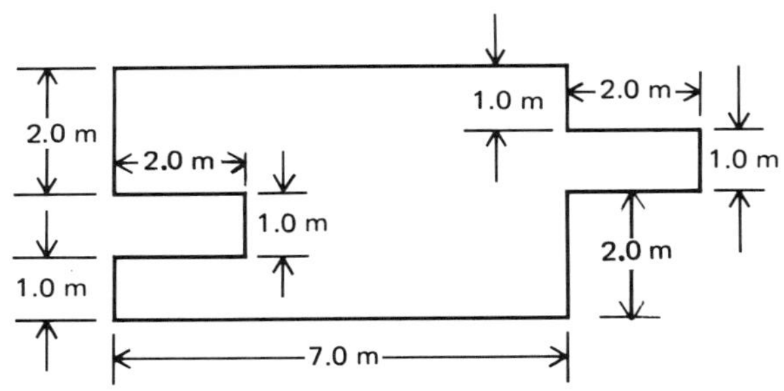

33. (nearest tenth for [a] ; nearest ten for [b])

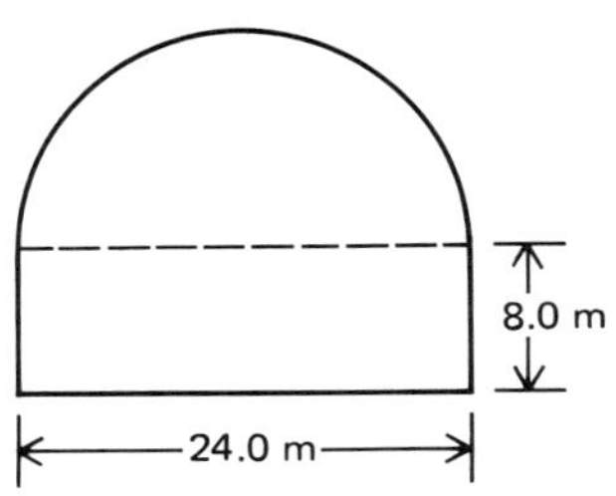

34. (nearest hundred)

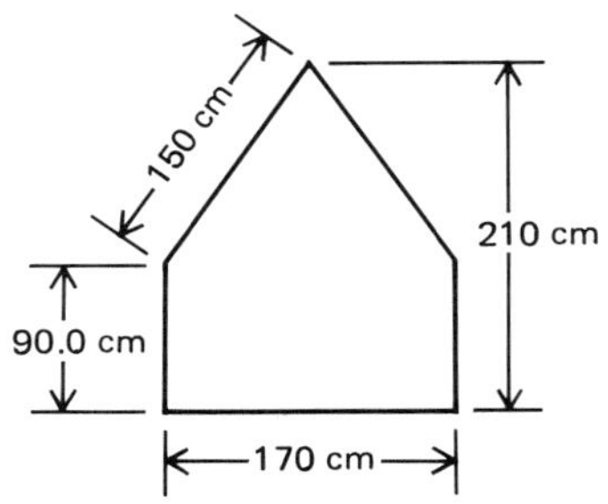

31. (a)

(b)

32. (a)

(b)

33. (a)

(b)

34. (a)

(b)

35. (a)

(b)

35. (nearest ten)

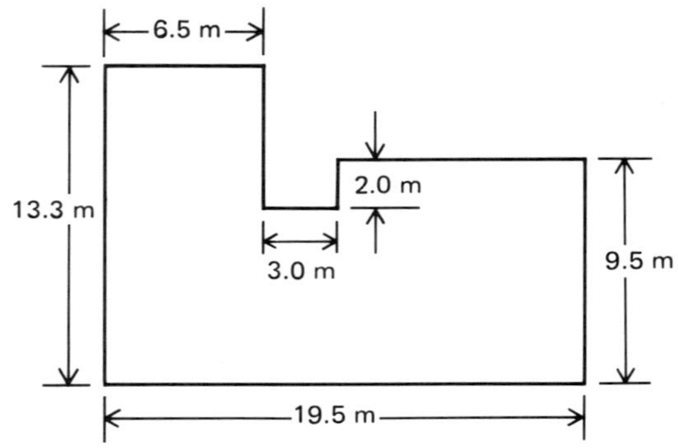

In exercises 36–40 find the area of the given figure, rounding off your answer as indicated. Use 3.14 for π.

36.

36. (nearest tenth)

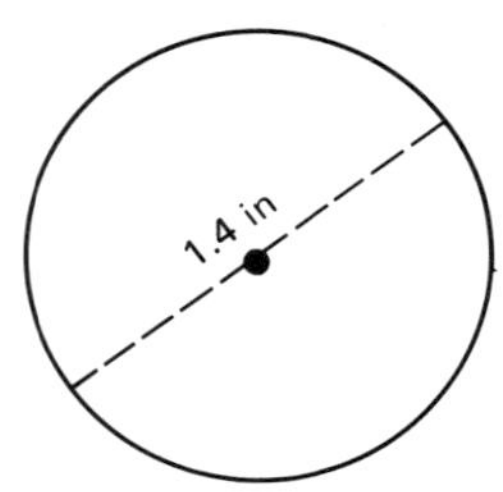

37.

37. (nearest ten)

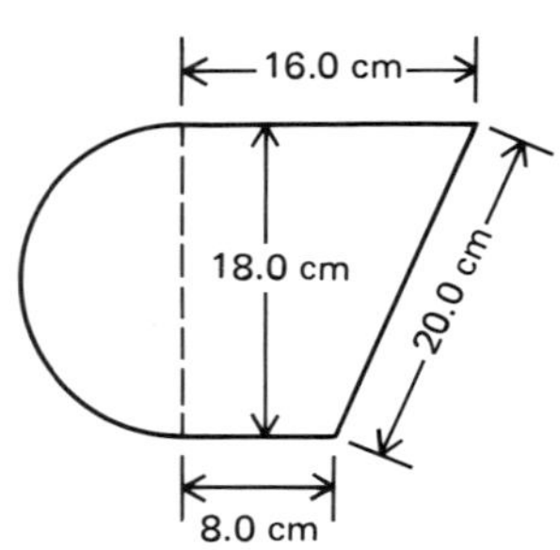

38.

38. (nearest ten)

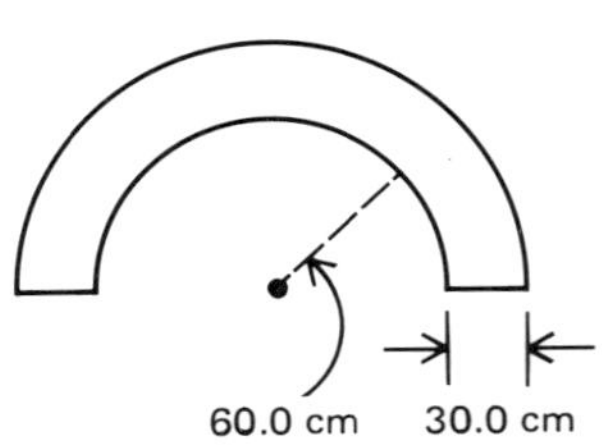

39. (nearest ten)

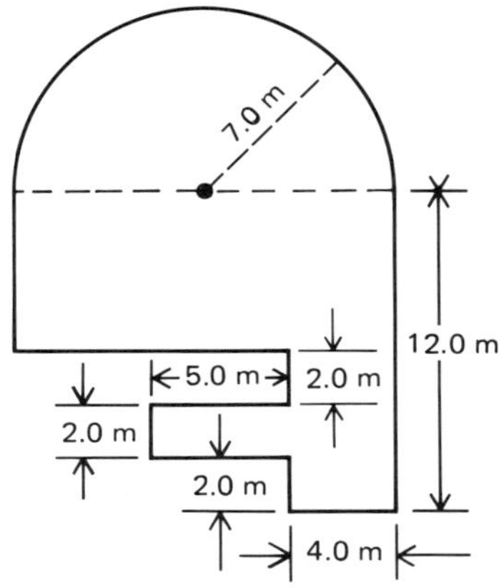

40. (nearest unit)

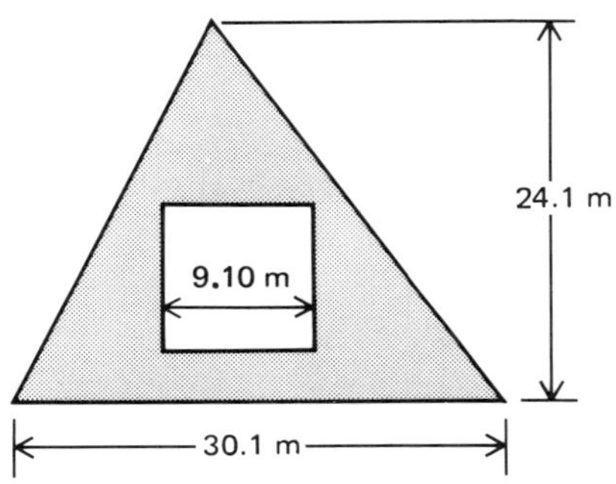

In exercises 41–58 round off your answer in a reasonable fashion.

41. The gymnasium floor at Kelly school is 72 feet wide and 95 feet long. Find (a) the area and (b) the perimeter of the floor.

42. How many square feet of carpet are required to cover a floor 33 feet long and 21 feet wide? How many square yards?

43. A new lawn is to be 12 meters long and 9 meters wide. If it requires a kilogram of lawn seed for every 30 square meters, how many kilograms of grass seed will be needed to sow the lawn?

39.

40.

41. (a)

(b)

42.

43.

44.

44. A garden is 18 m long and 15 m wide and is surrounded by a walk 3 m wide. What is the area of the walk?

45.

45. The stained glass in each of the four sides of a lamp shade is in the shape of a rectangle whose base is 24 cm and whose height is 32 cm. What is the cost of the glass in the shade if the glass sells for 50¢ per 100 cm^2?

46.

46. For a room 5 meters long and 3 meters wide, which will cost more and by how much–a 3-m by 4-m rug for \$196 or wall-to-wall carpeting at \$12 a square meter?

47.

47. A rectangle is 25 m by 12 m. Another rectangle has the same area as the first but its length is 20 m. What is its width?

48.

48. Two circular holes each 2.1 m in diameter are cut out of a 4.5-m by 2.7-m carpet. What area of carpet is left?

49.

49. Becky wishes to make a design in a triangular piece of leather. What is the cost of the piece of leather if it has an altitude of 24 cm and a base of 12 cm and leather costs \$2.22 for 100 cm^2?

50.

50. A square piece of plastic has an area of 36 cm^2. If cuts along both diagonals make four triangles of exactly the same size, what is the length of the base and the length of the altitude of each of these triangles?

51.

51. If a bicycle wheel is 71 cm in diameter, what is the distance it travels in one turn?

52. A rectangular field is 150 m by 100 m. A fence is placed from one corner A to the point M in the center of the longer opposite side as shown, dividing the original field into two smaller ones. Find the area of each of the smaller fields.

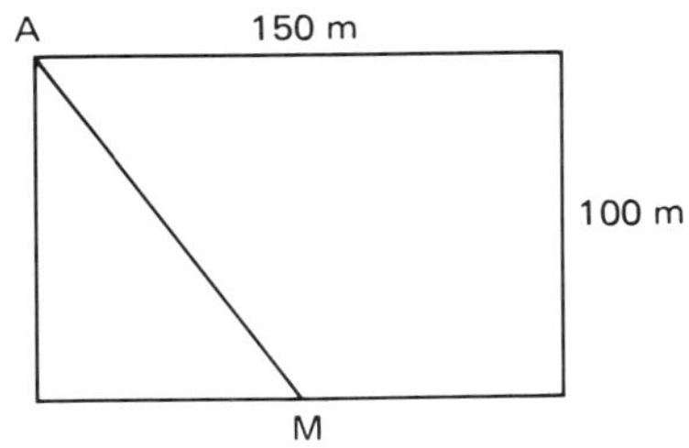

52.

53. A piece of wire 20.41 cm long is bent to form a circle. What is the diameter of the circle?

53.

54. A circular lampshade 36 cm in diameter needs new binding around the bottom. How long a strip of binding will be needed? (Disregard any need for overlap.)

54.

55. A merry-go-round on a playground has a radius of 4.5 m. If you sit on the edge of it, how far do you ride in ten turns around?

55.

56. The world's tallest living thing–a redwood tree in California–has a height of 367.8 feet and a girth (circumference) of 44 feet. What is its diameter?

56.

57. Which has the greater frying surface, a circular skillet 16 cm in diameter or a 14-cm by 14-cm square skillet?

57.

58. A square and a circle each have a perimeter of one meter. Which is greater, the area of the circle or the area of the square? How much greater?

58.

In exercises 59–64 find the volume of the given solid. Use 3.14 for π when needed and round off your answer as indicated.

59.

59. (nearest unit)

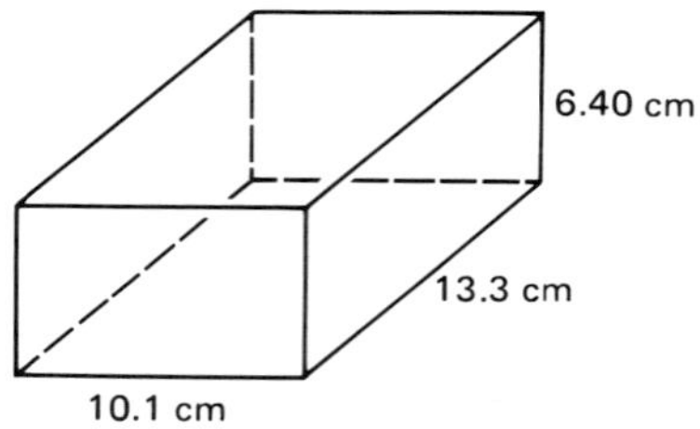

60.

60. (nearest ten)

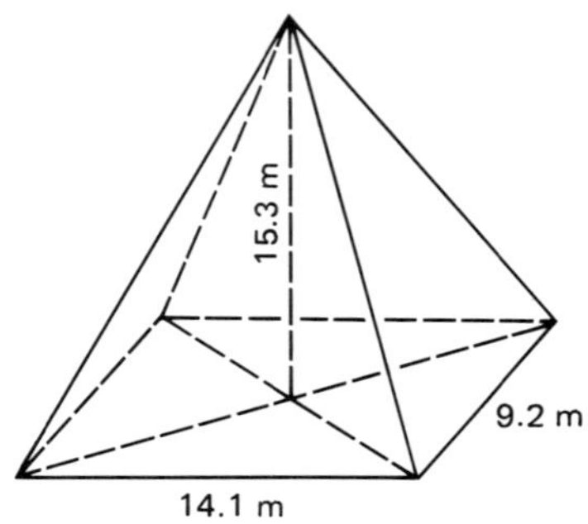

61.

61. (nearest hundred)

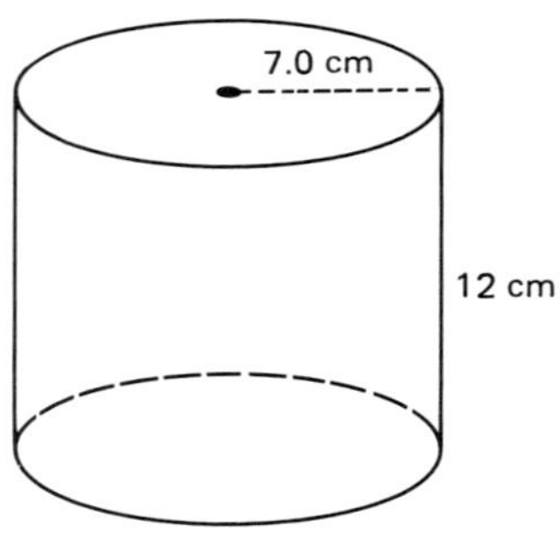

62.

62. (nearest hundred)

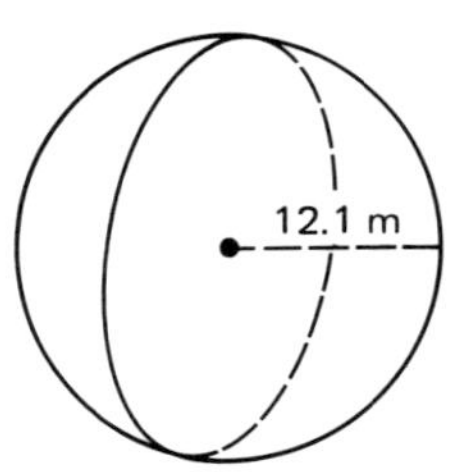

63. (nearest ten) 63.

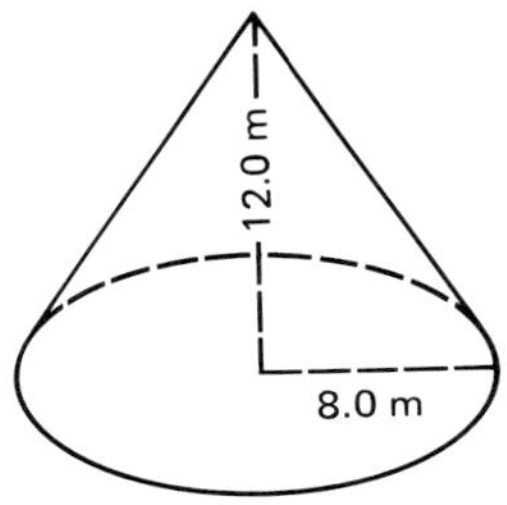

64. (nearest ten) 64.

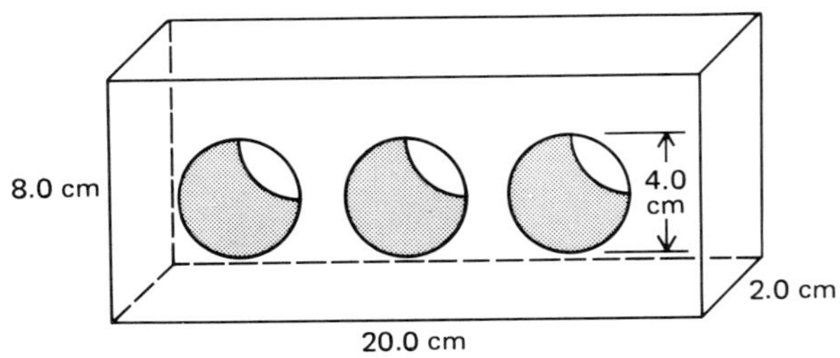

In exercises 65–73 round off your answer in a reasonable fashion.

65. A hot water tank is 2 feet in diameter and 49 inches high. How many gallons of water will it hold? (1 gallon ≈ 231 cubic inches) 65.

66. What is the inside length of a cylindrical railroad tank whose volume is 45,780 liters if the inside diameter is 2.13 meters? 66.

67. How many cubic meters of sand must be brought in to fill a dried up well 24 m deep and 61 cm in diameter? 67.

68. A buoy is made from two right circular cones with their bases fastened together. Each cone has a base of radius 2 m and a height of 2 m. What is the volume of the buoy? 68.

69.

69. A manufacturer of bolts orders the following: 1000 bars of 2-cm diameter steel, 4 m long; 1200 bars of 13-mm square steel, 3 m long; and 2000 bars of 3-cm square steel, 5 m long. If the density of steel is 7.79 g/cm^3, how much will the stock for the bolts cost at $75 per metric ton?

70.

70. How does the area of a rectangle change if both its length and its width are doubled?

71.

71. If both the base and the altitude of a triangle are doubled, what is the ratio of the area of the new triangle formed to the area of the given triangle?

72.

72. Answer the question in exercise 71 if the base and altitude are both tripled.

73.

73. What is the radius of a circle if the number of units in the circumference is the same as the number of units in the area?

Just for Fun

A. PATTERNS AND SEQUENCES

Write the next five terms of each sequence.

1. $\frac{2}{4}$, 6, $\frac{3}{4}$, 7, $\frac{4}{4}$, 8, $\frac{5}{4}$, ______, ______, ______, ______, ______

2. 6400, 3200, 1600, ______, ______, ______, ______, ______

3. 1, 2, 4, 8, 16, ______, ______, ______, ______, ______

4. A, D, G, J, ______, ______, ______, ______, ______

B. MORE PATTERNS

For those items indicated by an asterisk (*) you are to examine the pattern indicated by the previous computations and then write what you think is the result.

5.
$(1 \times 9) - 1 =$
$(21 \times 9) - 1 =$
$(321 \times 9) - 1 =$
*$(4321 \times 9) - 1 =$
*$(54321 \times 9) - 1 =$
*$(654321 \times 9) - 1 =$
*$(7654321 \times 9) - 1 =$

6.
$(1 \times 8) - 1 =$
$(21 \times 8) - 1 =$
$(321 \times 8) - 1 =$
*$(4321 \times 8) - 1 =$
*$(54321 \times 8) - 1 =$
*$(654321 \times 8) - 1 =$
*$(7654321 \times 8) - 1 =$

C. FIND THE MISSING DIGITS

In the following problems each letter represents a distinct digit. Find the numbers.

7. Add

```
      S E N D
(+)   M O R E
    ---------
    M O N E Y
```

8. Subtract

```
      H A M
(−)   A N D
    -------
      E G G
```

D. TRY THESE

9. A bookworm starts at page 1 of Volume I and bores its way in a straight line to the last page of Volume II. If the books are in order next to each other on the shelf, each cover is $\frac{1}{4}$ inch thick, and each book without the cover is 2 inches thick, how far does the bookworm travel?

10. Using the whole numbers 1 through 9, how many different ways can a sum of 15 be obtained using just three different numbers at a time? One example: $9 + 1 + 5 = 15$

11. Study the following pattern:

$$1 + 2 = 3$$
$$4 + 5 + 6 = 7 + 8$$
$$9 + 10 + 11 + 12 = 13 + 14 + 15$$

Verify each of the above equalities and write the next two equalities in the pattern. Will this pattern always hold?

E. MAGIC SQUARE

12. Complete this magic square:

$1\frac{1}{3}$	$\frac{1}{6}$	$\frac{1}{4}$	
$\frac{5}{12}$			$\frac{2}{3}$
$\frac{3}{4}$		$\frac{1}{2}$	
$\frac{1}{3}$	$1\frac{1}{6}$	$1\frac{1}{4}$	

F. TRIANGLE PUZZLE

13. Place the numbers 1 through 10 in the circles (using each one only once) so that the sum of rows of three are the same and the sum of rows of four are the same.

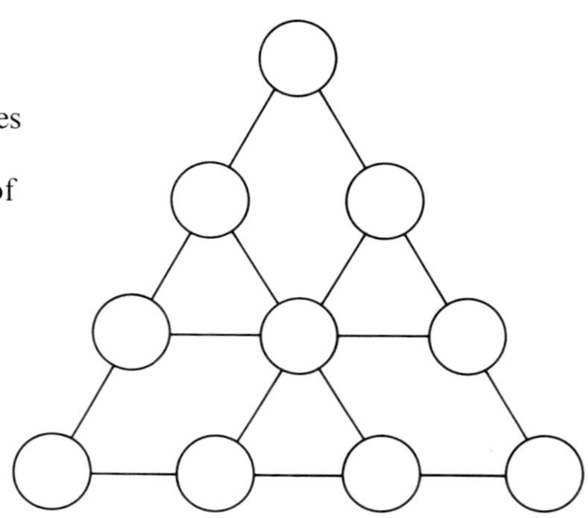

G. CALCULATOR GAME: "Divide and Conquer"

14. Play this game on your calculator.

	Sample Game:	*Display:*
1. Player A secretly multiplies a series of one- or two-digit numbers until a six- or seven-digit number appears in the display. Now he shows the calculator to B.	A enters 2 × 5 × 18 × 26 × 45	210600
2. B scores 1 point for each number he can divide into the number shown. If his division produces a decimal, the previous number is reset in the display and he loses 1 point. B continues until a prime number appears in the display.	B divides 210600 by 100 (210600 ÷ 100) and scores 1 point. (Note: He could have scored more points by dividing first by 2 or 5 rather than by 100.)	2106
If a prime number appears, he calls prime and scores 2 points. Player A may challenge the statement that it is a prime number.	B: 2106 ÷ 2 His score is now 2.	1053
If such a challenge is made and it turns out that the number is not prime, then player B loses 2 points and the round is over.	B: 1053 ÷ 9 (He could have scored more points by dividing by 3 instead of 9.) His score is now 3.	117
	B: 117 ÷ 3 Total score is now 4.	
If the challenger is not successful, he loses two points.	B: 39 ÷ 3 B calls prime. A does not challenge because he knows 13 is prime. B now gains 3 points and his total score is 7.	13
Player B now secretly multiplies a series of one- and two-digit numbers until a six- or seven-digit number appears in the display and play reverts to A.		

H. THE TALKING CALCULATOR

15. Perform the following calculation and use your calculator to find the word for a garland of flowers worn around the neck in Hawaii.

$$\left(\frac{21.3 \times 0.054}{97.4 \times 3.80}\right) + \left(\frac{560{,}000 \times 0.0045 \times 12{,}500}{1{,}050{,}000 \times 0.072}\right) - 279.66976$$

Mathematics and Athletics

BASEBALL, BASKETBALL, FOOTBALL

In reports of sports statistics, many times you find pairs or triples of numbers listed. Usually whenever a pair of numbers is listed, the first number indicates the number made or completed and the second number the number attempted.

Example: The report of a Seattle–Portland basketball game lists Walton: (12–23) and (2–3). This means Walton completed 12 out of 23 field goal attempts and 2 out of 3 free throw attempts.

Example: The report of a Buffalo–Detroit basketball game lists, for Detroit, the totals (47 12–15 106). This means Detroit made 47 field goals and scored on 12 out of 15 free throws for a total of 106 points.

In football, many times in a pair of numbers the first number indicates the number of attempts or times the ball is carried and the second number indicates the number of yards gained.

Example: The report of a Minnesota–Cincinnati football game lists, for Minnesota,

Rushes	27–69
Punts	8–41
Passes	25–43–3

In the case of rushes it means that 27 times the ball was carried for gains totaling 69 yards; for punts the 8–41 means they kicked 8 times for an average of 41 yards each time, while the 25–43–3 for passes indicated they completed 25 passes out of 43 attempts and had 3 passes intercepted.

Now consult the reports of games given on page 403 to do the following exercises.

A. BASKETBALL

1. In the Detroit–Buffalo basketball game, which team scored the highest percentage of free throws? What was the percentage of each team?

2. In the Chicago–Kansas City basketball game, what percentage of free throws did each team make?

3. In the Portland–Seattle basketball game, which player or players had the best percentage for free throws? What was it?

4. In the Portland–Seattle basketball game, which player or players had the best field goal percentage? What was it?

5. What was the team field goal percentage and the team free throw percentage for both Portland and Seattle?

6. Study the won-lost record of the NBA Western Conference and see if you can determine how the "PCT" and "GB" columns are calculated. Both these figures are determined by the won-lost record. PCT means *percentage of games won* while GB means *games behind the leader or first-place team.*) Note: For Portland, the newspaper report for the PCT was in error; it should be .701.) Compute the PCT and GB columns for the NBA Eastern Conference.

NBA EASTERN CONFERENCE

ATLANTIC DIVISION	W	L	PCT	GB
Philadelphia	18	12	______	______
Boston	16	13	______	______
N.Y. Knicks	15	15	______	______
Buffalo	13	19	______	______
N.Y. Nets	12	19	______	______
CENTRAL DIVISION				
Cleveland	19	11	______	______
Houston	17	10	______	______
New Orleans	17	16	______	______
San Antonio	16	16	______	______
Washington	13	16	______	______
Atlanta	12	22	______	______

B. FOOTBALL

7. Calculate the percentage of passes completed by the following teams: Minnesota, Cincinnati, New Orelans, Miami, San Francisco, Oakland.

8. Choose three teams from the six listed and complete the following table.

Team	Percentage of passes completed	Percentage of passes intercepted	Average yards per completed pass	Average yards per rush	Total yards punting	Average yards per penalty
(a)						
(b)						
(c)						

C. BASEBALL

9. In the Houston–Montreal baseball game, which player had the best batting average? What was it?

10. In the California–Cleveland game, which player had the best batting average? What was it?

11. The National League had seven players that batted .312 or above, while the American League had ten players batting .312 or above. What was the batting average of each of the first five players in each league?

12. Which team in each baseball league had the best batting average? What was it?

13. Study the won-lost record of the American League West teams and determine how the PCT and GB statistics are calculated. Then choose one of the other three divisions and calculate each team's PCT and GB statistics.

FOOTBALL

Cincinnati's defense flattens Vikings, 26-7

San Francisco	0	0	0	0	— 0
Oakland	13	10	7	3	— 33

Oak — FG Mann 24
Oak — FG Benirschke 47
Oak — Casper 20 pass from Stabler (Mann kick)
Oak — Davis 2 run (Benirschke kick)
Oak — FG Benirschke 38
Oak — Casper 13 pass from Stabler (Mann kick)
Oak — FG Mann 27

	49ers	Raiders
First downs	5	29
Rushes-yards	20-44	56-221
Passing yards	40	201
Return yards	24	56
Passes	5-15-0	17-30-0
Punts	9-35	3-39
Fumbles-lost	2-1	0-0
Penalties-yards	5-40	7-50

INDIVIDUAL LEADERS

RUSHING — San Francisco: Jackson 10-23, Ferrell, 3-15; Oakland: vanEeghen 12-55, Davis 13-54, Robiskie 10-35.

PASSING — San Francisco: Plunkett 5-15-0, 87 yards. Oakland; Stabler 13-23-0, 184; Rae 4-7-0, 47.

RECEIVING — San Francisco: Ferrell 2-11, Obradovich 1-30. Oakland: Casper 5-91, Biletnikoff 3-30, Davis 2-26.

A — 51,728

New Orleans	0	7	7	3	— 17
Miami	0	3	7	0	— 10

NO — Muncie 6 run (Szaro kick)
Mia — FG Yepremian 43
NO — Muncie 8 run (Szaro kick)
Mia — L. Harris 2 run (Yepremian kick)
NO — FG Szaro 39

	Saints	Dolphins
First downs	18	16
Rushes-yards	38-126	31-130
Passing yards	219	132
Return yards	79	132
Passes	15-21-0	15-27-4
Punts	4-28	1-38
Fumbles-lost	2-2	3-2
Penalties-yards	3-30	4-20

INDIVIDUAL LEADERS

RUSHING — New Orleans: Muncie 12-24, Galbreath 11-41. Miami: Davis 9-52, Malone 8-33.

PASSING — New Orleans: Scott 15-21-0-219. Miami: Griese, 6-12-1-62, Strock 9-15-3-70.

RECEIVING —New Orleans: Thaxton 5-131, Galbreath 3-31. Miami: N. Moore 4-32, D. Harris 1-25.

A — 31.704

Minnesota	0	7	0	0	— 7
Cincinnati	0	9	10	7	— 26

Cinn — FG Bahr 48
Minn — White 21 pass from Lee (Danmeier kick)
Cinn — 57 pass interception (kick failed)
Cinn — Davis 4 run (Bahr kick)
Cinn — FG Bahr 41
Cinn — Anderson 95 pass interception (Bahr kick)

	Vikings	Bengals
First Downs	19	14
Rushes-yards	27-69	41-189
Passing yards	181	21
Return yards	154	205
Passes	25-43-3	9-22-3
Punts	8-41	7-43
Fumbles-lost	3-2	2-1
Penalties-yards	6-54	10-84

INDIVIDUAL LEADERS

RUSHING — Minnesota: Kellar 15-36, McClanahan 6-19. Cincinnati: Griffin 12-64, Clark 9-51, Davis 9-28.

PASSING — Minnesota: Lee 13-25, 163, Kramer 7-18, 94. Cincinnati: Anderson 5-13, 19, Reaves 4-9, 18.

RECEIVING — Minnesota: Rashad 4-48, Kellar 6-35, Hartwig 2-35. Cincinnati: Curtis 4-27, Clark 2-7.

A — 52,942.

BASKETBALL

NBA standings

WESTERN CONFERENCE

Pacific Division

	W	L	Pct.	GB
PORTLAND	22	9	.710	—
Los Angeles	19	13	.594	3½
Golden State	15	14	.517	6
Phoenix	13	15	.464	7½
Seattle	15	18	.455	8

Midwest Division

	W	L	Pct.	GB
Denver	21	9	.700	—
Detroit	19	14	.576	3½
Indiana	15	18	.455	7½
Kansas City	14	19	.424	8½
Chicago	11	16	.407	7½
Milwaukee	7	25	.219	15

EASTERN CONFERENCE

Atlantic Division

	W	L
Philadelphia	18	12
Boston	16	13
NY Knicks	15	15
Buffalo	13	19
NY Nets	12	19

Central Division

	W	L
Cleveland	19	11
Houston	17	10
New Orleans	17	16
San Antonio	16	16
Washington	13	16
Atlanta	12	22

Braves, 115-106

BUFFALO (AP) — Ernie DiGregorio ignited an eight-point Buffalo burst late in the fourth quarter that carried the Braves to a 115-106 National Basketball Association victory over the Detroit Pistons.

DETROIT 106 — Carr 23, H. Porter 8, Lanier 15, K. Porter 5, Ford 19, Eberhard 4, Douglas, Simpson 22, Money 10, Barnes, Sellers. Totals — 47 12-15 106.

BUFFALO 115 — Dantley 30, Gianelli 2, Shumate 17, DiGregorio 16, Smith 24, Adams 16, Williams 6, Foster 2, Abdul Aziz, Terry, Gerard 1 0-0 2. Totals — 46 23-26 115.

Detroit	25	30	28	23—106
Buffalo	35	25	23	32—115

Fouled out — Carr. Total fouls — Detroit 25, Buffalo 15, A — 6,141.

CHICAGO 96 — May 18, M. Johnson 8, Gilmore 8, Van Lier 9, Holland 22, Pondexter 12, Mengelt 15, Boerwinkle 2, Laskowski 2, Starr, Kropp. Totals 35 26-33 96.

KANSAS CITY 91 — Washington 20, Wedman 12, Lacey 21, Boone 6, Barr 6, Taylor 12, Robinzine 6, Eakins 8, Hansen, O. Johnson. Totals 36 19-21 91.

Chicago	20	28	25	23 — 96
Kansas City	20	23	25	23 — 91

Fouled out — M. Johnson, Gilmore. Total fouls — Chicago 23, Kansas City 27. Technical fouls — Robinzine, Eakins, Kansas City coach Johnson. A — 9,576

SEATTLE (95)

	MP	FG	FT	R	A	P	TP
Bantom	31	5-9	2-3	11	2	4	12
Seals	15	0-4	0-0	0	1	1	0
Burleson	26	3-12	4-4	7	0	3	10
Oleynick	34	9-14	1-1	1	1	0	19
Watts	40	4-13	1-2	3	6	2	9
Norwood	32	8-17	2-2	4	2	3	6
Wilkerson	17	2-4	2-2	4	2	3	6
Green	24	3-6	2-2	6	1	2	8
Weatherspoon	16	6-11	1-2	5	0	1	13
Johnson	5	0-0	0-0	0	0	1	0
Team rebounds				5			
Totals	240	40-90	15-18	48	14	19	95

PORTLAND (110)

	MP	FG	FT	R	A	P	TP
Gross	29	5-7	2-2	10	6	2	12
Lucas	41	10-16	4-4	14	4	3	24
Walton	35	12-23	2-3	17	2	4	26
Hollins	30	2-12	0-1	4	6	5	4
Twardzik	23	2-5	0-0	4	5	3	4
Davis	19	3-9	0-0	4	3	1	6
Calhoun	3	0-1	0-0	0	0	0	0
Gilliam	24	6-11	0-0	2	3	0	12
Neal	20	5-11	4-9	4	1	1	14
Steele	15	4-6	0-0	0	3	2	8
Walker	1	0-0	0-0	0	0	0	0
Team rebounds				5			
Totals	240	49-101	12-19	64	33	21	110

Seattle	24	29	21	21— 95
Portland	17	29	37	27—110

Team turnovers — Seattle 21. Portland 19.
Technical fouls — Lucas Changing on rim
Attendance — 12,480
Officials — Powers, Rooney

BASEBALL

American League

WEST

	W	L	Pct.	GB
Kansas City	79	54	.594	—
Chicago	74	57	.565	4
Texas	74	59	.556	5
Minnesota	75	61	.551	5½
California	62	69	.473	16
Oakland	52	80	.394	26½
Seattle	54	83	.394	27

EAST

	W	L	Pct.	GB
New York	82	52		
Boston	77	56		
Baltimore	76	56		
Detroit	65	68		
Cleveland	63	72		
Milwaukee	58	83		
Toronto	45	86		

HOUSTON	ab	r	h	bi	MONTREAL	ab	r	h	bi
Puhl lf	4	0	0	0	Cash 2b	4	1	2	0
Cabell 3b	4	0	0	0	Crmrte lf	3	1	1	0
Cedeno cf	3	1	1	0	Unser lf	1	0	0	0
JCruz rf	4	0	0	0	Dawson cf	4	0	0	0
Fergsn c	4	0	2	1	Perez 1b	4	0	1	1
Watson 1b	4	0	2	0	Carter c	3	2	2	1
Howe ss	3	0	1	0	Mejias rf	4	0	0	0
Fschln ss	1	0	0	0	Parrish 3b	2	1	2	3
JGnzlz 2b	3	0	0	0	Speier ss	3	0	1	0
Banistr p	2	0	0	0	Rogers p	3	0	0	0
Pentz p	0	0	0	0					
Boswell ph	1	0	0	0					
Andiar p	0	0	0	0					
Total	33	1	6	1	**Total**	31	5	9	5

Houston	000 100 000	— 1
Montreal	200 201 00x	— 5

E—Ferguson. DP—Houston 2. LOB—Houston 6, Montreal 4. 2B—Cash, Perez, Cedeno, Carter, Parrish. 3B—Ferguson. HR—Parrish (10). SB—Cedeno, Parrish.

	IP	H	R	ER	BB	SO
Bannistr (L,5-8)	5 1-3	9	5	5	1	4
Pentz	1 2-3	0	0	0	0	0
Anduiar	1	0	0	0	1	0
Rogers (W,15-13)	9	6	1	1	0	4

HBP—By Rogers (Cedeno). T—2:08. A—16,222.

By The Associated Press
Complete through games of Friday

AMERICAN LEAGUE
TEAM BATTING

	AB	R	H	HR	RBI	Pct
New York	4605	694	1303	150	653	
Minnesota	4726	746	1329	109	690	.281
Chicago	4534	699	1270	168	674	.280
Boston	4496	693	1254	183	670	.279
Kansas City	4545	655	1235	115	616	.272
Cleveland	4585	581	1244	94	543	.271
Texas	4600	619	1223	105	569	.266
Detroit	4551	592	1194	147	562	.262
Milwaukee	4783	554	1235	102	515	.258
California	4366	565	1126	109	535	.258
Toronto	4403	509	1119	84	465	.254
Seattle	4552	510	1157	109	487	.254
Baltimore	4440	556	1128	112	522	.254
Oakland	4292	482	1017	92	434	.237

INDIVIDUAL BATTING
275 or more at bats

	AB	R	H	HR	RBI	Pct
Carew Min	522	107	196	12	83	
Bostock Min	504	93	170	11	77	
Rivers NY	475	64	157	10	61	
Singleton Bal	425	72	140	19	78	
LeFlore Det	538	87	172	16	51	
Bailor Tor	435	59	139	5	29	.320
Rice Bos	535	82	170	33	90	.318
LJohnson Chi	296	41	94	16	54	.318
Hargrove Tex	436	78	136	13	53	.312
GBrett KC	459	81	143	17	67	.312

National League

EAST

	W	L	Pct.	GB
Philadelphia	59	27		—
Pittsburgh	48	38		
New York	48	45		
St. Louis	40	48		
Chicago	36	53		
Montreal	26	57		

WEST

	W	L	Pct.	GB
Cincinnati	57	34		—
Los Angeles	50	41		
Houston	47	45		
San Diego	45	47		
Atlanta	42	48		
San Francisco	39	54		

CALIFORNIA	ab	r	h	bi	CLEVELAND	ab	r	h	bi
Bosley cf	5	1	2	0	Dade rf	3	0	0	0
Remy 2b	4	0	2	1	Kuiper 2b	3	0	0	0
Bonds rf	4	0	0	0	BBell 3b	3	0	2	0
Solaita 1b	3	0	1	0	Thortn 1b	4	0	0	0
Briggs 1b	0	0	0	0	Bochte lf	3	0	1	0
Baylor dh	4	0	2	0	Maning cf	0	0	0	0
Mlinks ss	4	0	1	0	Carty dh	4	0	0	0
Chalk 3b	4	1	1	0	Pruitt lf	3	1	1	0
Flores lf	3	1	2	0	Blanks ss	3	1	1	1
Hmphy c	2	0	0	0	Duffy ss	1	0	0	0
Gurero ph	1	0	1	0	Kendall c	1	0	1	1
Echbrn c	0	0	0	1	JNorris ph	1	0	0	0
Total	34	3	12	2	**Total**	29	2	6	2

California	000 000 201	— 3
Cleveland	020 000 000	— 2

E—Bosley, Humphrey, Garland, Kendall. DP—Cleveland 2. LOB—California 7, Cleveland 8. 2B—Baylor, BBell, Chalk. SB—Dade, Blanks, Remy, Duffy. S— BBell, Flores. SF—Kendall, Etchbrren.

	IP	H	R	ER	BB	SO
Ryan (W,18-13)	9	6	2	0	6	7
Garland (L10-17)	9	12	3	2	1	2

T—2:30. A—11,054.

NATIONAL LEAGUE
TEAM BATTING

	AB	R	H	HR	RBI	Pct
Philadelphia	4500	677	1247	156	638	
Cincinnati	4605	684	1260	159	640	.274
St. Louis	4539	592	1234	75	550	.272
Pittsburgh	4717	623	1282	117	578	.272
Chicago	4567	594	1232	97	561	.270
Los Angeles	4606	640	1230	154	608	.267
Montreal	4671	554	1214	113	518	.260
SanFrancisco	4578	562	1164	107	517	.254
Atlanta	4531	544	1132	112	512	.250
Houston	4542	538	1135	91	505	.250
San Diego	4685	592	1167	102	561	.249
New York	4452	466	1079	82	48	.242

INDIVIDUAL BATTING
275 or more at bats

	AB	R	H	HR	RBI	Pct
Parker Pgh	543	93	189	18	81	
Stennett Pgh	453	53	152	5	51	
Simmons StL	430	69	142	17	81	
Griffey Cin	493	95	158	10	47	
Tmpleton StL	509	77	163	6	58	
Smith LA	418	91	132	25	75	.316
Luzinski Phl	455	78	142	3mp	110	.312

7

More on Computations with Whole Numbers and Decimals

7.1 Scientific Notation

Scientific notation is useful for many purposes and, indeed, is a built-in feature of higher priced calculators. For example, such a (eight-place display) calculator will record the answer to

$$29{,}328 \times 3642$$

as

$$1.0681258 \times 10^8$$

instead of simply indicating an overflow. This means that

$$29{,}328 \times 3642 \approx 1.10681258 \times 100{,}000{,}000 = 106{,}812{,}580$$

When 106,812,580 is written as 1.0681258×10^8 it is said to be written in *scientific notation*. (Note: $29{,}328 \times 3642 = 106{,}812{,}576$.)

In general, when a number is written as the product of

(1) a number that is greater than or equal to 1 and less than 10, and

(2) a power of 10

it is said to be written in scientific notation. (Any whole number or decimal that is greater than or equal to 1 and less than 10 is already written in scientific notation—as, for example, 2.67.) Here are some other examples:

Given number	In scientific notation
8.79	8.79
879	8.79×10^2
26.7	2.67×10
2157.0002	2.1570002×10^3

Do exercises 1 through 10 in the adjoining column.

OBJECTIVES

You should:

(a) Be able to use scientific notation.

(b) Understand the rules for multiplying powers of 10.

****

In exercises 1–10 write the given number in scientific notation.

1. 595.58

2. 37.089

3. 2581.5

4. 4.365

5. 506.27

6. 324.783

7. 14.523

8. 9.89

9. 69,916

10. 772.15

STOP

In exercises 11–20 write the given number in scientific notation.

11. 0.7

12. 0.07

13. 0.007

14. 0.421

15. 0.0421

16. 0.00421

17. 0.000421

18. 0.3756

19. 0.00016877

20. 0.0000000588

STOP

In exercises 21–24 write the given expression as a power of 10.

21. $10^3 \times 10^2$

22. $10^4 \times 10$

23. $10^3 \times 10^4$

24. $10^5 \times 10^3$

STOP

All of our examples so far have dealt with numbers greater than 1. What about a number less than 1, such as 0.3? We know, of course, that

$$0.3 = \frac{3}{10}$$

but we can also use the definition that

$$10^{-1} = \frac{1}{10}, \qquad 10^{-2} = \frac{1}{10^2}, \qquad 10^{-3} = \frac{1}{10^3}$$

etc., to write

$$0.3 = \frac{3}{10} = 3 \times \frac{1}{10} = 3 \times 10^{-1}$$

$$0.034 = \frac{3.4}{100} = \frac{3.4}{10^2} = 3.4 \times \frac{1}{10^2} = 3.4 \times 10^{-2}$$

$$0.000572 = \frac{5.72}{10{,}000} = \frac{5.72}{10^4} = 5.72 \times \frac{1}{10^4} = 5.72 \times 10^{-4}$$

Then the scientific notation

for 0.3 is 3×10^{-1},

for 0.034 is 3.4×10^{-2},

for 0.000572 is 5.72×10^{-4},

etc.

Do exercises 11 through 20.

In the next sections we will find it useful to know rules for multiplying and dividing powers of 10. Considering the following examples should enable you to determine such rules. First consider

$$10^2 \times 10^3 = (10 \times 10) \times (10 \times 10 \times 10) = 10^5 = 10^{2+3}$$

$$10 \times 10^5 = 10^1 \times 10^5 = 10 \times (10 \times 10 \times 10 \times 10 \times 10)$$

$$= 10^6 = 10^{5+1}$$

Do exercises 21 through 24.

Next consider

$$10^{-2} \times 10^{-3} = \frac{1}{10^2} \times \frac{1}{10^3} = \frac{1}{10^2 \times 10^3} = \frac{1}{10^5}$$

$$= 10^{-5} = 10^{-(2+3)}$$

Do exercises 25 through 30.

In exercises 25–30 write the given expression as a power of 10.

25. $10^{-3} \times 10^{-2}$

26. $10^{-5} \times 10^{-3}$

27. $10^{-1} \times 10^{-1}$

28. $10^{-10} \times 10^{-6}$

29. $10^{-9} \times 10^{-2}$

30. $10^{-7} \times 10^{-4}$

STOP

Continuing our examples of multiplication of powers of 10, we have

$$10^5 \times 10^{-3} = 10^5 \times \frac{1}{10^3} = \frac{10^5}{10^3} = \frac{10 \times 10 \times 10 \times 10 \times 10}{10 \times 10 \times 10}$$

$$= 10 \times 10 = 10^2 = 10^{5-3}$$

$$10^{-5} \times 10^3 = \frac{1}{10^5} \times 10^3 = \frac{10^3}{10^5} = \frac{10 \times 10 \times 10}{10 \times 10 \times 10 \times 10 \times 10}$$

$$= \frac{1}{10 \times 10} = \frac{1}{10^2} = 10^{-2} = 10^{-(5-3)}$$

$$10^{-3} \times 10^3 = \frac{1}{10^3} \times 10^3 = \frac{10^3}{10^3} = 1$$

Note that in the last example, we can also write our answer as

$$10^{3-3} = 10^0$$

if we define 10^0 as 1.

Do exercises 31 through 38.

In exercises 31–38 write the given expression as a power of 10.

31. $10^3 \times 10^{-2}$

32. $10^7 \times 10^{-3}$

33. $10^5 \times 10^{-2}$

34. $10^{13} \times 10^{-9}$

35. $10^{-5} \times 10^2$

36. $10^{-4} \times 10^3$

37. $10^{-9} \times 10^5$

38. $10^9 \times 10^{-7}$

STOP

For division we can simply note that we can convert any division problem involving powers of 10 into a multiplication problem if we notice that, just as

$$10^{-1} = \frac{1}{10}, \quad 10^{-2} = \frac{1}{10^2}, \quad 10^{-3} = \frac{1}{10^3},$$

etc., so

$$\frac{1}{10^{-1}} = \frac{1}{\frac{1}{10^1}} = 1 \div \frac{1}{10^1} = 1 \times \frac{10^1}{1} = 10$$

In exercises 39–46 write the given expression as a power of 10.

39. $10^2 \div 10^1$

40. $10^4 \div 10^2$

41. $10^7 \div 10^3$

42. $10^7 \div 10^5$

43. $10^{-7} \div 10^7$

44. $10^3 \div 10^{-5}$

45. $10^{-3} \div 10^{-5}$

46. $10^{-3} \div 10^{-3}$

STOP

Similarly,

$$\frac{1}{10^{-2}} = 10^2, \quad \frac{1}{10^{-3}} = 10^3$$

etc. Thus, for example,

$$10^5 \div 10^3 = \frac{10^5}{10^3} = 10^5 \times 10^{-3} = 10^2$$

$$10^3 \div 10^5 = \frac{10^3}{10^5} = 10^3 \times 10^{-5} = 10^{-2}$$

$$10^2 \div 10^{-3} = \frac{10^2}{10^{-3}} = 10^2 \times 10^3 = 10^5$$

$$10^{-2} \div 10^{-3} = \frac{10^{-2}}{10^{-3}} = 10^{-2} \times 10^3 = 10^1 = 10$$

Do exercises 39 through 46.

Exercises 7.1

Name ______________________

Class ______________________

In exercises 1–18 write the given number in scientific notation.

1. 689
2. 5319
3. 259,000
4. 2 million
5. 3,240,000
6. 1 billion
7. 19,400,000
8. 67,300
9. 0.0403
10. 0.0033
11. 0.00067
12. 0.00000093
13. 0.15
14. 0.6304
15. 15 thousandths
16. 0.005
17. 0.00105
18. 0.0607

1.
2.
3.
4.
5.
6.
7.
8.
9.
10.
11.
12.
13.
14.
15.
16.
17.
18.

In exercises 19–35 write the given number in decimal notation.

19.	19. 6.5×10^4
20.	20. 1.92×10^3
21.	21. 6.5×10^{-4}
22.	22. 1.92×10^{-3}
23.	23. 7.01×10^3
24.	24. 3.106×10^6
25.	25. 2.54×10^5
26.	26. 4.061×10^{-4}
27.	27. 2.636×10^{-2}
28.	28. 1.0×10^{-2}
29.	29. 3.047×10^4
30.	30. 6.25×10^7
31.	31. 2.104×10^{-3}
32.	32. 8.009×10^{-2}
33.	33. 9.9999×10^8
34.	34. 2.54×10^1
35.	35. 6.6×10^{-2}

In exercises 36–53 write the given expression as a power of 10.

36. $10^2 \times 10^3$

37. $10^3 \times 10^5$

38. $10^1 \times 10^0$

39. $10^4 \times 10^3$

40. $10^{-2} \times 10^3$

41. $10^3 \times 10^{-2}$

42. $10^{-2} \times 10^{-1}$

43. $10^4 \times 10^{-5}$

44. $10^5 \times 10^7$

45. $10^4 \times 10^{-3}$

46. $10^{12} \times 10^{13}$

47. $10^{12} \times 10^{-3}$

48. $10^{-12} \times 10^{13}$

49. $10^{-12} \times 10^{-3}$

50. $10^7 \times 10^5$

51. $10^7 \times 10^{-5}$

52. $10^{-7} \times 10^5$

53. $10^{-7} \times 10^{-5}$

In exercises 54–61 change any numbers in decimal notation to scientific notation and change any numbers in scientific notation to decimal notation.

54. Some computers can perform several addition problems in 4.5×10^{-6} seconds.

55. The diameter of the sun is 8.64×10^5 miles.

56. One foot of steel pipe will increase about 0.000011 foot for a one-degree Celsius rise in temperature.

57. A steel rod will expand 0.000024 inch for each inch of length when the temperature is increased one degree Fahrenheit.

58. It is estimated that 7,500,000,000,000,000 BTU's of energy were used in the United States in 1900.

59. In 1976 Americans used 63,300,000,000,000,000 BTU's of energy.

36.	37.
38.	39.
40.	41.
42.	43.
44.	45.
46.	47.
48.	49.
50.	51.
52.	53.
54.	
55.	
56.	
57.	
58.	
59.	

60.

60. The amount of surface water on the earth provided by the oceans is 3.487×10^{20} gallons.

61.

61. The amount of surface water on the earth provided by the atmosphere is 3.41×10^{17} gallons.

7.2 Handling Overflow in Multiplication

In section 1.6 we discussed how overflow in a multiplication problem can be handled on a simple HHC. Now on an HHC with scientific notation, overflow is handled directly by simply getting an approximation for such products. Thus, for example, where we obtained the answer of 127,176,588 to the multiplication

$$85{,}239 \times 1492$$

an HHC with scientific notation and an eight-digit display would record

$$1.2717658 \times 10^8$$

Since

$$1.2717658 \times 10^8 = 127{,}176{,}580$$

we see that this is an approximation to 85,239 × 1492 that differs from the exact answer by only 8.

Without an HHC with scientific notation, it is still possible to obtain this approximate answer of 127,176,580 with little effort by using scientific notation to re-write the problem. Thus we write

$$85{,}329 \times 1492 = (8.5329 \times 10^4) \times (1.492 \times 10^3)$$

which, by the commutative and associative properties of multiplication, is equal to

$$(8.5329 \times 1.492) \times (10^4 \times 10^3)$$

Now, using an HHC to multiply 8.5329 by 1.492, we have, since $10^4 \times 10^3 = 10^7$,

$$85{,}239 \times 1492 \approx 12.717658 \times 10^7 = 127{,}176{,}580$$

(Note: Since 127,176,588 rounds off to 127,176,590 to the nearest ten, one would expect to get 1.2717659 whether or not the HHC has scientific notation. Indeed, on some HHC's the latter answer will be obtained; as mentioned before, the differing internal logic of HHC's can easily result in differences in the last digit of an eight-digit answer.)

Do exercises 1 and 2 in the adjoining column.

OBJECTIVE

You should be able to find approximate answers to multiplication problems that cause your HHC to overflow.

****

1. To find an approximation to

   984,502 × 67,474

   on an HHC without scientific notation, we write (fill in the boxes and blanks):

   $984{,}502 \times 67{,}474 =$

   $(9.84502 \times 10^{\square})$

   $(6.7474 \times 10^{\square})$

   Then, since an HHC gives

   $9.84502 \times 6.7474 \approx$

   ____

   we have

   $984{,}502 \times 67{,}474$

   $\approx (\_\_\_\_ \times 10^{\square}) \times 10^{\square}$

   $\approx \_\_\_\_\_\_ \times 10^{\square}$

   $\approx$ ____________

2. Use scientific notation and the procedure shown in the text and exercise 1 to find an approximate answer to the multiplication:
   325,337 × 65,201

STOP

Exercises 7.2

Name ______________________

Class ______________________

In these exercises use scientific notation to find an approximate answer to the problem.

Example:

$$20{,}512 \times 314{,}269 = (2.0512 \times 10^4) \times (3.14269 \times 10^5)$$
$$= (2.0512 \times 3.14269) \times (10^4 \times 10^5)$$
$$\approx 6.4462857 \times 10^9 = 6{,}446{,}285{,}700$$

1. 205,123 × 31,426

2. 66,051 × 74,702

3. 40,076 × 768,953

4. 36,604 × 361,006

5. 658,135 × 398,808

6. 110,231,542 × 2917

7. 45,508 × 8,240,013

8. 353,227 × 42,614,326

1.

2.

3.

4.

5.

6.

7.

8.

9.

9. 94,223 X 60,124 X 1567

10.

10. 3146 X 14,142 X 17,321

11.

11. The circumference of the earth at the equator is 24,901 miles. What is this circumference in feet? (1 mile = 5280 feet)

12.

12. The area of the earth is about 196,938,800 square miles. What is its area in acres? (1 square mile = 640 acres)

13.

13. In 1972 in the U.S.A., 2,378,500 new housing units were begun. If the average cost of each housing unit was $20,846, what was the value of all of the new housing?

14.

14. In 1972 in the U.S.A., 8,823,938 new cars were sold at an average price of $2621. Find the total value of the new car sales in that year.

15.

15. Light that left the star Arcturus in 1893 was used to operate a photo-electric cell and turn on the lights for the World's Fair in Chicago in 1933. If light travels at 186,000 miles per second, how far away is Arcturus from the earth? (Assume 365 days in each year.)

7.3 Some Peculiarities in Computations with Decimals on an HHC

We said previously that, except for pressing the decimal point key and sometimes getting an excess number of decimal places in the answer, there is no real difference between working with decimals and working with whole numbers on an HHC. This is really not completely true, although it has been true of the computations with decimals that we have considered so far.

Do exercise 1 in the adjoining column.

Unless your HHC possesses scientific notation, it will have recorded an answer of 0 to exercise 1! Why?

Do exercise 2.

For the first multiplication in exercise 2 you should have obtained an answer of 0.0276; for the second, 0.000276; for the third, 0.0000027 ("losing" the "6"); and for the fourth, 0.0000002 (losing all but the "2"). Clearly, all that has happened in exercise 1 is that the HHC has run out of space!

What to do? We simply write

$$0.000012 \times 0.0023 = (1.2 \times 10^{-5}) \times (2.3 \times 10^{-3})$$

$$= (1.2 \times 2.3) \times (10^{-5} \times 10^{-3})$$

$$= 2.76 \times 10^{-8}$$

for the answer in scientific notation (and this would be the form of the answer shown on an HHC with scientific notation if the problem were done directly on it). Written as a decimal, our answer would be 0.0000000276.

Do exercises 3 and 4.

OBJECTIVES

You should:

(a) Understand some of the limitations of the HHC in working with decimals.

(b) Know how to overcome some of these limitations.

****

1. Attempt to compute, on your HHC,

   0.000012×0.0023

STOP

2. Perform the following sequence of multiplications on your HHC:

   $0.12 \times 0.23 =$ ____________

   $0.012 \times 0.023 =$ ____________

   $0.0012 \times 0.0023 =$ ____________

   $0.0012 \times 0.00023 =$ ____________

STOP

In exercises 3–4 use the procedure shown in the text to perform the indicated multiplication. Express your answer in both scientific notation and as a decimal.

3. $0.00000475 \times 0.0000827$

4. $0.00004704 \times 0.0009053$

STOP

Division can also cause problems.

Do exercise 5.

5. Attempt to compute, with your HHC,

   0.0000013 ÷ 264

STOP

Again, unless your HHC has scientific notation, you will have obtained an answer of 0 for exercise 5.

There are several ways to obtain a non-zero answer in such situations. One way, which will produce the same answer as would be obtained by the use of an HHC with scientific notation and an eight-digit display, is illustrated by the following examples:

1. $0.0000013 \div 264 = (1300 \times 10^{-9}) \div 264$

   $= (1300 \div 264) \times 10^{-9}$

   $= 4.9242424 \times 10^{-9}$

   $= 0.0000000049242424$

2. $0.0000000142 \div 0.29 = (142 \times 10^{-10}) \div (29 \times 10^{-2})$

   $= (142 \div 29) \times (10^{-10} \div 10^{-2})$

   $= 4.8965517 \times 10^{-8}$

   $= 0.000000048965517$

Note that in each problem both the number being divided (the *dividend*) and the divisor have been written as whole numbers times a power of ten, where the number of digits in the whole number part of the dividend is one more than the number of digits in the whole number part of the divisor. Thus

Whole number part of dividend	Whole number part of divisor
1300 (4 digits)	264 (3 digits)
142 (3 digits)	29 (2 digits)

Do exercises 6 through 8.

In exercises 6–8 use the procedure described in the text to perform the indicated division.

6. 0.0000092 ÷ 756

7. 2376 ÷ 0.00000532

8. 0.000000158 ÷ 0.92

STOP

There are other less significant differences in computations with decimals as compared to computations with whole numbers.

Do exercise 9.

9. (a) Compute

   23.46987 × (10.295 × 15.392)

   (Remember that this means first to multiply 10.295 by 15.392 and then multiply 23.46987 by the result.)

   (b) Compute

   (23.46987 × 10.295) × 15.392

   (Remember that this means first to multiply 23.46987 by 10.295 and then multiply the result by 15.392.)

STOP

If you did exercise 9 correctly, you may have found that the answers to the two parts differ in the last decimal place whereas, by the associative property of multiplication, one would expect the two answers to be the same.

What has happened is that, in both parts of exercise 9, an HHC needs to round off the exact answer of 3719.0506209168, and the rounding off procedure can differ slightly from one HHC to another.

Exercises 7.3

Name ____________

Class ____________

In exercises 1–20 find an answer to the problem (exact or approximate) by using your HHC and scientific notation.

1. $0.001176 \times 0.0001877$

2. 0.364×0.00000489

3. 0.00185×0.0516

4. $0.00089 \div 0.000448$

5. $0.0000408 \div 0.00000582$

6. $0.007194 \div 0.00000002103$

7. $0.0000000828 \div 0.00075$

8. $0.000009742 \div 0.00000093$

9. $(0.000614 \times 0.00236) \div 0.000035$

10. $(0.0000101 \div 0.00294) \times 0.009048$

1.

2.

3.

4.

5.

6.

7.

8.

9.

10.

11.

11. $0.00159 \div (0.0693 \times 0.00000474)$

12.

12. $0.0000981 \div (0.000818 \div 0.02961)$

13.

13. $0.0921 \times (0.0039 \times 0.0000931)$

14.

14. $0.0064 \times (0.0023 \div 0.00176)$

15.

15. $0.0000258 \times (0.000026 \div 0.00000789)$

16.

16. $(0.00056 \times 0.003 \times 500) \div (0.00012 \times 0.028 \times 200)$

17.

17. $(650{,}000 \times 24{,}000) \div (0.00013 \times 0.0002)$

18.

18. $(1{,}760{,}000 \times 95{,}000{,}000) \div 0.000088$

19.

19. $(6{,}300{,}000 \times 0.00045) \div (0.0003 \times 0.07)$

20.

20. $(144{,}000 \times 0.00036) \div (0.00072 \times 960{,}000)$

21. (a) Compute

$8.143272 \times (2.314876 + 5.92313)$

(b) Compute, without using the distributive property,

$(8.143272 \times 2.314876) + (8.143272 \times 5.92313)$

By the distributive property, the answers to (a) and (b) should be the same. Are they? Discuss!

In exercises 22–25 use scientific notation to solve each problem.

22. Wrought iron expands 0.00001166 inch per inch for each degree Celsius the temperature increases. Find the expansion in a 12-foot piece of wrought iron when the temperature increases 11.8° C.

23. How many tons of sea water would be needed to produce one ounce of gold if sea water is $2.45 \times 10^{-10}\%$ gold?

24. The unit strain on a structural member is defined as the total elongation (or shortening) of the member divided by its original length. Find the unit strain on an aluminum wire 48 inches long that stretches 1.2×10^{-2} inches when subjected to a pull of 60 pounds.

25. A concrete pier is subjected to a load that produces a unit strain (see exercise 24) of 8×10^{-4}. By how much is this pier shortened if the length before the load was applied was 22 feet?

21. (a)

(b)

22.

23.

24.

25.

7.4 Squares and Square Roots

Do exercise 1 in the adjoining column.

We call 1764 the *square* of 42 since $42^2 = 1764$ and 1.5129 the square of 1.23 since $1.23^2 = 1.5129$.

Now let us look at the reverse situation: What number gives, when squared, the number 1764 and what number gives, when squared, the number 1.5129? We call the (positive) number whose square is 1764, the *square root* of 1764 and the number whose square is 1.5129, the square root of 1.5129. Thus the square root of 1764 is 42 because $42^2 = 1764$, and the square root of 1.5129 is 1.23 because $1.23^2 = 1.5129$. We write

$$\sqrt{1764} = 42, \qquad \sqrt{1.5129} = 1.23$$

and, in general, write, for positive numbers a and b,

$$\sqrt{a} = b \quad \text{if} \quad b^2 = a$$

Do exercises 2 through 6.

Given a number, it is easy to find its square—only a multiplication is involved. But how can we find the square root of a number? (Some HHC's have a square root key, but we are concerned here only with those that do not.)

Suppose, then, that our HHC does not have a square root key and that we do not know that $\sqrt{1764} = 42$. How could we find $\sqrt{1764}$? What if we were to guess that $\sqrt{1764} = 20$. Is this guess a correct one? We calculate that $20^2 = 400$ and conclude that $\sqrt{1764} \neq 20$. Indeed, 20 is clearly too small and so we might try 50. But $50^2 = 2500$ and so 50 is certainly too large. Trying various numbers between 20 and 50 would narrow the search until we eventually would try 42 and find that $42^2 = 1764$, so that $\sqrt{1764} = 42$.

Note that when we get around to trying 42, we have

$$1764 \div 42 = \frac{1764}{42} = 42$$

Thus another way of describing the square root of a number a is to say that it is a number b such that $a \div b = b$. That is, if

$$\sqrt{a} = b, \quad \text{then} \quad a \div b = \frac{a}{b} = b$$

and if

$$a \div b = \frac{a}{b} = b, \quad \text{then} \quad \sqrt{a} = b$$

OBJECTIVES

You should:

(a) Know the meaning of squares and square root.

(b) Be able to calculate approximations to square roots.

****

1. Calculate 42^2 and 1.23^2.

STOP

In exercises 2–5 determine by squaring whether the given equality holds.

2. $\sqrt{26} = 5.09$

3. $\sqrt{169} = 13$

4. $8.944 = \sqrt{80}$

5. $\sqrt{20.25} = 4.5$

6. Show by squaring that none of 1.7, 1.73, 1.732, 1.7321, 1.73205, and 1.7320508 is equal to $\sqrt{3}$ but that each, in turn, is a better approximation.

$1.7^2 =$ ______________

$1.73^2 =$ ______________

$1.732^2 =$ ______________

$1.7321^2 =$ ______________

$1.73205^2 =$

____

$1.7320508^2 =$

____

STOP

Let us now apply this way of looking at square roots to see if we can find $\sqrt{3}$. From exercise 6 we know that $1.7 \neq \sqrt{3}$ because $1.7^2 \neq 3$. But, by the preceding remarks, we also know that $1.7 \neq \sqrt{3}$ because

$$\frac{3}{1.7} = 1.7647058 \neq 1.7$$

Similarly, $1.7647058 \neq \sqrt{3}$ because

$$\frac{3}{1.7647058} = 1.7 \neq 1.7647058$$

Thus 1.7 is too small and 1.7647058 is too large to be equal to the square root of 3. Is it not natural, then, to assume that the average of 1.7 and 1.7647058, i.e.,

$$\frac{1.7 + 1.7647058}{2} = 1.7323529$$

will be a better approximation to $\sqrt{3}$ than either 1.7 or 1.7647058? Indeed this is the case, since

$$1.7^2 = 2.89 \qquad \text{and} \qquad 1.7647058^2 = 3.1141865$$

whereas

$$1.7323529^2 = 3.0010465$$

Do exercise 7.

This process (the *divide-and-average process*) can always be repeated. Thus a still better approximation to $\sqrt{3}$ is

$$\frac{1.7323529 + \dfrac{3}{1.7323529}}{2} = 1.7320508$$

the average of 1.7323529 and $\dfrac{3}{1.7323529}$. Indeed, as you saw in exercise 6,

$$1.7320508^2 = 2.9999999$$

so that we have found the closest approximation to $\sqrt{3}$ possible with an 8-digit display HHC. (Note that the quickest way of computing the average above is to first compute

$$\frac{3}{1.7323529}$$

and then add it to 1.732529.)

Do exercise 8.

7. In what follows replace the □'s by the correct numbers to find a second approximation to $\sqrt{3.2}$: If we were were to take 1.8 as our first approximation to $\sqrt{3.2}$, we could get a better approximation by taking the average of 1.8 and

$$\frac{3.2}{\square}$$

That is, our second approximation would be

$$\frac{1.8 + \dfrac{3.2}{\square}}{\square} = 1.7888888$$

STOP

8. Continue exercise 7 by filling in the blank: A still better approximation to $\sqrt{3.2}$ than 1.7888888 is

$$\frac{1.7888888 + \dfrac{3.2}{\square}}{2} = 1.7888543.$$

Finally, check that if we use 1.7888543 in the divide and average process, we again obtain 1.7888543, and so conclude that we have obtained as close an approximation to $\sqrt{3.2}$ as is possible with our HHC.

STOP

Although the divide-and-average process provides us with a way of obtaining better and better approximations to $\sqrt{3}$, it turns out that there is no decimal exactly equal to $\sqrt{3}$. Indeed, it is rather rare for the square root of a number to be exactly equal to either a whole number or a decimal.

Do exercise 9.

9. Use the divide-and-average method as in exercises 7 and 8 to find as close an approximation to $\sqrt{10}$ as is possible with your HHC. Use 3 as your first approximation.

STOP

An important application of the finding of square roots is in connection with the *Pythagorean Theorem.* This theorem states that the sum of the squares of the lengths of the *sides* of a right triangle is equal to the square of the length of its *hypotenuse.* Thus

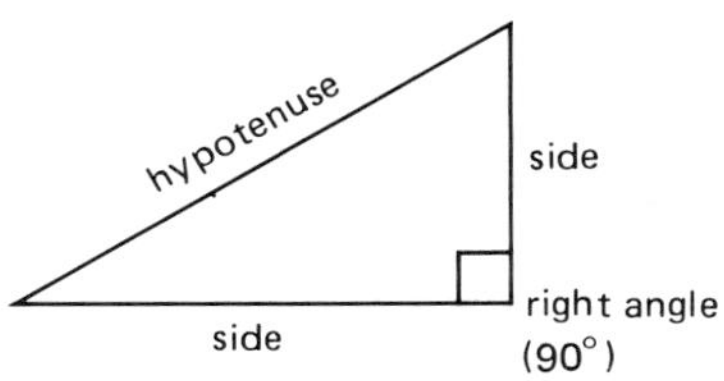

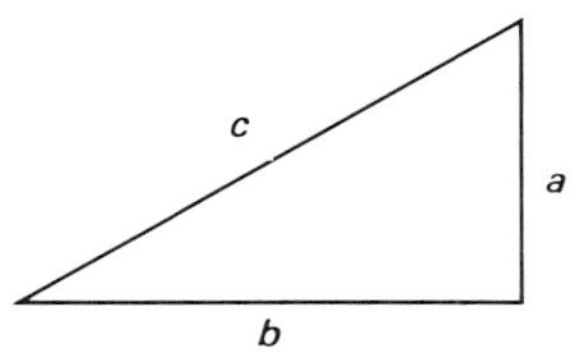

$$c^2 = a^2 + b^2 \quad \text{and so} \quad c = \sqrt{a^2 + b^2}$$

For example, if the lengths of the sides of a right triangle are 3 inches and 4 inches, the length of its hypotenuse, in inches, is

$$\sqrt{3^2 + 4^2} = \sqrt{9 + 16} = \sqrt{25} = 5$$

Do exercise 10.

10. Find, to the nearest tenth of a foot, the length of the hypotenuse of a right triangle whose sides are of lengths 3.2 feet and 4.6 feet.

STOP

Exercises 7.4

Name ____________

Class ____________

In exercises 1-20 use the divide and average method to find the decimal approximation to the given square root to the nearest ten thousandths.

1. $\sqrt{11}$

2. $\sqrt{24}$

3. $\sqrt{38}$

4. $\sqrt{76}$

5. $\sqrt{1.6}$

1.

2.

3.

4.

5.

6.

6. $\sqrt{250}$

7.

7. $\sqrt{26.3}$

8.

8. $\sqrt{2.1}$

9.

9. $\sqrt{2.57}$

10.

10. $\sqrt{59.3}$

11. $\sqrt{79.21}$

11.

12. $\sqrt{8.37}$

12.

13. $\sqrt{5.42}$

13.

14. $\sqrt{45,897}$

14.

15. $\sqrt{17,840}$

15.

16.

17.

18.

19.

20.

16. $\sqrt{34.7}$

17. $\sqrt{\dfrac{1.316}{0.016}}$

18. $\sqrt{\dfrac{32 \times 36}{6.65}}$

19. $\sqrt{\dfrac{280 \times 3.1416}{57}}$

20. $\sqrt{\dfrac{4.3 \times 7.8}{1.6}}$

21. Do you think that $\sqrt{a} + \sqrt{b} = \sqrt{a + b}$ for all numbers a and b? If your answer is yes, give at least four examples where the equality holds. If your answer is no, give an example to justify your answer. 21.

22. Proceed as in exercise 21 in regard to the validity of the equality 22.

$$\frac{\sqrt{a}}{\sqrt{b}} = \sqrt{\frac{a}{b}}$$

for all numbers a and b with $b \neq 0$.

23. Proceed as in exercise 21 in regard to the validity of the equality 23.

$$\sqrt{a} \times \sqrt{b} = \sqrt{a \times b}$$

for all numbers a and b.

Use the formula $c = \sqrt{a^2 + b^2}$ to find c in exercises 24–31, correct to the nearest hundredth.

24. $a = 6,\ b = 8$ 24.

25. $a = 5,\ b = 12$ 25.

26. $a = 40,\ b = 9$ 26.

27.

27. $a = 37,\ b = 47$

28.

28. $a = 10.32,\ b = 7.52$

29.

29. $a = 408,\ b = 523$

30.

30. $a = 1.06,\ b = 5.25$

31.

31. $a = 27.6,\ b = 85.7$

In exercises 32–46 use a right triangle and the Pythagorean Theorem to solve the problem. Express each answer to the nearest hundredth.

32. A baseball diamond is in the form of a square whose sides are 90 feet long. How far is it from first base to third base?

32.

33. The pitcher's box in a baseball diamond is 60 feet from home plate. (a) How far is it from second base? (b) How far is it from third base?

33. (a)

(b)

34. A shortstop fields a batted ball at a point one-third of the way between second and third base. How far must he throw the ball to make an out at first base?

34.

35. Find the length of a diagonal brace for a rectangular gate 4 feet high and 8 feet long.

35.

36. Find the diagonal of a square whose perimeter is 40 centimeters.

36.

37.

37. Find the diagonal of a cube whose edge is 10 inches.

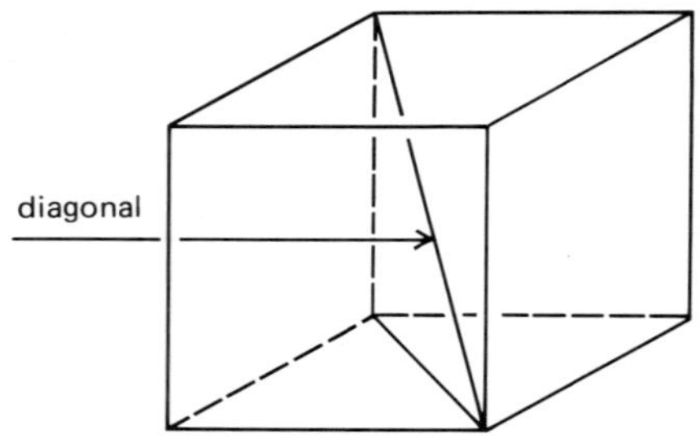

38.

38. Find the longest diagonal of a box that is 3 meters by 4 meters by 2 meters. (This is the shortest distance that a fly could travel to get from one lower corner to the opposite upper corner.)

39.

39. Find the length of a wire from the top of a vertical pole 25 feet tall to a point on the ground 8 feet from the pole.

40.

40. An 18-foot ladder just reaches a window on the side of a house. If the ladder is 3 feet from the side of the house, how high is the window?

41. At noon a ship sailed due east from Key West, Florida traveling at 15 knots (nautical miles per hour). Two hours later a second ship left Key West and sailed due south at 18 knots. How far apart were the two ships at 6:00 P.M.?

41.

42. What must be the least diameter of round stock steel so that a square bolt head $1\frac{7}{8}$ inch on a side can be milled from it?

42.

43. A radio antenna tower is to be fastened by 3 guy wires that reach the ground 50 feet from the foot of the tower. How much guy wire will be needed if the wire is attached to the tower at a point 70 feet above the ground?

43.

In exercises 44 and 45 refer to the following diagram, which shows a spacecraft at P, at an altitude of h above a planet. The distance to the horizon A is d, and the radius of the planet is given as r.

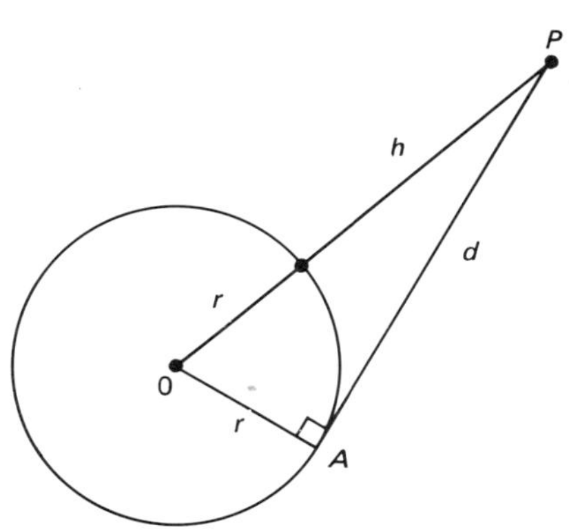

44.

44. Find the distance to the horizon if a spacecraft is 100 miles above the earth. (Radius of the earth is 3960 miles.)

45.

45. How far is it to the horizon on the moon if a spacecraft is 70 miles above its surface? (The moon's diameter is 2160 miles).

46.

46. A tree was broken off 12 feet above the ground and fell so that its top lay 47 feet, 3 inches from the foot of the tree, the end where it was broken resting on the stump. What was the height of the tree?

7.5 Compound Interest

In our work with interest in Chapter 5 we considered only what is called *simple* interest where the principal (amount of money drawing interest) stays the same. For example, suppose that \$100 is invested at 6% per year for three years. Then the interest per year on the \$100 would be \$6 ($100 \times 0.06 = 6$) and the total interest received in three years would be three times that or \$18. Thus the \$100 investment would have increased in value to \$118.

Do exercises 1 and 2 in the adjoining column.

Suppose now that the interest of \$6 after one year were added to the principal to give a new principal of \$106. Then, by exercise 1, the interest earned in the second year would be \$6.36, and if this interest were added to the \$106, the new principal, in dollars, would be

$$106 + 6.36 = 112.36$$

So, by exercise 2, the interest earned in the third year would be \$6.74 (to the nearest cent) and if this interest were added to \$112.36 the value, in dollars, of the \$100 at the end of three years would be

$$112.36 + 6.74 = 119.10$$

Thus the interest earned in three years in this way would be \$19.10 as compared to the \$18 earned with simple interest.

Do exercise 3.

OBJECTIVES

You should:

(a) Understand the concept of compound interest.

(b) Be able to solve problems involving compound interest.

****

1. What interest will be earned in one year on an investment of \$106 if the interest rate is 6% per year?

2. What interest will be earned in one year on an investment of \$112.36 if the interest rate is 6% per year?

STOP

3. If \$100 is invested at $7\frac{1}{2}\%$ per year simple interest, what interest would be earned at the end of three years?

STOP

Do exercises 4 and 5.

4. Suppose that \$100 was invested at $7\frac{1}{2}\%$ for three years with the interest for each year added to the principal (as in the text example with 6% just discussed). Find to the nearest cent the interest earned at the end of three years.

5. How much more interest is earned under the conditions of exercise 4 than under the conditions of exercise 3?

STOP

When the interest earned over a certain period is added to the principal on a regular basis, we have *compound* interest. Thus in the example of \$100 invested at 6% we would say that the interest was *compounded annually* at 6% and that the compound interest for the three-year period was \$19.10.

In our example we had, in dollars, at the end of the first year,

$$\begin{aligned} 100 + (0.06 \times 100) &= 100 \times (1 + 0.06) \\ &= 100 \times 1.06 \end{aligned}$$

So, at the end of two years we will have 100×1.06 dollars plus 6% of 100×1.06 dollars; that is,

$$\begin{aligned} 100 \times 1.06 + 0.06 \times (100 \times 1.06) &= (100 + 0.06 \times 100) \times 1.06 \\ &= 100 \times (1 + 0.06) \times 1.06 \\ &= 100 \times 1.06 \times 1.06 \\ &= 100 \times (1.06)^2 \end{aligned}$$

dollars.

A continuation of this process would show us that if \$100 were compounded annually at 6% interest for n years, the amount of money at the end of n years would be

$$100 \times (1.06)^n$$

Thus, for example, the amount of money at the end of 15 years would be

$$100 \times (1.06)^{15}$$

Do exercises 6 and 7.

6. Enter 1.06 and then ⊗ on your HHC. Now press the ⊜ key. If the display reads 1.1236 (which is 1.06^2), press the ⊜ key again (without clearing the display) and record your answer.

7. Compare your result in exercise 6 with the direct computation of $(1.06)^3$:

1.06	×	1.06	×	1.06	=

STOP

The results of exercises 6 and 7 may or may not be the same, depending on the make and model of your HHC. If they are the same, they indicate a quick method of calculating such powers as $(1.06)^{15}$. That is, to find $(1.06)^{15}$ we simply enter

1.06	×

and then press ⊜ 14 times. (If your HHC does not have this feature, we suggest that you borrow one that has to do such calculations.)

Do exercise 8.

In general, if P dollars are invested at $R\%$ compounded annually, the amount at the end of n years will be

$$A = P \times \left(1 + \frac{R}{100}\right)^n$$

For example, if \$240 is invested at 5% compounded annually, the amount in dollars at the end of 12 years will be

$$240 \times \left(1 + \frac{5}{100}\right)^{12} = 240 \times (1.05)^{12}$$

$$= 240 \times 1.7958558$$

$$= 431.00539$$

or \$431.01 to the nearest cent.

Do exercises 9 and 10.

The compounding of interest need not take place on a yearly basis; it may be semi-annually, quarterly, monthly, weekly, or even daily.

If a bank says it will pay 6% compounded quarterly on savings accounts, this means that interest will be calculated and compounded four times a year (every three months) at the rate of $(6 \div 4)\% = 1.5\%$. Thus at the end of three years we would have $3 \times 4 = 12$ periods so that \$100 would amount to

$$100 \times (1.015)^{12} \approx 100 \times 1.1956176$$

dollars which, to the nearest cent, is \$119.56. Note that this is a little more than the amount would be if the interest were only compounded annually: $100 \times (1.06)^3$ gives us \$119.10 to the nearest cent. In general, the more times interest is compounded, the greater the resulting amount for the same amount of time and the same annual rate of interest.

8. What will \$100 amount to in 15 years if it is invested at 6% compounded annually?

STOP

9. What will \$500 amount to in 12 years if it is invested at $7\frac{1}{2}\%$ compounded annually?

10. What will \$250 amount to in ten years if it is left in a savings account at $5\frac{1}{4}\%$ interest compounded annually?

STOP

As another example, suppose that \$100 is invested at 9% compounded monthly. To find the amount at the end of the year we calculate

$$\begin{aligned} 100 \times \left(1 + \frac{0.09}{12}\right)^{12} &= 100 \times (1 + 0.0075)^{12} \\ &\approx 100 \times 1.0938063 \\ &= 109.38063 \end{aligned}$$

and conclude that the amount would be \$109.38 to the nearest cent. Note that the interest is \$9.38 and we would also get this if \$100 were invested at 9.38% simple interest; 9.38% is called the *effective* rate of interest.

Do exercises 11 and 12.

11. Find what \$244 would amount to in 6 years if it is invested at 5% compounded quarterly (four times a year).

12. Find the effective rate of interest if money is invested at 6% compounded monthly.

STOP

Exercises 7.5

Name ____________

Class ____________

In exercises 1–20 find (a) the amount of money available at the end of the given time, (b) the amount of interest earned, (c) the effective rate of interest to the nearest hundredth of a percent.

1. $820 for 3 years at 6% compounded annually

2. $820 for 3 years at 6% compounded semi-annually

3. $600 for 5 years at 4% compounded annually

4. $600 for 5 years at 4% compounded semi-annually

5. $4000 for 2 years at 8% compounded annually

6. $4000 for 8 years at 2% compounded annually

7. $4000 for 2 years at 8% compounded quarterly

8. $500 for 1 year at 5% compounded monthly

9. $500 for 1 year at 5% compounded bimonthly (every 2 months)

10. $2600 for 4 years at 2% compounded annually

11. $2600 for 4 years at 4% compounded annually

12. $2750 for 9 years at 6% compounded annually

1. (a)
(b)
(c)

2. (a)
(b)
(c)

3. (a)
(b)
(c)

4. (a)
(b)
(c)

5. (a)
(b)
(c)

6. (a)
(b)
(c)

7. (a)
(b)
(c)

8. (a)
(b)
(c)

9. (a)
(b)
(c)

10. (a)
(b)
(c)

11. (a)
(b)
(c)

12. (a)
(b)
(c)

13. (a)
(b)
(c)

14. (a)
(b)
(c)

15. (a)
(b)
(c)

16. (a)
(b)
(c)

17. (a)
(b)
(c)

18. (a)
(b)
(c)

19. (a)
(b)
(c)

20. (a)
(b)
(c)

21.

22.

23.

24.

13. \$6500 for 6 years at 5% compounded semi-annually

14. \$4800 for 4 years at $5\frac{1}{2}$% compounded semi-annually

15. \$1900 for 6 months at 6.3% compounded monthly

16. \$500 for 24 years at 6% compounded annually

17. \$2000 for 10 years at 6% compounded semi-annually

18. \$750 for 10 years at 4% compounded quarterly

19. \$1200 for 5 years at 6% compounded monthly

20. \$10 for 1 year (365 days) at 5% compounded daily

Exercises 21–25 refer to the following situations: You deposit \$20,000 in a bank that pays 8% annual interest and make no deposits or withdrawals for 5 years.

21. If the interest were compounded annually, how much money would be in your account at the end of the 5-year period?

22. If the interest were compounded semi-annually, how much money would be in your account at the end of the 5-year period?

23. If the interest were compounded quarterly, how much money would be in your account at the end of the 5-year period?

24. How much more interest would you have if the interest were compounded quarterly rather than annually?

25. How much more interest would you have if the interest were compounded quarterly rather than semi-annually? 25.

26. If $500 were invested on the day you were born at 6% compounded annually, how much would it amount to on your eighteenth birthday? 26.

27. What is the difference in interest earned for each $100 invested when interest is compounded at 5% for 5 years semi-annually instead of annually? 27.

28. Will $100 compounded annually for 4 years at 8% amount to the same as $100 compounded annually for 8 years at 4%? If not, which amount is greater and how much greater? 28.

29. If the population of the United States in 1975 was 213 million and it increased at a rate of 2% per year, what would the population be in the year 2000? 29.

30. If the national debt in 1975 was 500 billion dollars and it increased at a rate of 20% per year, what would be the debt in the year 2000? 30.

31. Using the information in exercises 29 and 30, find how much the national debt would increase per person from 1975 to 2000 under those conditions. 31.

32. If a loaf of bread cost 59 cents in 1975 and the rate of inflation is 6% each year, what would you expect to pay for a loaf of bread in the year 2000? 32.

33. A savings bank pays its depositors 5% interest compounded quarterly. On the day of his son's birth, a man wished to deposit a sum of money that will accumulate to $5000 on the boy's 20th birthday. How much money must he deposit? 33.

34. Your grandson is 10 years old. If you wish to give him a present of $1000 on his 21st birthday, how much must you deposit now in a savings and loan association that pays 6% interest compounded semi-annually? 34.

35. How much money will it take to amount to $15,000 at the end of 10 years if it is invested at 4% compounded quarterly? 35.

36. How long will it take $1000 to double itself if it is invested at 6% compounded annually? 36.

37.

37. How long will it take $1000 to triple itself if it is invested at 6% compounded annually?

*38.

*38. If you could invest $1000 each year in an investment that yields 6% and you reinvest the interest at the same rate each time it is paid, how much will you have at the end of 20 years?

*39.

*39. If a man begins investing $500 annually at 5% when he is 25 years of age and continues this investing until he is 60, keeping all interest reinvested, how much will he have accumulated?

Review Exercises 7

Name ______________________

Class ______________________

In exercises 1–16 express the given number in scientific notation.

1. 37,650,000
2. 379,000,000
3. 857,200,000,000
4. 0.000187
5. 0.000000007396
6. 0.00000892
7. 0.00031
8. 60.098
9. 48,876.42
10. The half-life of uranium 235 is 710,000,000 years.
11. When the temperature increases one degree Celsius, one foot of a certain kind of steel pipe increases 0.000011 foot in length.
12. The wave length of yellow light is 0.00000059 meter.
13. Television signals travel at the rate of 30,000,000,000 centimeters per second.
14. The weight of the earth is 6,660,000,000,000,000,000,000 tons.

1.
2.
3.
4.
5.
6.
7.
8.
9.
10.
11.
12.
13.
14.

15.	15. The wave length of some x-rays is about 0.00000095 centimeter.
16.	16. An oil film on water is about 0.0000002 inch thick.
	In exercises 17–30 write the given number in decimal notation.
17.	17. 4.67×10^{6}
18.	18. 2.367×10^{-6}
19.	19. 9.40×10^{3}
20.	20. 7.108×10^{-3}
21.	21. 8.5×10^{-8}
22.	22. 3.1853×10^{9}
23.	23. The cross sectional area of a certain wire is 4.35×10^{-3} square centimeter.
24.	24. The average distance from the sun to the earth is 9.29×10^{7} miles.
25.	25. The mass of a proton is 1.67×10^{-24} gram.
26.	26. A certain electrical resister has a resistance of 4.5×10^{3} ohms.
27.	27. Some computers can perform an arithmetic operation in 1.5×10^{-6} second.
28.	28. There are about 1.85×10^{5} centimeters in one mile.
29.	29. The escape velocity of a rocket is 2.5×10^{4} miles per hour.
30.	30. Atmospheric pressure is about 1.013×10^{6} dynes per square centimeter.

In exercises 31–40 perform the indicated computation using scientific notation.

31. $8{,}007{,}000{,}000 \times 0.000127$ — 31.

32. 0.0000456×0.00959 — 32.

33. $0.000068317 \times 621{,}000{,}000{,}000$ — 33.

34. $783{,}042{,}000{,}000 \div 26{,}742$ — 34.

35. $0.00000000128 \div 0.000374$ — 35.

36. $365{,}742{,}850{,}000 \div 0.0000578$ — 36.

37. $0.000175 \times (7{,}800{,}000{,}000 \div 54{,}400{,}000)$ — 37.

38. $(237{,}687{,}000{,}000 \div 0.00615) \times 0.0000411$ — 38.

39. $\dfrac{0.00028 \times 357{,}000{,}000}{43{,}850{,}000 \times 0.0887}$ — 39.

40. $\dfrac{844{,}730{,}000{,}000 \times 136{,}000{,}000}{621{,}000{,}000 \times 948{,}000{,}000{,}000}$ — 40.

41.

41. An angstrom is a unit of length equal to 10^{-8} centimeter. If the wave length of yellow light is 0.00000059 meter (100 centimeters = 1 meter), what is the wave length in angstroms?

42.

42. The coefficient of linear expansion of a solid is the increase in unit length per unit of increase in temperature. For a certain kind of steel, the coefficient is 24×10^{-6} foot per degree Fahrenheit. If a steel rod is 32 feet at 65° Fahrenheit in the morning, find the length of the rod when the temperature reaches 94° Fahrenheit in the afternoon.

43.

43. The coefficient of cubical expansion of petroleum of a certain grade is 90×10^{-4} gallon per degree Celsius. If 5000 gallons of this petroleum were measured at 0° Celsius, what would be the number of gallons at 30° Celsius?

44.

44. Steel rails (see exercise 42) are laid when the temperature is 50° Fahrenheit and, at that temperature, they measure 34 feet. If the maximum temperature expected is 122° Fahrenheit, how much space should be left between the rails to allow for expansion?

*45.

*45. If a computer can print 150 digits (0, 1, 2, . . . , 9) to a line and 5000 lines a minute, how long would it take the computer to print out the numeral for a googol? (Googol = 10^{100}.) Scientists estimate the earth is 5 billion years old (5×10^9). Compare this time with the time it would take the computer to write a numeral for the googol. What conclusion can you draw?

In exercises 46–55 find the square root of the given number to the nearest ten thousandth by the divide-and-average method.

46.

46. 73

47.

47. 96

48. 24

49. 0.124

50. 0.2815

51. 0.5177

52. 1535

53. 5177

54. 97.327

55. 71.62

48.

49.

50.

51.

52.

53.

54.

55.

In exercises 56–60 use the Pythagorean Theorem to solve the given problem. Give your answer to the nearest hundredth.

56. 56. A wire is to be stretched from the top of a pole 75 feet high to a point 35 feet from the foot of the pole. How long a wire will be needed?

57. 57. A road rises 120 feet for every 2200 feet of horizontal travel. What is the length of the road if the rise is 120 feet?

58. 58. Will a 48-inch umbrella fit in a trunk that is 3 feet long, 2 feet wide, and 2 feet high?

59. 59. The span of a roof is 32 feet and its rise at the center of the span is 7.5 feet. How long must the rafters be?

60. 60. If a kite is flying 100 feet high at a distance on the ground of 300 feet from the person flying it, how long must the string be?

In exercises 61–70 calculate the amount of money available at the end of the given time if the interest is compounded as indicated.

		Principal	*Rate*	*Compounded*	*Time*
61.	61.	\$1000	4%	annually	3 years
62.	62.	\$2000	6%	semi-annually	2 years
63.	63.	\$5000	3%	quarterly	2 years

	Principal	*Rate*	*Compounded*	*Time*	
64.	\$10,000	8%	semi-annually	$2\frac{1}{2}$ years	64.
65.	\$795	$5\frac{1}{4}$%	quarterly	$3\frac{1}{2}$ years	65.
66.	\$1580	$8\frac{3}{4}$%	quarterly	2 years	66.
67.	\$1360	$7\frac{1}{2}$%	monthly	6 months	67.
68.	\$2250	5%	annually	7 years	68.
69.	\$1670	6%	annually	10 years	69.
70.	\$1728.25	7%	semi-annually	3 years	70.

71. If you were to deposit \$1000 in a savings account with interest being credited every six months at a 4% annual rate, how much money would you have at the end of five years, assuming no withdrawals or additional deposits? 71.

72. What is the effective rate of interest corresponding to 4% compounded annually? 72.

73. A savings bank offers $4\frac{1}{2}$% compounded annually and another bank offers 4% compounded semi-annually. Which is the better investment? Find how much \$100 will amount to if invested in each bank for five years. 73.

74. What is the effective rate of interest corresponding to 4% compounded semi-annually? 74.

75.

75. A father invests \$50 at 5% compound annual interest on the day his daughter is born and the same amount on each birthday. What will these investments amount to when the daughter is 12 years old? When the daughter is 18 years old?

76.

76. How long will it take \$500 to double itself at 7% interest compounded annually?

77.

77. What is the effective rate of interest corresponding to 8% compounded quarterly?

78.

78. How much more interest is earned at the end of three years on \$5000 when it is compounded quarterly rather than annually at 8%?

79.

79. How much would you have to invest now at 8% compounded quarterly in order to have a total of \$5000 five years from now?

Just for Fun

A. PATTERNS AND SEQUENCES

Write the next five terms of each sequence.

1. 1.3, 1.9, 2.5, ______, ______, ______, ______, ______.

2. $1, \frac{4}{3}, \frac{13}{9}, \frac{40}{27}$, ______, ______, ______, ______, ______.

3. $0, \frac{1}{2}, \frac{5}{6}, \frac{4}{3}, \frac{5}{3}$, ______, ______, ______, ______, ______.

4. 1, 5, 17, 53, 161, ______, ______, ______, ______, ______

B. MORE PATTERNS

For those items indicated by an asterisk (*) you are to examine the pattern indicated by the previous computations and then write what you think is the result.

5.
$1 \times 9 + 2 =$
$12 \times 9 + 3 =$
$123 \times 9 + 4 =$
$*1234 \times 9 + 5 =$
$*12345 \times 9 + 6 =$
$*123456 \times 9 + 7 =$
$*1234567 \times 9 + 8 =$

6.
$1 \times 8 + 1 =$
$12 \times 8 + 2 =$
$123 \times 8 + 3 =$
$*1234 \times 8 + 4 =$
$*12345 \times 8 + 5 =$
$*123456 \times 8 + 6 =$
$*1234567 \times 8 + 7 =$

C. FIND THE MISSING DIGITS

7. Subtraction

$$\begin{array}{cccc} 5 & \_ & 7 & \_ \\ \_ & 1 & \_ & 4 \\ \hline 3 & 2 & 7 & 9 \end{array}$$

8. Multiplication

$$\begin{array}{cccccc} & & & 4 & \_ & 6 \\ & & & \_ & 4 & \_ \\ \hline & & & \_ & \_ & 2 \\ & & \_ & \_ & 0 & 4 \\ \_ & \_ & \_ & \_ & & \\ \hline 1 & 4 & 5 & 6 & 9 & 2 \end{array}$$

D. TRY THESE

9. A man weighing 100 kilograms has two sons each weighing 50 kilograms. They cross a deep river in a boat which can carry a maximum load of 100 kilograms. How do they do it?

10. A man is $\frac{3}{8}$ of the way across a trestle when he hears a train coming. If the train is traveling at 40 miles per hour, how fast does the man run, if by running he can make it to either end of the trestle just before the train reaches that end?

11. (a) Try $\left(\left\{[(1 \div 2) + 1] \div 2 + 1\right\} \div 2 + 1\right) \div \cdots$ on your HHC. What happens if you continue to divide by 2, add 1, divide by 2, add 1, . . . ?

    (b) Try this with some other pairs of numbers other than 1 and 2.

    (c) Can you explain what happens?

E. MAGIC SQUARE

12. Complete this magic square.

17			8	15
	5		14	
4		13	20	22
	12	19	21	3
	18			9

F. THE MAGIC CUBE

13. Place the numbers 1 through 8 in each of the circles so that the four numbers on each face of the cube (front, back, top, bottom, and both sides) have the same sum. Use each number only once.

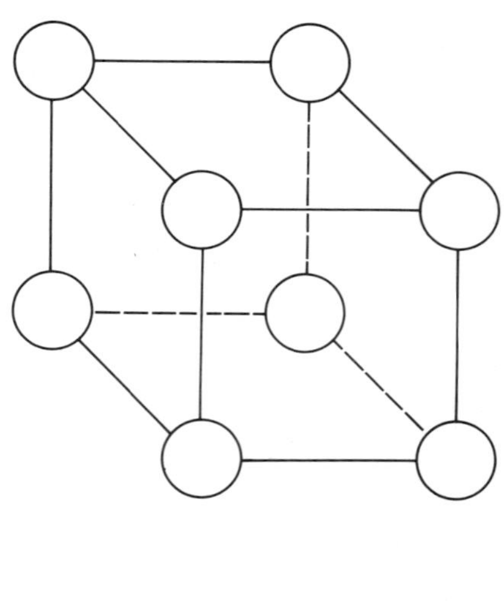

G. CALCULATOR GAME: "HIT THE TARGET"

14. Play this game on your calculator.

(a) Each player enters a six-digit number with the first number divided by the second number.

(b) The first two digits on the left of the quotient is the target and the third digit is the operator.

(c) Using the number chosen as the operator and only the keys +, −, ×, ÷, and =, each player strives to obtain the target number.

(d) Each player counts the number of keys pressed in obtaining the target number. The winner is the one with the least number of keys pressed. His score is the difference in the number of keys pressed by each player. If a decimal appears in the display, that player loses, and the other player scores 10 points.

(e) Five rounds are played, and the player with the highest score wins.

Sample game:	*Display:*
A enters 217080.	217080
B divides by 964182.	0.2251442
Target: 22	
Operator: 5	

A enters 5 and proceeds as follows:

Key	5	÷	5	+	=	+	5	+	5	+	5	+	5	+
Display	5	5	5	1	2	2	5	7	5	12	5	17	5	22

B enters 5 and proceeds as follows:

Key	5	5	÷	5	+	=
Display	5	55	55	5	11	22

Note: Not all calculators will give the above sequences because of differences in internal logic.

A pressed 14 keys.

B pressed 6 keys.

B is the winner.

B's score is 8. (14 − 6)

H. THE TALKING CALCULATOR

15. Perform the following calculation to find what the confederate of a gambler is called.

$$(4 \times 25) + (318 \times 145) + 27{,}085 + (275 \times 14)$$

Mathematics and Your Dog and Cat

Use the information given on page 458 to do the following exercises.

1. Compare the costs of the six dog and cat foods listed by completing the following table:

	A	B	C	D	E	F
Amount in ounces						
Amount in pounds						
Total cost in dollars						
Total calories						
Cost per ounce in cents						
Cost per pound in cents						
Cost per 100 calories in cents						
Calories per dollar						

2. If a 7-pound cat eats 5 ounces of Brand A food each day, how many ounces of protein and how many ounces of fat does the cat eat each day?

3. If a cat's diet should contain 1.25 ounces of fat, how much additional fat should be fed to the cat of exercise 2?

4. What would be the cost per day of feeding the cat of exercise 2?

5. What would it cost to feed a 15-pound, moderately active dog per week if (a) he is fed Brand B? (b) he is fed Brand D?

6. What would it cost per week to feed a 42-pound inactive dog

   (a) Brand B dog food?

   (b) Brand D dog food?

7. What would it cost per day to feed a 55-pound hunting dog

   (a) Brand B dog food?

   (b) Brand D dog food?

8. A 23-pound, moderately active dog is usually fed Brand B dog food. If the dog has a three-degree fever, how many ounces of this food should the dog eat per day?

9. If a 45-pound, inactive dog is overweight and its food must be reduced 20%, how many pounds of food should it be fed each day if it is fed

   (a) Brand B dog food?

   (b) Brand D dog food?

10. (a) A $14\frac{1}{4}$-pound, moderately active dog would eat how much more or less crude fiber each day using Brand B dog food compared to Brand D dog food? (b) What is the ratio of fiber in Brand B to Brand D?

11. Information on the package of dry dog food indicates that 1 cup of food should be used for every 3 pounds of body weight. Does this meet the specifications given in the table for the number of required calories if 1 cup of dry food weighs 4 ounces? If it does not, by how much does it differ?

A. **PET FOODS**
Bonus Buy!
CAPITOL DOG or CAT, 15-oz
SAVE 24c
7 for $1.

GUARANTEED ANALYSIS:
PROTEIN (CRUDE) .... 10.0% MIN
FAT (CRUDE) 3.0% MIN
FIBER (CRUDE) 1.5% MAX.
MOISTURE 75.0% MAX
ASH 2.5% MAX

B. **DOG FOOD** 15 oz or more **4 FOR $1**

C. **Dog Food** ASSORTED VARIETIES .... **3** 14½ oz TINS **$1**

FOR

COMPLETE 100% NUTRITION
NET WT. 15 OZ.

FOOD NEEDED BY DOGS AND CATS
KITTENS—1 oz/lb/day; CATS—¾ oz/lb/day of canned food
INACTIVE DOG—40 cal/lb/day; MODERATELY ACTIVE DOG—add 20%
SICK DOG—add 7% for each degree of fever
HUNTING DOG—add 60%

Canned food gives about 500 calories per pound
Dry food gives about 1500 calories per pound

D. **Dog Food** $1.50 Off Label **50** lb Bag **$7.98**

E. **DOG FOOD** 50 LB BAG **$6.99**

GUARANTEED ANALYSIS:

Crude Protein	(Minimum)	28.0%
Crude Fat	(Minimum)	8.0%
Crude Fiber	(Maximum)	4.0%
Moisture	(Maximum)	10.0%
Calcium	(Minimum)	1.1%
Phosphorus	(Minimum)	0.9%
Vitamin A	(Minimum)	5000 IU per Kg
Vitamin D	(Minimum)	500 IU per Kg
Vitamin E	(Minimum)	50 IU per Kg

F. 50 LB BAG **$10.89**

Appendix A

Use of a Memory Key

The instructions for the use of memory keys on HHC's vary somewhat, so do consult the manual accompanying your particular HHC for details. Basically two keys are involved, one to enter the number to be stored and one to recall it when needed. Typically, the entering key will be labeled MS (memory storage) and the recall key MR (memory recall). Then, for example, to compute

$$3 \times 4 + 7 \times 8$$

we proceed as follows:

Enter	3	×	4	=	MS	C	7	×	8	+	MR	=
Display	3.	3.	4.	12.	12.	0.	7.	7.	8.	56.	12.	68.

Here MS stored the number 12 which is 3 × 4 and MR recalled it to be added to 56, the product of 7 and 8.

If a number is entered into the memory it will, in some HHC's, be removed from the memory by pressing the [C] key. On the other hand, some HHC's have a clear memory key ([MC]) so that, until this key is pressed (or the HHC turned off), a number will stay in the memory. This feature is useful if we need to do a sequence of operations where one of the numbers stays the same. For example, to compute the products

$$12 \times 53$$
$$22 \times 53$$
$$29 \times 53$$

we can proceed as follows on such an HHC:

Enter	53.	MS	×	12	=	C	MR	×	22	=	C	MR	×	29	=
Display	53.	53.	53.	12.	636.	0.	53.	53.	22.	1166.	0.	53.	53.	29.	1537.

Appendix B

Rules for Rounding Off

As we have pointed out in Section 7.1, all measurements are approximate. In any measurement, we speak of the *greatest possible error,* which is always one-half of the unit of measure used in making the measurement.

For example, if we have a ruler that is marked off only in centimeters (i.e., no markings for millimeters), we could still expect to make the measurement to the nearest centimeter. The greatest possible error we would have (assuming we were not careless!) would be $\frac{1}{2}$ cm. Similarly, in using a ruler marked off in millimeters, the greatest possible error is $\frac{1}{2}$ mm.

When two measurements are compared for their *precision,* the more *precise* of the two measurements is the one for which the greatest possible error is the smaller. Thus if measurements of the length of a given line segment are taken first with a ruler marked only in centimeters, and then with one marked in millimeters, the second measurement will be the more precise.

It is important that measurements be stated in a way which shows clearly how precise they are. One way of doing this is to make certain agreements about what is meant when we write the number for a measure in decimal form. The decimal form is the one most often used in scientific and technical measurements. For example, when we write 3.42 cm for a certain measurement, we understand that the measurement has been made correct to the nearest 0.01 cm and that the greatest possible error is $\frac{1}{2}$ (0.01 cm), or 0.005 cm. This is equivalent to writing (3.42 ± 0.005) cm for the measurement. By this agreement each of the digits in 3.42 is *significant.* That is, each digit in 3.42 tells us something about the measurement. Nonzero digits in a measure are always significant. (Note that we refer to the number of units in a measurement as the *measure.* That is, if a measurement is 3.42 cm, then the measure is 3.42.)

On the other hand, if one or more zeros appear in a measure, they may or may not be significant. If one or more zeros come between two significant digits or are ending zeros to the right of a decimal point, they are significant. But zeros that are to the left of all the nonzero digits are not significant.

Example:

Measure	*Significant digits (in order)*
4.25	4, 2, 5
2041	2, 0, 4, 1
0.0027	2, 7
12.60	1, 2, 6, 0
30.08	3, 0, 0, 8

In the numeral 2041, the "0" is between the significant digits 2 and 4 and is itself significant. In the numeral 0.0027, all of the zeros are to the left of the nonzero digits 2 and 7 and are not significant. Indeed, the first "0" in the numeral 0.0027 is used simply to emphasize the decimal point and the other two "0's" are used only to locate the decimal point; hence, none of them are significant. In the numeral 12.60, the "0" (an ending zero to the right of the decimal point) is significant. It serves to tell us that the measurement is precise to the nearest hundredth of a unit. In the numeral 30.08, both "0's" are significant because they come between the two significant digits, 3 and 8.

Another way of looking at the question of whether or not a zero is significant is to ask what happens when the units are changed. For example, suppose that we have a measurement of 3.01 m. Since there are 100 cm in a meter, this could be represented as 301 cm; the zero stays! But if we have 0.03 m, this could be represented as 3 cm; the zeros disappear!

Suppose the distance between two objects is measured and the measurement is recorded as 360 m. If there is nothing to indicate otherwise, we assume that the "0" in the numeral 360 is not significant. Thus the measurement of 360 m has two significant digits (3 and 6) in its measure. The measurement is precise to the nearest 10 m and the greatest possible error is $\frac{1}{2}$ (10 m), or 5 m.

On the other hand, if the measurement of 360 m is precise to the nearest meter, we may indicate this by writing $36\underline{0}$ m. Note that the zero has been underlined to indicate that it is significant. Thus the measure $36\underline{0}$ has three significant digits (3, 6, and 0). The greatest possible error in the measurement of $36\underline{0}$ m is $\frac{1}{2}$(1 m), or 0.5 m. Similarly, if the measurement of 93,000,000 miles is precise to the nearest 100,000 miles, we may indicate this by writing $93{,}\underline{0}00{,}000$ miles for the measurement. The zero (in the hundred thousands place) following the digit 3 has been underlined to indicate that it is significant.

Still another way of indicating that certain zeros to the left of a decimal point are significant is to use scientific notation. (See Section 7.1.) When a number is expressed in scientific notation we assume that all the digits in the first factor are significant. Thus

9.3×10^7 miles has two significant digits (9 and 3)

9.30×10^7 miles has three significant digits (9, 3, and 0)

4.8×10^5 grams has two significant digits (4 and 8)

4.800×10^5 grams has four significant digits (4, 8, 0, and 0)

We may summarize our discussion of significant digits as follows: If a measure is recorded as a decimal numeral, then we can say

1. Every nonzero digit in the numeral is significant.
2. The zeros are significant if
   (a) they appear between two nonzero significant digits (as in 20.03)
   (b) they appear as ending zeros to the *right* of a decimal point (as in 2.300)
   (c) they are underlined or in some way stated to be significant (as in $23\underline{0}$)
3. The zeros are not significant if they are to the *left* of all the nonzero digits (as in 0.0034).

What about computations with numbers involving measurements? For example, suppose that we have two metal rods; the length of one is measured carefully as 2.63 m whereas the length of the other is measured less carefully as 6.8 m. What is the most correct measure of the total length of rod available?

The measurement of 6.8 m is precise to the nearest tenth of a meter. Thus the length of this rod in meters is somewhere between 6.75 (= 6.8 − 0.05) and 6.85 (= 6.8 + 0.05). Another way of expressing this is to write

$$6.75 < \text{length} < 6.85$$

which combines the two statements "length > 6.75" and "length < 6.85." Similarly, for the length of the other rod in meters we can write

$$2.625 < \text{length} < 2.635$$

So we have

Sum of smallest possible measures	*Sum of largest possible measures*
6.75	6.85
2.625	2.635
9.375	9.485

Hence the sum of the two measures is somewhere between 9.375 and 9.485. That is,

$$9.375 < \text{sum} < 9.485$$

Note that if we add the two measures as they were recorded, we have

$$\begin{array}{r} 6.8 \\ 2.63 \\ \hline 9.43 \end{array}$$

Are we then to say that the sum of the two measures is 9.43 meters? The chances seem to be about 1 in 100 that 9.43 would be the total length of rod available to the nearest hundredth of a meter.

To say that

$$6.8 \text{ m} + 2.63 \text{ m} = 9.43 \text{ m}$$

implies that the measurement of 6.8 m has the same precision as a measurement of 6.80 m. This is certainly not true. The measurement of 6.8 m is precise only to the nearest tenth of a meter, whereas a measurement of 6.80 m is precise to the nearest hundredth of a meter.

These observations lead to the following rule for adding or subtracting measures:

To add or subtract measures, round each measure to the same precision as the least precise measure involved, and then perform the addition or subtraction.

Thus in our example we would round 2.63 to 2.6 and write

$$6.8 \text{ m} + 2.6 \text{ m} = 9.4 \text{ m}$$

Similarly, for the difference we would write

$$6.8 \text{ m} - 2.6 \text{ m} = 4.2 \text{ m}$$

Now let us consider multiplication and division with measures. For example, suppose we need to find the area of a rectangle whose dimensions are 72 cm by 42 cm, where the base and the altitude have been measured precise to the nearest centimeter. That is,

$$\text{base} = (72 \pm \tfrac{1}{2}) \text{ cm} = (72 \pm 0.5) \text{ cm}$$

and

$$\text{altitude} = (42 \pm \tfrac{1}{2}) \text{ cm} = (42 \pm 0.5) \text{ cm}$$

Recall that the area of a rectangle is found by using the formula

$$A = b \times h$$

where b is the measure of the base, h is the measure of the altitude, and A is the measure of the area. Are we to write

$$A = 72 \times 42 = 3024$$

and say that the area of the rectangle is 3024 cm^2? If the measurements of the sides of a rectangle are 72 cm and 42 cm, then the area of the rectangle lies between (71.5×41.5) cm^2 and (72.5×42.5) cm^2. If we perform the indicated multiplications, we see that

$$2967.25 < A < 3081.25$$

The figure below illustrates this situation. The shaded part of the figure shows the difference between the smallest possible area and the largest possible area of the rectangle. There is a difference of 114 cm^2 in the two calculations for the area. Since half of 114 is 57, the computed area of 3024 cm^2 is about 57 cm^2 greater than the smallest possible area and about 57 cm^2 less than the greatest possible area.

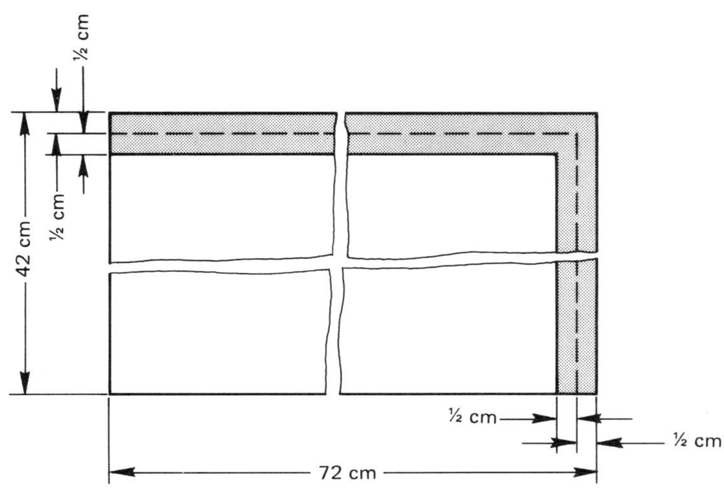

One way of expressing the approximate area, then, is to write

$$3024 \pm 57 \text{ cm}^2$$

We see that the measure of the computed area cannot be more accurate than two significant digits and may not even be that accurate. We also note that there are two significant digits in each of the measures of the sides of the rectangle. It should seem reasonable, then, to round the measure of the area of the rectangle (and hence the product of 72 and 42) to the same accuracy as the least accurate of the measures used in obtaining the measure of the area. In this instance, since there are two significant digits in each of the measures 72 and 42, we round the measure of the area to two significant digits and write $3\underline{0}00$ cm^2 for the area, where the first zero is underlined to show that it is a significant digit. Alternatively, we could write

$$3.0 \times 10^2 \text{ cm}^2$$

The following rule is frequently used when multiplying numbers obtained from measurements or from approximations of any kind:

When multiplying two or more numbers obtained as approximations, keep as many significant digits in the product as there are in the factor with the fewest significant digits.

Example 1: Find the area of a rectangle 3.8 m by 0.9 m.

Solution: $A = b \times h$

$$A = 3.8 \times 0.9 = 3.42 \approx 3 \text{ (to one significant digit)}$$

The area is 3 m^2.

Although 3.8 and 0.9 both have the same precision of measurement (tenth of an inch), 0.9 has only one significant digit and thus, according to our rule, our product should have one significant digit. We use the same rule for dividing numbers obtained as approximations. That is:

When dividing two or more numbers obtained as approximations, keep as many significant digits in the quotient as there are in the number with the fewest significant digits.

The following examples show what happens when we compute with measures that involve approximations or rounding of numbers.

Example 2: Find the area of a rectangle whose base is 10.2 cm and whose altitude is 6.7 cm.

Solution: We multiply the measures 10.2 and 6.7, obtaining 68.34. Since the least precise measurement, 6.7, has two significant digits, we round the answer to two significant digits and write 68 cm^2 for the area.

Should we have rounded the measure 10.2 to two significant digits first, and then multiplied? Had we done so, we would have obtained 10×6.7 cm^2, or 67 cm^2 for the area. But the actual area of the rectangle is between 10.15×6.65 cm^2 and 10.25×6.75 cm^2. Or, stating this in another way, we may write

$$10.15 \times 6.65 = 67.4975 < \text{measure of area} < 69.1875 = 10.25 \times 6.75$$

We see that the measure 68 lies in this interval, whereas the measure 67 does not. Therefore 68 cm^2 is a better measurement of the area than is 67 cm^2. For this reason, we use the following rule when multiplying or dividing two numbers arising from approximations and when one of the two numbers has more significant digits than the other:

If one of two numbers in a product or a quotient has more significant digits than the other, round off, if necessary, the one with more significant digits so that it has only one more significant digit than the least precise number. Then multiply or divide and, finally, keep as many significant digits in the answer as there are in the number with the fewer significant digits.

Example 3: Find the circumference of a circle whose diameter is 4.3 inches.

Solution: The circumference of a circle is found by using the formula $c = \pi d$ where d is the measure of the diameter and $\pi \approx 3.14159$. Since the measure of the diameter is given to two significant digits, we round π to 3.14 (three significant digits). The circumference is found by multiplying 3.14 by 4.3 and rounding the product to two significant digits. Thus $3.14 \times 4.3 = 13.502$ and the circumference is 14 inches to two significant digits.

Appendix C

U.S. Customary Measures and Metric Equivalents

In the following tables, some of the equivalents have been rounded off.

Length Equivalents

U.S. to metric	Metric to U.S.
1 in = 2.54 cm	1 cm = 0.394 in
1 ft = 0.305 m	1 m = 3.28 ft
1 yd = 0.915 m	1 m = 1.09 yd
1 mi = 1.61 km	1 km = 0.62 mi

Area Equivalents

U.S. to metric	Metric to U.S.
1 in^2 = 6.45 cm^2	1 cm^2 = 0.155 in^2
1 ft^2 = 0.093 m^2	1 m^2 = 10.76 ft^2
1 yd^2 = 0.836 m^2	1 m^2 = 1.196 yd^2
1 acre = 0.405 ha	1 ha = 2.47 acres

Volume Equivalents

U.S. to metric	Metric to U.S.
1 in^3 = 16.387 cm^3	1 cm^3 = 0.06 in^3
1 ft^3 = 0.028 m^3	1 m^3 = 35.315 ft^3
1 qt = 0.946 l	1 l = 1.06 qt
1 gal = 3.785 l	1 l = 0.264 gal

Mass Equivalents

U.S. to metric	Metric to U.S.
1 oz = 28.35 g	1 g = 0.035 oz
1 lb = 0.454 kg	1 kg = 2.205 lb

Answers Section

Margin Exercises Introduction

3. Correct 4. (a) 133,147 (b) 625 5. Correct 6. (a) 95,592 (b) 15,857 7. Correct
8. (a) 20,764,868 (b) 3887 (c) 28,728 9. Correct 10. (a) 78 (b) 12,276 11. Correct
12. Answers will vary. Some will, some will not.

CHAPTER 1

Margin Exercises 1.1

1. (a) Correct (b) 14,843 2. 14,833 3. 14,833 4. 14,833 5. 30,460 6. 183 7. 286 8. 300
9. 7400 10. 729 11. Associative property of addition 12. Commutative property of addition
13. Associative property of addition 14. Answers will vary. 15. Answers will vary. 16. 14,795 17. 147,950,000
18. 13,670 19. 139,913,885 20. \$1023 21. \$544 22. 38,172,000 cords

Exercises 1.1

1. Correct 3. 33,483 5. 14,986 7. 117 9. 186 11. 270 13. 1200 15. 320 17. 11,445,377
19. 8,907,097 21. 56,735,710 23. 916,728 sq miles 25. \$92,400,000,000 27. \$22,706 29. 104,071 bu
31. 96,095 bu 33. \$12,139,000,000 35. \$9,940,000,000
37. Answers will vary. One example is 2408 (Seattle–New York) + 1092 (New York–Miami) + 1397 (Miami–Omaha) + 1094 (Omaha–Philadelphia) + 1089 (Philadelphia–New Orleans) + 1051 (New Orleans–Minneapolis) + 290 (Minneapolis–Omaha) + 1369 (Omaha–Seattle) = 9790 mi.
39. Answers will vary.

Margin Exercises 1.2

1. Two hundred eighty-four thousand, nine hundred sixty
2. One hundred seventy-two million, four hundred fifty-six thousand, three hundred sixty-seven
3. One hundred six thousand, seventy 4. 9,304,750 5. 35,679 6. 174,026
7. (b) $1 \times 1000 + 8 \times 100 + 7 \times 10 + 5$ (c) $1 \times 10^3 + 8 \times 10^2 + 7 \times 10^1 + 5$
8. (b) $3 \times 100{,}000 + 6 \times 10{,}000 + 8 \times 1000 + 4 \times 100 + 9 \times 10 + 2$
   (c) $3 \times 10^5 + 6 \times 10^4 + 8 \times 10^3 + 4 \times 10^2 + 9 \times 10^1 + 2$
9. (a) 70 (b) 70, 700 (c) 700, 7000 (d) 7000, 70,000 (e) 70,000, 700,000 (f) 700,000, 7,000,000 (g) 7,000,000, 70,000,000
10. Ten 11. Hundred 12. Hundred thousand 13. Seven hundred 14. Seventy 15. Seven thousand

Exercises 1.2

1. One (1) 3. Hundred (100) 5. One (1) 7. Thousand (1000) 9. Million (1,000,000) 11. 30 (thirty)
13. 3 (three) 15. 3000 (three thousand) 17. 30 (thirty) 19. 300,000 (three hundred thousand)
21. (b) $3 \times 1000 + 5 \times 100 + 6 \times 10 + 7 \times 1$ (c) $3 \times 10^3 + 5 \times 10^2 + 6 \times 10^1 + 7 \times 10^0$
23. (b) $2 \times 100{,}000 + 8 \times 10{,}000 + 4 \times 1000 + 9 \times 100 + 6 \times 10 + 0 \times 1$
    (c) $2 \times 10^5 + 8 \times 10^4 + 4 \times 10^3 + 9 \times 10^2 + 6 \times 10^1 + 0 \times 10^0$
25. (b) $1 \times 100{,}000 + 0 \times 10{,}000 + 8 \times 1000 + 0 \times 100 + 4 \times 10 + 0 \times 1$
    (c) $1 \times 10^5 + 0 \times 10^4 + 8 \times 10^3 + 0 \times 10^2 + 4 \times 10^1 + 0 \times 10^0$
27. (b) $6 \times 10{,}000 + 0 \times 1000 + 0 \times 100 + 9 \times 10 + 0 \times 1$ (c) $6 \times 10^4 + 0 \times 10^3 + 0 \times 10^2 + 9 \times 10^1 + 0 \times 10^0$
29. Seven thousand, eight hundred, forty-nine 31. Seventy-one thousand, seventy-one
33. Seventy million, six hundred three thousand, forty-nine 35. Sixty billion, four million, eighty-four thousand, six hundred
37. 16,304 39. 300,016 41. 7,042,033 43. 80,090,040,020 45. 16,032,229,800,010 47. 9,876,543,210
49. 456, 465, 546, 564, 645, 654 51. 101 53. 999,999 55. 1000 57. 102 59. 10^{10}
61. Quadrillion, quintillion, sextillion, septillion, octillion

Margin Exercises 1.3

1. Incorrect (9081) **2.** Correct **3.** 18,179 **4.** 3670 **5.** 9 **6.** 2665 **7.** 44,973 **8.** 39,240 **9.** 39,240
10. 6143 **11.** 6143 **12.** 5196 – 3124 + 132 = 2204 **13.** 80,144 – 75,386 + 40,129 = 44,887
14. 6142 – 4324 – 1609 = 209 **15.** 17,969 – 6143 – 8965 = 2861 **16.** 4329 + 7812 – 1321 – 8692 = 2128
17. 14,129 + 29,207 – 17,283 – 6128 = 19,927 **18.** 3296 – 1324 – 1396 + 969 = 1545
19. 60,368 – 5364 – 50,680 + 10,823 = 15,147 **20.** –1805 **21.** Positive: yards gained; negative: yards lost
22. 632 **23.** 10,829 **24.** –2281 **25.** \$3 **26.** 248 feet **27.** \$1050

Exercises 1.3

1. 5213 **3.** 68,095 **5.** 2977 **7.** 187,157 **9.** 171,539 **11.** –7337 **13.** 67,253 **15.** 3697
17. 16,338 **19.** –3221 **21.** 865 **23.** 6570 **25.** 714,841 **27.** 25 ft wider, 500 ft higher **29.** 500,700 lb
31. \$79 **33.** 55 ft **35.** \$411 **37.** 1960-1970, \$1744 **39.** 1957-1958, \$22 **41.** 47,361,000,000,000,000 BTU
43. 1960-1970, 10,934,000,000,000,000 BTU **45.** 3,784,000,000,000,000 BTU **47.** 1,931,398 **49.** 552,129
51. **(a)** 6317 – 3499 = 2818, **(b)** 110,064 – 10,098 = 99,966

53.

18	(15)	12	26
(13)	25	19	14
24	(10)	(17)	(20)
16	21	23	11

55. There is more than one way. One example is:

6	1	8
7	5	3
2	9	4

Margin Exercises 1.4

1. Incorrect, 254,939 **2.** Correct **3.** 357,948; 357,948; 489 **4.** 1,044,288; 1,044,288 ÷ 84 = 12,432
5. 1,648,704; 1,648,704 ÷ 124 = 13,296 **6.** 690,000 **7.** 96,700 **8.** 322,000 **9.** 192,000 **10.** Commutative
11. Associative **12.** Associative **13.** Commutative **14.** Commutative **15.** Answers will vary.
16. **(a)** 54,625,000,000 **(b)** 57,010,800,000 **(c)** 1,839,330,000 **17.** 4^5 **18.** 5^6 **19.** 6^7 **20.** 336 cans
21. \$738 **22.** \$192 **23.** 420 miles

Exercises 1.4

1. 3,876,094 **3.** 4,922,368 **5.** Incorrect (28,500) **7.** Correct **9.** 16,851,000,000 **11.** 116,886,000,000
13. 140,055,300,000 **15.** 6200 **17.** 7500 **19.** 637,000 **21.** 485,000 **23.** 846,000 **25.** 362,500
27. 625 **29.** 3375 **31.** 5082 **33.** –802 **35.** 1 **37.** 5,764,801 **39.** 657 **41.** 375 owls
43. 648 cans **45.** 85,500,000 eggs **47.** 1,331,000 g **49.** 1,475,000 g **51.** 368,500,000,000 gal **53.** \$6600
55. \$3060 **57.** 1344 students **59.**

```
  427
   39
 ----
 3843
1281
-----
16653
```

61.

```
   4067
    825
   ----
  20335
  8134
32536
-------
3355275
```

63. Bell (7738)

Margin Exercises 1.5

1. Correct **2.** Incorrect (6166.139) **3.** 234; 234; 22,932 **4.** 2132; 2132 × 112 = 238,784
5. 546; 927 × 546 = 506,142 **6.** 8124; 8124 × 6923 = 56,242,452 **7.** 24; 1 **8.** 35; 576 **9.** 234
10. 234 **11.** 19,092 **12.** 19,092 **13.** Answers will vary. **14.** 0, 0, 0, 0; $0 \div a = 0$ if $a \neq 0$ **15.** 91
16. 225 mph **17.** 79 **18.** \$560 **19.** 19 bottles **20.** \$1235 **21.** \$1625 **22.** \$11 **23.** 75 wpm

Exercises 1.5

1. 56 **3.** 507 **5.** 42,860 **7.** **(a)** 1 **(b)** 2916 **9.** **(a)** 6962 **(b)** 2 **11.** **(a)** 3842 **(b)** 26
13. **(a)** 383,856 **(b)** 383,856
15. The order by which you divide and multiply a given number by the same two numbers does not change the answer to the computation.
17. \$205 **19.** 29 months **21.** 50° F **23.** 25 tickets **25.** 35 miles **27.** 475 mph **29.** 2640 steps
31. \$3.25 **33.** 23 **35.** 3 **37.** 35 min at 5.7 mph **39.** 4,500,000 **41.** 104,980,000
43. Yes. Example: $1 \div 1 = 1 \div 1$
45. The number formed is always divisible by 3, 7, 13, and 37. Examples: $323{,}232 \div 3 = 107{,}744$; $107{,}744 \div 7 = 15{,}392$; $15{,}392 \div 13 = 1184$; $1184 \div 37 = 32$.
47. Beige (36138)

Margin Exercises 1.6

1. 124,768 **2.** 353,574 **3.** $8288 - (13 \times 572) = 852$ **4.** $(128 \times 14) + 69 - (18{,}208 \div 32) = 1292$
5. $(437 \times 56) + (537 \times 172) + (38 \times 59) = 119{,}250$ **6.** 256 **7.** 6900 **8.** 36,400 **9.** 624 **10.** 2,222,158
11. 2,222,158 **12.** 211,945 **13.** 170 **14.** 6340 **15.** 627,228 **16.** 627,228 **17.** 232,324 **18.** 129,921
19. 2,921,670 **20.** 1,414,200 **21.** 1130 **22.** 1130 **23.** Ex. 21 **24.** 136,822 **25.** 136,822 **26.** Ex. 24
27. 42 **28.** 42 **29.** Ex. 27 **30.** 287,000 **31.** $1262 - 997 + 431 = 696$ **32.** $5369 - 573 - 89 = 4707$
33. $469 + 872 - 891 - 67 = 383$ **34.** $468 - 322 - 189 + 72 = 29$ **35.** 542 **36.** 544
37. $(68 \times 27) + 392 = 2228$ **38.** $(68 + 112) \times 92 = 16{,}560$ **39.** $(215 + 1122) \times 89 = 118{,}993$
40. $(118 - 105) \times 97 = 1261$ **41.** $(41 - 39) \times 52 = 104$ **42.** $(2832 + 360) \div 24 = 133$ **43.** $(672 - 462) \div 46 = 5$
44. Answers will vary. **45.** 932; 567; 528,444; 1000; 163,296; 296,000; 528,444; 824,444; 163,824,444
46. Correct **47.** 29,186 **48.** Correct **49.** 54,199 **50.** 91
51. Estimate: 35,000,000; Computation: 36,951,436 **52.** 75,000,000; 79,116,288 **53.** 1,500,000; 1,273,788
54. 829 **55.** 829 **56.** They are the same. **57.** 347 **58.** 45 **59.** 10,057
60. Key items are: one, \$16, 18, \$17, how much did he receive altogether; the first sentence is not relevant.
61. $16 + (17 \times 18) =$ total sales **62.** $(17 \times 18) + 16 =$ total sales **63.** 320 **64.** 322 **65.** Altogether he makes \$322.
66. Key items are: \$265; savings account; 39 hours; \$4 per hour, deducts; \$35; deposits; receives for his work; how much will he then have in his savings account.
67. $265 + (39 \times 4 - 35) =$ amt. in savings **68.** $(39 \times 4) - 35 + 265 =$ amt. **69.** \$390 **70.** \$386
71. He will have \$386 in his savings account. **72.** 1215 lb

Exercises 1.6

1. Correct **3.** Correct **5.** −4447 **7.** 2626 **9.** 71,367 **11.** 2471 **13.** 1449 **15.** 1009
17. $12 \times (9 + 8) = 204$ **19.** None needed **21.** $12 \times (9 + 4) \times 13 = 2028$ **23.** $12 \times (9 - 4) \times 13 = 780$
25. $23 + 16 \times (15 + 12) = 455$ **27.** $(55 + 80) \times 864 = 116{,}640$ **29.** $(268 - 243) \times 542 = 13{,}550$
31. $(10{,}904 - 6768) \div 188 = 22$ **33.** $(187 - 45) \times 232 = 32{,}944$ **35.** $(56{,}325 + 27{,}900) \div 75 = 1123$
37. 2684 **39.** 7 **41.** 768,930 **43.** 2602 **45.** 255,245 **47.** $(3 \times 45) + (7 \times 17) =$ T.S.; \$254
49. $10{,}470 - (12 \times 234) =$ T.H.P.; \$7662 **51.** $(215 - 169) \times 35 = P$; \$1610
53. $1110 + (30 - 10) \times 2 = s$; 1150 f.p.s. **55.** $(55 - 43) \times 12 = s$; \$1.44 **57.** $3 \times 129 + 250 =$ T.C.; \$637
59. $9000 + 6 \times 150 = A$; $8400 + 6 \times 250 = A$; both \$9900
61. $186{,}262 \times 31{,}536{,}000 = 5{,}873{,}958{,}432{,}000$ miles
63. Yes; $\left(\frac{6N + 12}{3} - 2\right) \div 2 - N + 9 = 10$
65. Sleigh (461375)

Review Exercises for Chapter 1

1. 3700 **3.** 62,988,209 **5.** 72,137 **7.** 9702 **9.** 51,710,208 **11.** –41 **13.** –20,454 **15.** 325
17. $4 \times 10^4 + 2 \times 10^3 + 8 \times 10^2 + 3 \times 10^1 + 2 \times 10^0$
19. $6 \times 10^6 + 3 \times 10^5 + 3 \times 10^4 + 2 \times 10^3 + 7 \times 10^2 + 0 \times 10^1 + 9 \times 10^0$ **21.** 10 **23.** 100 **25.** 10,000,000
27. 700 **29.** 70,000 **31.** Twenty-seven thousand, one hundred fifty-five **33.** Twenty-five thousand thirteen
35. Seventy thousand ninety-one **37.** 8072 **39.** 27,009 **41.** $165 \times 52 = A$; \$8580
43. $2324 \times 2 \times 5 \times 4 = A$; 92,960 parts **45.** $1{,}219{,}680 \div 1386 = W$; \$880
47. $7834 - (84 \times 74) = A$; \$1618 more needed **49.** $6 \times 55 = D$; 330 miles
51. $(1{,}225{,}465 + 65{,}750 + 104{,}785) - 700{,}000 = C$; \$696,000
53. $(1490 + 4278 + 4678) \times 1{,}000{,}000 = V$; \$10,446,000,000
55. $[(5 \times 778) - (5 \times 408)] \times 1{,}000{,}000 = V$; \$1,850,000,000
57. $[(3 \times 140) + (10 \times 300) + 388] \div 14 = s$; \$272 **59.** $(68 + 75 + 83 + 77 + 79 + 72 + 78) \div 7 = A$; 76
61. $(170 + 190) - (160 + 180) = A$; \$20 gain **63.** 1070 **65.** 111,111 + 11,111 + 1111 + 111 + 11 + 1 = 123,456
67. 64

Just for Fun—Chapter 1

1. 162; 486; 1458; 4374; 13,122 **3.** 13, 21, 34, 55, 89 **5.** 1, 11, 21, 31, 41, 51, 61, 71, 81, 91
7. 638 + 475 + 253 = 1366 **9.** $4 \times 30 \times 2224 = D$; 266,880 yards **11.** Answers will vary; see example.

13.

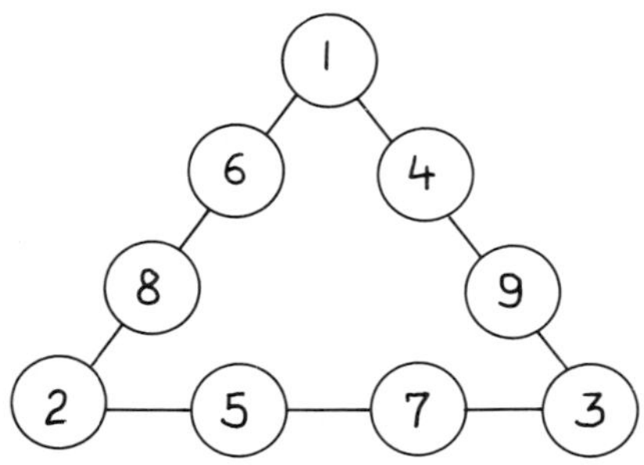

15. Oboe; 3080

Mathematics and Recreation

Bowling: **3.** 176 **5.** 300
Golf: **1.** 1 → 4, 2 → 4, 3 → 5, 4 → 3, 5 → 4, 6 → 4, 7 → 5, 8 → 5, 9 → 3, 10 → 4, 11 → 5, 12 → 3, 13 → 4, 14 → 4, 15 → 5, 16 → 3, 17 → 5, 18 → 4
3. First 9: 3170 yd; second 9: 3390 yd; total 6560 yd **5.** greater **7.** 4

CHAPTER 2

Margin Exercises 2.1

1. 10; 10; 100; 100; 1000; 1000; 10,000; 10,000; 100,000 **2.** 0.1; 0.1; 0.01; 0.01; 0.001; 0.001; 0.0001; 0.0001; 0.00001
3. **(a)** 2000 + 300 + 40 + 6 + 0.5 + 0.07 + 0.009 **(b)** $2 \times 1000 + 3 \times 100 + 4 \times 10 + 6 \times 1 + 5 \times 0.1 + 7 \times 0.01 + 9 \times 0.001$
4. One thousand **5.** Three thousand **6.** One hundredth **7.** Seven hundredth
8. Two thousand three hundred forty-six and five hundred seventy-nine thousandths **9.** 27; 270; 2.7; 270; 2.7; 27; 270
10. 2560; 19.23; 1143.29; 82,600 **11.** 2.7; 0.27; 0.027; 0.027; 0.00027
12. 0.256; 0.0256; 0.011349; 0.000321
13. $\frac{45}{10}$; $\frac{56}{100}$; $\frac{21}{1000}$; $\frac{257}{100{,}000}$ **14.** 70 **15.** 500 **16.** 51,000 **17.** 69,370 **18.** 750,400 **19.** 19,000
20. 76,100 **21.** 27,569,000 **22.** 27,570,000 **23.** 27,569,550 **24.** 299.584 **25.** 299.58 **26.** 299.6
27. 300 **28.** 300 **29.** 300 **30.** 100.00

Exercises 2.1

1. (a) $100 + 30 + 8 + 0.2 + 0.03$ (b) $1 \times 100 + 3 \times 10 + 8 \times 1 + 2 \times 0.1 + 3 \times 0.01$
3. (a) $300 + 7 + 0.02 + 0.0005$ (b) $3 \times 100 + 0 \times 10 + 7 \times 1 + 0 \times 0.1 + 2 \times 0.01 + 0 \times 0.001 + 5 \times 0.0001$
5. (a) $200 + 3 + 0.008$ (b) $2 \times 100 + 0 \times 10 + 3 \times 1 + 0 \times 0.1 + 0 \times 0.01 + 8 \times 0.001$
7. 49.716 9. 73.543 11. 33,030.03 13. Fifty-seven hundredths
15. Twenty-one and one hundred thirty-six thousandths
17. Six thousand seven hundred fifteen and eight thousand six hundred nine ten-thousandths
19. 13.08 21. 503.047 23. 0.000024 25. One 27. Tenths 29. Hundredths 31. Eight-tenths
33. Eight-hundredths 35. Eight-thousandths 37. 0.35 39. 0.053 41. 0.0073 43. $\frac{19}{100}$ 45. $\frac{405}{10,000}$
47. $\frac{231}{100}$ 49. 6.104 51. 8.01 53. 26.67 55. Answers will vary. 57. Answers will vary.
59. Answers will vary.

Margin Exercises 2.2

1. 15.905 2. 36.296 3. 0.1870128 4. 0.4432456 5. 24.666666 mpg 6. 81.7 cm^2 (81.6714)
7. $5552.76 (5552.7567) 8. 1035; 103.5; 10.35; 10.35; 1.035; 0.1035; 0.001035; 0.0001035
9. 0.08; 0.0008; 0.0006; 0.63; 0.00026; 0.001204 10. 0 11. 0 12. 0 13. 0 14. No
15. 0.0001; 0.0001 16. –0.0001; $0.302 > 0.3019$ 17. 4.93, 4.8, 4.72, 4.65 18. 0.0001, 0.001, 0.01, 0.1
19. 0.47, 0.465, 0.4053, 0.405

Exercises 2.2

1. 226.3594 3. 63.000153 5. 0.015447 7. Correct 9. 0.0015 11. 0.00014 13. 0.072 15. 5.4
17. 2.2, 0.31, 0.301, 0.29 19. 0.0001, 0.001, 0.01, 0.1 21. 0.70, 0.683, 0.6126, 0.61
23. 0.007, 0.5483, 0.603, 6.02 25. 0.983, 0.98, 0.9, 0.89 27. 12.4, 12.3567, 12.35, 12.3498, 12.3 29. 2.07 inches
31. 33.75 inches 33. $11.37 (11.3742) 35. $5.41 (5.4057142) 37. 68.9¢ (0.6885057) 39. $27
41. 1357.63 acres 43. 18.626 inches 45. 1.66 mpg decrease 47. 65,686,401 cars 49. 0.8 gal 51. 35 miles
53. Guess at a multiplication—keep trying—finally $16 \times 34 = 544$. Thus $544 \div 16 = 34$. 55. $1 + 2 + 3 = 1 \times 2 \times 3$

Margin Exercises 2.3

1. 0.329042 2. 0.1895199 3. 0.0256055 4. 0.0000494 5. 0.0333177 6. 59.619622 7. 26.659845
8. 16.964750 9. $1.48 (1.482) 10. 22.6 mpg (22.601449) 11. 34.0501 12. 29.21510
13. 40 pieces (40.145833)

Exercises 2.3

1. 5.769 3. 7.187 5. 25.206 7. 21.41335 9. 80.609786 11. 9.3642583 13. 1067
15. 2.9727623 17. 53.006173 19. 7.4620207 21. $12 \times 1.37 = R$, 16.44 inches
23. $24.18 \times 0.6 = M$, 14.508 miles 25. $47.6 \div 1.5 = B$, 31 boxes (31.733333) 27. $47,869 \div 1.25 = B$, 38,295.2 bu
29. $(39.5 + 36.75 + 42.12 + 37.87 + 40.01) \div 5 = A$, 39.25 inches
31. $(13,536.2 - 13,243.7) \div 15 =$ mpg, 19.5 mpg 33. $21.25 \div (0.035 + 0.050) = N$, 250 sheets
35. $25 + [(3.70 - 2.50) \div 0.08] = N$, 40 pieces 37. $(42 \times 294.99 + 8 \times 224.99) - (50 \times 156.32) = P$, $6373.50
39. $(48 \times 5.25) \div 37 = R$, 7 rolls (6.8108108) 41. $100 \times 3 \div 5280 \div 9.1 \times 3600 = s$, 22.48 mph (22.477521)
43. Begs (5638)

Review Exercises for Chapter 2

1. **(a)** 90 + 4 + 0.5 + 0.05 **(b)** 9 × 10 + 4 × 1 + 5 × 0.1 + 5 × 0.01
3. **(a)** 0.4 + 0.03 + 0.001 + 0.0009 **(b)** 4 × 0.1 + 3 × 0.01 + 1 × 0.001 + 9 × 0.0001
5. **(a)** 0.009 + 0.00002 + 0.0000009 **(b)** 9 × 0.001 + 0 × 0.0001 + 2 × 0.00001 + 0 × 0.000001 + 9 × 0.0000001
7. Two and four hundred eighty-one thousandths **9.** Eight thousand seven hundred sixty-seven hundred thousandths
11. 70.146 **13.** 8998.037 **15.** 0.000025 **17.** Hundred **19.** Thousandths **21.** Seven-hundredths
23. Seven-hundredths **25.** Seven hundred **27.** 190 **29.** 0.964 **31.** 8.97 **33.** 2.0 **35.** 29.005
37. 0.01161 **39.** 10.042 **41.** 0.3608 **43.** 1899 **45.** 5668 **47.** 24.689021 **49.** 2.2160168
51. 1.209785 **53.** 69.952967 **55.** 68.368 **57.** 14,132.366 **59.** 1139.3109
61. 0.037 × 367 × 40 × 50 = A, $27,158 **63.** (8.16 − 7.37) ÷ 2 = T, 0.395 cm
65. 2.15 + [(25 − 3) × 0.07] = C, $3.69 **67.** (300 ÷ 3.2) × 8 = R, 750 miles
69. 8.25 + 9.09 + 7.00 + 6.51 = L; 31 feet (30.85); 0.15 feet
71. 4013.2 − 3827.4 = D, 185.8 miles; 185.8 ÷ 6 = S, 31 mph (30.96); 185.8 ÷ 11.2 = mpg, 16.6 mpg (16.589285)
73. (6.59 + 2.47) ÷ 3 = W, 3.02 g **75.** 100 ÷ 0.625 = N, 160 bolts
77. 174 ÷ (678 × 2) = C, 12.8¢ per mile (0.1283185) **79.** $1.65

Just for Fun—Chapter 2

1. 32, 65, 128, 256, 512 **3.** 22, 29, 37, 46, 56 **5.** 81; 9801; 998,001; 99,980,001; 9,999,800,001; 999,998,000,001
7. 100,064 − 10,098 = 99,966 **9.** 11 days, 13 hr, 46 min, 40 sec; 6.9 weeks (6.9444444)
11. Answers will vary. One example is: 1 + 2 + (3 × 4) − 5 − 6 + 7 + 8 − 9 = 10.

13. Answers will vary. One example is given:

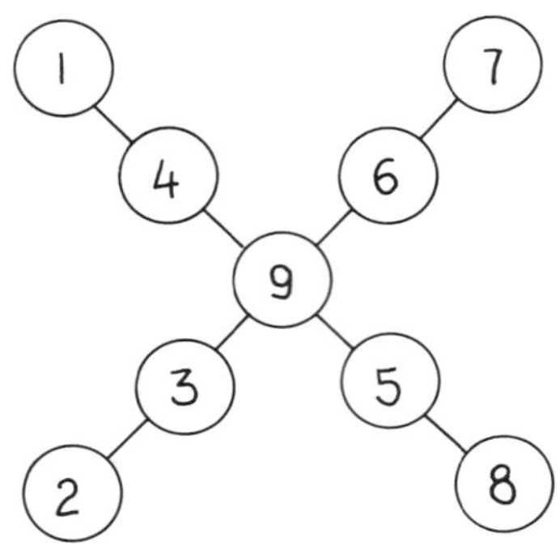

15. Helios (501734)

Mathematics and Nutrition

1. **(a)** 425 calories **(b)** 31 g carbohydrate **3.** **(a)** 1075 calories **(b)** 89 g carbohydrate
5. Answers will vary. One example is given:
(a) Calories needed: 2565

Breakfast		*Lunch*		*Dinner*	
$\frac{1}{2}$ grapefruit	45 calories	3 oz hamburger	245 calories	Baked potato	90 calories
2 eggs	160	2 bread	130	2 pork chops	520
2 bacon	90	French fries	310	1 cup peas	115
2 toast	130	Cola	145	Apple pie	350
Black coffee	0	Apple	70	1 cup milk	160
	425 calories		900 calories		1235 calories

Total calories: 425 + 900 + 1235 = 2560 ≈ 2565

(b) Calories needed: 1845

Breakfast		*Lunch*		*Dinner*	
1 egg	80 calories	Beef stew	210 calories	Macaroni and cheese	430 calories
2 toast	130	Milk	160	$\frac{1}{2}$ cup peas	58
$\frac{1}{2}$ grapefruit	45	Pie	350	Beans	230
Black coffee	0	Orange	65	Apple	70
	255 calories		785 calories		788 calories

Total calories: 255 + 785 + 788 = 1828 ≈ 1845

7. (a) Answers will vary. One example is given:

Breakfast		*Lunch*		*Dinner*	
1 egg	T g	3 oz hamburger	0 g	Beef stew	15 g
1 toast	10 g	2 bread	20 g	1 banana	23 g
$\frac{1}{2}$ grapefruit	11 g	1 apple	18 g		38 g
	21 g		38 g		

Total carbohydrate: $21 + 38 + 38 = 97$ g $\approx$ 100 g

(b) Answers will vary. One example is given:

Breakfast		*Lunch*		*Dinner*	
1 egg	T g	4 oz hamburger	0 g	Pork chops	0 g
2 bacon	T g	1 bread	10 g	$\frac{1}{2}$ cup peas	9.5 g
$\frac{1}{4}$ grapefruit	5.5 g	$\frac{1}{2}$ banana	12 g	$\frac{1}{2}$ baked potato	14 g
	5.5 g		22 g		23.5 g

Total carbohydrate: $5.5 + 22 + 23.5 = 51$ g

9. 272 calories per day 11. Answers will vary.

CHAPTER 3

Margin Exercises 3.1

1. No 2. Yes 3. No 4. No 5. Yes 6. 56 × 109 7. 227 × 318 8. 5842 × 1 9. 1 × 10,756
10. 981 × 52,314 11. 2, 5, 10 yes 12. 2, 5, 10 no 13. 2 yes; 5, 10 no 14. 5 yes; 2, 10 no
15. 2 yes; 5, 10 no; even 16. 5 yes; 2, 10 no; odd 17. 2, 5, 10 no; odd 18. 2, 5, 10 yes; even
19. 3, 4, 6 yes; 9 no 20. 3 yes; 4, 6, 9 no 21. 3, 4, 6, 9 yes

Exercises 3.1

1. 2, 5, 10 no 3. 2, 5, 10 yes 5. 2, 5, 10 yes 7. 5 yes; 2, 10 no 9. 2, 5, 10 yes 11. 3, 6, 9 13. 3, 4, 6
15. 3, 4, 6, 9 17. 3, 6 19. 3 21. 1, 2, 4, 8 23. 1, 3, 5, 15 25. 1, 2, 4, 8, 16, 32 27. 1, 3, 17, 51
29. 1, 67
31. (a) 4: 1, 2, 4; 9: 1, 3, 9; 14: 1, 2, 7, 14; 16: 1, 2, 4, 8, 16; 18: 1, 2, 3, 6, 9, 18; 21: 1, 3, 7, 21
(b) 2: 1, 2; 3: 1, 3; 5: 1, 5; 7: 1, 7; 11: 1, 11; 13: 1, 13
(c) (1) Three or more factors for each number (2) Only two factors
33. Answers will vary. One example is: 400,005. 35. Answers will vary. One example is: 333,333. 37. 1, 5, 7, 11
39. Answers will vary. One example is: $736 - 637 = 99$, $99 \div 11 = 9$.
41. Answers will vary. Two examples are: $13 \times (13^2 + 1) = 2210$, $2210 \div 2 = 1105$; $12 \times (12^2 + 1) = 1740$, $1740 \div 2 = 870$.
43. Answers will vary. One example is: $(21 \times 22 \times 23 \times 24) \div 4 = 63{,}756$.
45. 1243; 1342; 2134; 2431; 3124; 3421; 4213; 4312 47. No. One example is: $(5 + 9) \div 3 = 14 \div 3 = 4.\overline{6}$ 49. 25
51. See *Mathematics Teacher,* April 1968, pp. 370-373, published by the National Council of Teachers of Mathematics, Reston, Virginia.

Margin Exercises 3.2

1. 1:1; 2: 1, 2; 3: 1, 3; 4: 1, 2, 4; 5: 1, 5; 6: 1, 2, 3, 6; 7: 1, 7; 8: 1, 2, 4, 8; 9: 1, 3, 9; 10: 1, 2, 5, 10
2. 11, 13, 17, 19, 23 3. 12, 14, 15, 16, 18 4. 2, 3, 5, 7, 11, 13, 17, 19, 23, 29, 31, 37, 41, 43, 47
5. Philosophically–yes; physically–no 6. Yes

Exercises 3.2

1. 2, 3, 5, 7, 11, 13, 17, 19, 23, 29, 31, 37, 41, 43, 47, 53, 59, 61, 67, 71, 73, 79, 83, 89, 97 **3.** More composite
5. False. 15 is both odd and composite. **7.** True **9.** 293 **11.** C **13.** C **15.** C **17.** P **19.** P **21.** P
23. C **25.** C **27.** C **29.** C
31. **(a)** Answers will vary. One example is:
$11 + 1 = 12$; $2 \times 6 = 12$
$29 + 1 = 30$; $6 \times 5 = 30$
$53 + 1 = 54$; $6 \times 9 = 54$
(b) Answers will vary. One example is:
$13 - 1 = 12$; $2 \times 6 = 12$
$31 - 1 = 30$; $6 \times 5 = 30$
$73 - 1 = 72$; $6 \times 12 = 72$
33. **(a)** 29, 31, 37, 41, 43, 47 **(b)** Answers will vary. **35.** 1, 3, 7, 9 **37.** No; none
39. Answers will vary. One example is: $2^2 - 2 = 2$; $3^2 - 2 = 7$; $7^2 - 2 = 47$; $9^2 - 2 = 79$; $13^2 - 2 = 167$; $15^2 - 2 = 223$
41. Answers will vary. One example is: $2 \times 2 + 1 = 5$; $2 \times 3 + 1 = 7$; $2 \times 5 + 1 = 11$
43. 5, 7; 11, 13; 17, 19; 29, 31; 41, 43; 59, 61; 71, 73
45. Answers will vary. One example is: Even: $4 = 2 + 2$; $6 = 3 + 3$; $8 = 5 + 3$; $10 = 7 + 3$; $12 = 5 + 7$
Odd: $9 = 5 + 2 + 2$; $11 = 5 + 3 + 3$; $13 = 5 + 5 + 3$; $15 = 5 + 5 + 5$; $17 = 5 + 5 + 7$
47. 2 and any other prime that gives a remainder of 1 when divided by 4 (5, 13, 17, 29, 37, . . .)
49. Answers will vary. One example is: 24, 25, 26, 27, 28.

Margin Exercises 3.3

1. Correct **2.** Incorrect **3.** Incorrect **4.** Correct **5.** Correct **6.** Incorrect **7.** Correct **8.** Prime
9. $2 \times 3 \times 7 \times 31$ **10.** Prime **11.** $2 \times 19 \times 41$ **13.** $19 \times 19 \times 23$

14. Answers will vary. One example is:

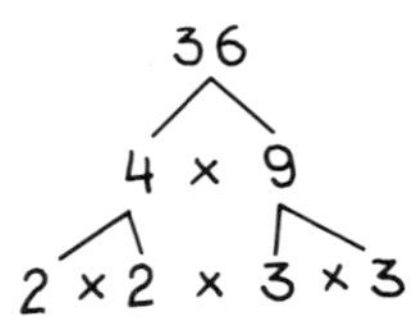

Exercises 3.3

1. 700 **3.** 2700 **5.** 240 **7.** 298,116 **9.** 44,955,000
11. Answers will vary. One example is:

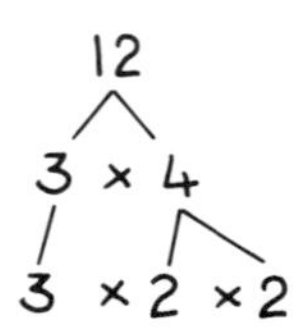

13. Answers will vary. One example is:

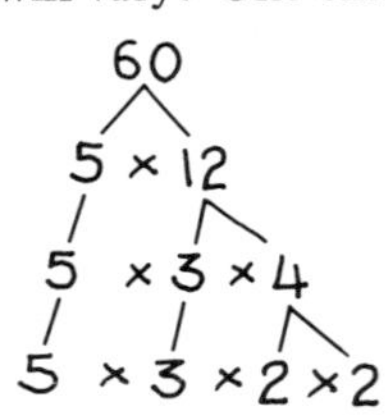

15. Answers will vary. One example is:

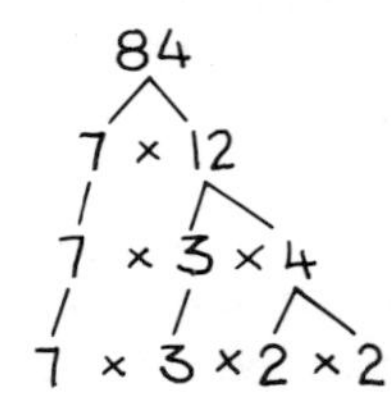

17. $2^2 \times 3 \times 13$ **19.** $2^5 \times 3^2$ **21.** $2^3 \times 3 \times 13$ **23.** $3 \times 5 \times 23$ **25.** $2^3 \times 211$ **27.** 5×691
29. $3 \times 5 \times 11 \times 17$ **31.** 7 **33.** 5 **35.** 101 **37.** 197 **39.** 173 **41.** 1, 2, 3, 6, 9, 27, 54
43. 1, 2, 3, 4, 5, 6, 8, 12, 15, 20, 30, 40, 60, 120 **45.** 1, 2, 3, 4, 5, 6, 10, 12, 15, 20, 25, 30, 50, 75, 100, 150, 300
47. 1 and the prime number
49. $2^{2^1} + 1 = 4 + 1 = 5$; $2^{2^2} + 1 = 16 + 1 = 17$; $2^{2^3} + 1 = 256 + 1 = 257$; $2^{2^4} + 1 = 65{,}536 + 1 = 65{,}537$

Margin Exercises 3.4

1. 3, 5, 7, 1, 2, 2, 1, 2, 2, 3675 **2.** 6; 2×3 **3.** 4; 2^2 **4.** 3 **5.** 12; $2^2 \times 3$ **6.** Relatively prime
7. Not relatively prime **8.** Not relatively prime **9.** Relatively prime **10.** Relatively prime
11. Not relatively prime **12.** Relatively prime **13.** 3 **14.** 1 **15.** 1 **16.** 7 **17.** 1 **18.** 180; $2^2 \times 3^2 \times 5$
19. 144; $2^4 \times 3^2$ **20.** 378; $2 \times 3^3 \times 5 \times 7$ **21.** 2250; $2 \times 3^2 \times 5^3$ **22.** 5928; $2^3 \times 3 \times 13 \times 19$ **23.** a; 1
24. a; b

Exercises 3.4

1. 6, 12, 18, 24, 30, 36, 42, 48, 54, 60 **3.** 7, 14, 21, 28, 35, 42, 49, 56, 63, 70
5. 12, 24, 36, 48, 60, 72, 84, 96, 108, 120 **7.** 20, 40, 60, 80, 100, 120, 140, 160, 180, 200 **9.** 42 **11.** 12
13. 40 **15.** 30 **17.** $2^2 \times 3 = 4 \times 3 = 12$ **19.** $3 \times 5 \times 7 \times 11 = 1155$ **21.** $3^2 \times 11^2 = 9 \times 121 = 1089$
23. $2^3 \times 3 \times 5 = 8 \times 3 \times 5 = 120$ **25.** $2^3 \times 3 \times 7 = 168$ **27.** $2 \times 3^2 \times 5 \times 11 = 990$ **29.** $2^2 \times 3^3 \times 5 = 540$
31. $2^5 \times 3^2 \times 7 = 2016$ **33.** $2 \times 3 \times 5 \times 11 \times 19 = 6270$ **35.** $2^4 \times 3^4 \times 5 \times 17 = 110{,}160$ **37.** 52, 28; 4
39. 475, 75; 25 **41.** 300, 2250; 150 **43.** 360, 180, 300; 60 **45.** 975; 650; 38,025; 325 **47.** 6 **49.** 1
51. 7 **53.** 22 **55.** 5 **57.** 10:00 AM **59.** 1:12 AM **61.** Yes. Example, 8 and 15 **63.** If their GCD is 1
65. *b*

Margin Exercises 3.5

1. 211 **2.** 2311 **3.** $510{,}511 = 19 \times 26{,}869$ **4.** $9{,}699{,}691 = 347 \times 27{,}953$ **5.** 2309 **6.** $510{,}509 = 61 \times 8369$

Review Exercises for Chapter 3

1. $2 \times 3 \times 7$ **3.** $2^2 \times 3 \times 5$ **5.** $2^2 \times 3 \times 11$ **7.** $2^2 \times 3^2 \times 5$ **9.** $3^2 \times 5^2$ **11.** $2^5 \times 3 \times 5$ **13.** 2^{10}
15. $2^4 \times 3 \times 37$ **17.** 36 **19.** 30 **21.** 864 **23.** 195 **25.** 360 **27.** 6 **29.** 16 **31.** 6 **33.** 1
35. 6 **37.** 9 inches; 8 red, 5 white, 12 blue **39.** 60 ties **41.** False, 1 **43.** True **45.** False, 2
47. False, $2 + 3 = 5$ **49.** True **51.** True **53.** True **55.** 2 **57.** 4, 9 **59.** 210 **61.** 13, 31; 37, 73; 79, 97
63. Deficient **65.** Abundant **67.** Abundant **69.** Abundant **71.** Abundant
73. $1 + 3 + 5 + 7 + 15 + 21 + 27 + 35 + 45 + 63 + 105 + 135 + 189 + 315 = 975 > 945$

Just for Fun—Chapter 3

1. 7.0, 8.4, 9.8, 11.2, 12.6 **3.** 2.25, 3, 3.875, 4.875, 6
5. 81; 8811; 888,111; 88,881,111; 8,888,811,111; 888,888,111,111
7.

$$\begin{array}{r} 7853 \\ \times\ 608 \\ \hline 62824 \\ 4711800 \\ \hline 4774624 \end{array}$$

9. 1 hr, 23 min, 19 sec **11.** Answers will vary. One example is: $1 + 2 + 3 + 4 + 5 + 6 + 7 + (8 \times 9) = 100$

13. Answers will vary. One example is:

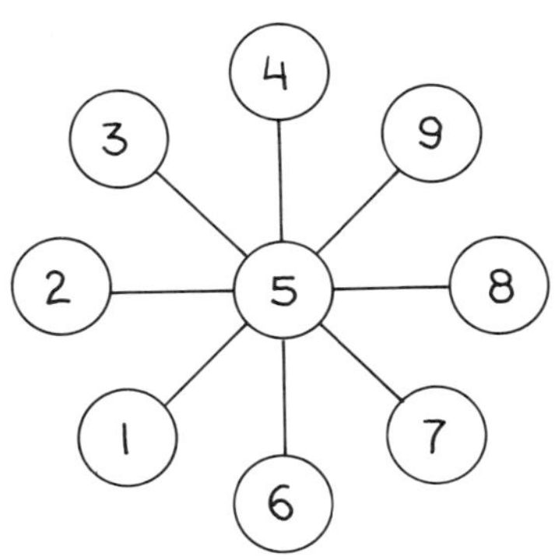

15. Isis (5151)

Mathematics and the Supermarket

1. Bacon: A. 8.25¢/oz B. 10.56¢/oz C. 8.17¢/oz. Cheese: A. 8.17¢/oz B. 9.92¢/oz C. 10.75¢/oz
Tide: A. 2.73¢/oz B. 2.36¢/oz C. 2.37¢/oz. Peanut butter: A. 3.58¢/oz B. 5.44¢/oz C. 4.21¢/oz
Lemonade: A. 2.08¢/oz B. 2.42¢/oz C. 2.38¢/oz
3. A: $35.62; B: $44.39; C: $37.87
5. Cost of items individually is $34.57. Store A: $35.62, more by $1.05; Store B: $44.39, more by $9.82; Store C: $37.87, more by $3.30
7. A: $30.00; B: $26.39; C: $26.00

CHAPTER 4

Margin Exercises 4.1

1. $\frac{1}{3}$ 2. $\frac{2}{4}$ 3. $\frac{3}{8}$ 4. $\frac{3}{6}$ 5. $\frac{1}{5}$ 6. $\frac{4}{5}$ 7. $\frac{8}{5}$ 8. $\frac{12}{5}$ 9. (a) 5 (b) 16 (c) Proper

10. (a) 13 (b) 8 (c) Improper 11. (a) 15 (b) 15 (c) Improper 12. (a) $\frac{637}{793}$ (b) $\frac{49}{61} = 0.8032786$; $\frac{637}{793} = 0.8032786$

13. (a) $\frac{69}{82}$ (b) $\frac{3933}{4674} = 0.8414634$; $\frac{69}{82} = 0.8414634$ 14. $\frac{5}{6}$ 15. $\frac{27}{64}$ 16. $\frac{4}{3}$

Exercises 4.1

1. $\frac{4}{6}$ 3. $\frac{3}{8}$ 5. Answers will vary. One example is:

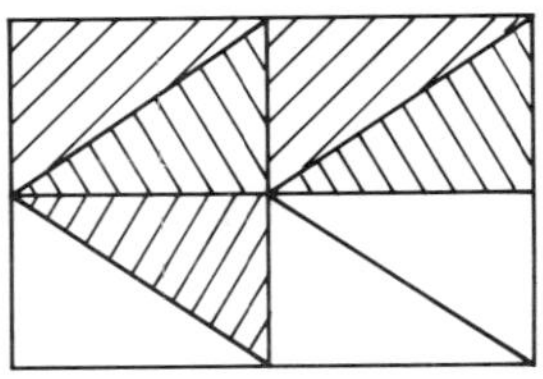

7. Answers will vary. One example is:

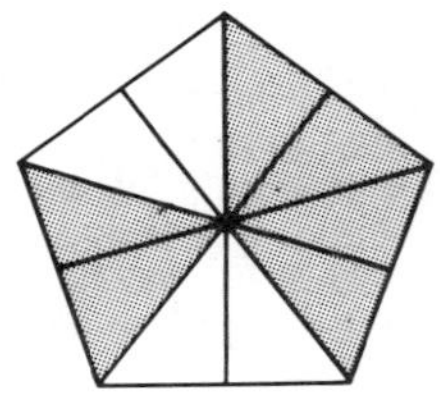

9. Answers will vary.

11. (a) $\frac{90}{133}$ (b) $\frac{43}{133}$ 13. (a) $\frac{216}{1164}$ (b) $\frac{240}{1164}$ (c) $\frac{108}{1164}$ 15. Improper 17. Improper 19. Proper

21. 2, 3, 5, 6, 12 23. 10, 20, 35, 60, 75 25. 26, 95, 143, 247, 351

27. Answers will vary. One example is: $\frac{3}{2}, \frac{6}{4}, \frac{12}{8}, \frac{15}{10}, \frac{18}{12}$

29. Answers will vary. One example is: $\frac{214}{206}, \frac{321}{309}, \frac{428}{412}, \frac{535}{515}, \frac{642}{618}$ 31. $\frac{7}{9}$ 33. $\frac{9}{16}$ 35. $\frac{5}{1}$ 37. $\frac{3}{4}$

39. $\frac{5}{4}$ 41. $\frac{3}{4}$ 43. $\frac{25}{3}$ 45. $\frac{23}{24}$ 47. $\frac{55}{226}$ 49. Answers will vary.

51. (a) $\frac{2}{2}, \frac{2}{5}, \frac{2}{7}, \frac{2}{9}, \frac{5}{5}, \frac{5}{2}, \frac{5}{7}, \frac{5}{9}, \frac{7}{7}, \frac{7}{2}, \frac{7}{5}, \frac{7}{9}, \frac{9}{9}, \frac{9}{2}, \frac{9}{5}, \frac{9}{7}$ (b) Smallest: $\frac{2}{9}$; largest: $\frac{9}{2}$

(c) Less than 1: $\frac{2}{5}, \frac{2}{7}, \frac{5}{7}, \frac{5}{9}, \frac{7}{9}$ Greater than 1: $\frac{5}{2}, \frac{7}{2}, \frac{7}{5}, \frac{9}{2}, \frac{9}{5}, \frac{9}{7}$ Equal to 1: $\frac{2}{2}, \frac{5}{5}, \frac{7}{7}, \frac{9}{9}$

(d) $\frac{2}{2} = \frac{5}{5} = \frac{7}{7} = \frac{9}{9}$ (e) $\frac{2}{9}, \frac{2}{7}, \frac{2}{5}, \frac{5}{7}, \frac{5}{9}, \frac{7}{9}, 1 \left(\frac{2}{2}, \frac{5}{5}, \frac{7}{7}, \frac{9}{9}\right), \frac{9}{7}, \frac{7}{5}, \frac{9}{5}, \frac{5}{2}, \frac{7}{2}, \frac{9}{2}$

53. Answers will vary. One example is:

13: (3, 5, 7, 9) $\frac{3}{3}, \frac{3}{5}, \frac{3}{7}, \frac{3}{9}, \frac{5}{3}, \frac{5}{7}, \frac{5}{9}, \frac{7}{3}, \frac{7}{5}, \frac{7}{9}, \frac{9}{3}, \frac{9}{5}, \frac{9}{7}$

11: (1, 2, 3, 4) $\frac{1}{1}, \frac{1}{2}, \frac{1}{3}, \frac{1}{4}, \frac{2}{1}, \frac{2}{3}, \frac{3}{1}, \frac{3}{2}, \frac{3}{4}, \frac{4}{1}, \frac{4}{3}$

9: (3, 5, 6, 10) $\frac{3}{3}, \frac{3}{5}, \frac{3}{6}, \frac{3}{10}, \frac{5}{3}, \frac{5}{6}, \frac{6}{3}, \frac{6}{5}, \frac{10}{3}$

7: (2, 2, 3, 4) $\frac{2}{2}, \frac{2}{3}, \frac{2}{4}, \frac{3}{2}, \frac{3}{4}, \frac{4}{3}, \frac{4}{2}$,

5: (1, 1, 2, 4) $\frac{1}{1}, \frac{1}{2}, \frac{1}{4}, \frac{2}{1}, \frac{4}{1}$

3: (1, 1, 1, 2) $\frac{1}{2}, \frac{1}{1}, \frac{2}{1}$

1: (1, 1, 1, 1) $\frac{1}{1}$

Margin Exercises 4.2

1. $\frac{5}{3}$ 2. $\frac{7}{5}$ 3. $\frac{3}{2}$ 4. $\frac{5}{4}$ 5. $\frac{5}{4}$ 6. $\frac{3}{8}$ 7. $\frac{5}{6}$ 8. $\frac{19}{12}$ 9. $\frac{713}{91}$ 10. $\frac{19}{4}$ 11. $\frac{59}{36}$ 12. $\frac{169}{96}$ 13. $\frac{19}{12}$
14. $\frac{37}{40}$ 15. $\frac{31}{24}$ 16. $\frac{61}{80}$ 17. $\frac{155}{96}$ 18. $\frac{457}{160}$ 19. $\frac{15}{8}$ 20. $\frac{61}{32}$ 21. $\frac{243}{196}$ 22. $\frac{7}{2}$ 23. $\frac{87}{16}$
24. 261.34782, 0.34782, 7.99986, 8, $261\frac{8}{23}$ 25. $2\frac{11}{16}$ 26. $7\frac{3}{35}$ 27. $3\frac{5}{8}$ 28. $1\frac{134}{245}$ 29. $9\frac{1}{3}$ 30. $15\frac{1}{7}$
31. $9\frac{1}{16}$

Exercises 4.2

1. $\frac{26}{17}$ 3. $\frac{11}{24}$ 5. $\frac{56}{87}$ 7. $\frac{271}{288}$ 9. $\frac{59}{28}$ 11. $8\frac{3}{4}$ 13. $10\frac{5}{12}$ 15. $56\frac{4}{11}$ 17. $17\frac{31}{35}$ 19. $14\frac{44}{51}$
21. $\frac{69}{16}$ 23. $\frac{71}{12}$ 25. $\frac{495}{23}$ 27. $\frac{4121}{216}$ 29. $\frac{1475}{56}$ 31. $6\frac{8}{15}$ 33. $16\frac{1}{12}$ 35. $17\frac{19}{24}$ 37. $132\frac{13}{30}$
39. $21\frac{17}{78}$ 41. $3\frac{1}{2}+3\frac{1}{2}+1\frac{1}{2}+1\frac{1}{2}=D$, 10 miles 43. $\frac{5}{16}+\frac{3}{64}+\frac{3}{8}+\frac{1}{32}=W$, $\frac{49}{64}$ inches
45. $3\frac{3}{8}+1\frac{1}{4}+\frac{2}{3}=F$, $5\frac{7}{24}$ cups 47. $\frac{3}{64}+\frac{3}{64}+\frac{7}{32}+\frac{1}{8}=T$, $\frac{7}{16}$ inches
49. $6\frac{1}{4}+5\frac{3}{8}+4\frac{1}{64}+14\frac{1}{2}+16\frac{5}{8}+\frac{5}{8}=A$, $47\frac{25}{64}$ inches 51. $\frac{8}{5}+\frac{7}{6}=2\frac{23}{30}$ 53. (a) $1\frac{1}{2}$ (b) $1\frac{3}{4}$ (c) $1\frac{7}{8}$ (d) $1\frac{511}{512}$

Margin Exercises 4.3

1. $\frac{3}{11}$ 2. $\frac{2}{9}$ 3. $\frac{17}{20}$ 4. $\frac{11}{24}$ 5. $\frac{1}{20}$ 6. $\frac{1}{100}$ 7. $4\frac{1}{3}$ 8. $4\frac{1}{5}$ 9. $1\frac{5}{18}$ 10. $7\frac{13}{18}$ 11. $58\frac{7}{10}$

Exercises 4.3

1. $\frac{2}{7}$ 3. $\frac{1}{2}$ 5. $\frac{1}{15}$ 7. $\frac{453}{7564}$ 9. $\frac{1}{14}$ 11. $4\frac{3}{8}$ 13. $7\frac{39}{110}$ 15. $1\frac{13}{20}$ 17. $2\frac{7}{8}$ 19. $6\frac{5}{12}$ 21. $10\frac{19}{48}$
23. $14\frac{59}{126}$ 25. $4\frac{188}{549}$ 27. $12\frac{1}{2}$ 29. $9\frac{5}{48}$ 31. $30\frac{67}{210}$ 33. $13\frac{29}{48}$ 35. $20\frac{1}{2}$
37. $\frac{3}{4}-\frac{41}{64}=R$, $\frac{7}{64}$ inches 39. $\left(2\frac{1}{4}+3\frac{3}{4}\right)-3\frac{1}{2}=A$, $2\frac{1}{2}$ bushels 41. $3\frac{1}{2}-1\frac{5}{8}=A$; $1\frac{7}{8}$ yd; yes, $1\frac{7}{8}>1\frac{3}{4}$
43. $\frac{1}{4}+\frac{2}{5}+\frac{1}{10}+\frac{1}{12}+\frac{1}{10}=\frac{14}{15}$; no; $\frac{1}{15}$ not accounted for 45. $\frac{2}{5}-\frac{3}{8}=C$, $\frac{1}{40}$ inch
47. $\left(\frac{1}{16}+\frac{3}{64}+\frac{1}{32}\right)+\frac{1}{16}=L$, $\frac{13}{64}$ inch 49. $3\frac{7}{16}-2\frac{3}{4}=A$, $\frac{11}{16}$ inch 51. $\frac{12}{9}-\frac{10}{11}=\frac{14}{33}$

Margin Exercises 4.4

1. 0.04 2. 0.875 3. 0.65625 4. 0.95 5. $\frac{7}{10}$ 6. $\frac{546}{100}$ 7. $\frac{2431}{1000}$ 8. $\frac{23}{100{,}000}$
9. Terminating, factors of 2 10. Terminating, factors of 2, 5 11. Not terminating, factor of 3
12. Not terminating, factor of 3 13. Terminating, factors of 5 14. $\frac{75}{100}$, 0.75 15. $\frac{35}{100}$, 0.35 16. $\frac{1875}{10{,}000}$, 0.1875
17. 0.8333333 18. 0.8181818 19. 0.1428571 20. 0.6666666 21. 0.416666 22. 0.1351351
23. 0.5555555 24. 0.3535353 25. 0.4747474 26. 0.2132132
27. $\frac{7}{9}=0.\overline{7}$; $\frac{13}{99}=0.\overline{13}$; $\frac{28}{99}=0.\overline{28}$ 28. $0.\overline{7}$ 29. $0.\overline{23}$ 30. $1.\overline{36}$ 31. $0.\overline{12}$ 32. $0.\overline{157}$
33. 285,714; 428,571; 571,428; 714,285; 857,142 34. $\frac{58}{40}$ 35. $\frac{151}{105}$ 36. $\frac{58{,}199}{81{,}771}$ 37. $\frac{7}{20}$ 38. $\frac{123}{459}$
39. $\frac{12{,}235}{116{,}402}$ 40. $13\frac{219}{374}$ 41. $2\frac{1}{84}$ 42. $42\frac{1}{14}$

Exercises 4.4

1. 0.9   3. 0.375   5. 0.175   7. $\frac{481}{1000}$ 9. $\frac{10,096}{100,000}$ 11. Terminating, only factors of 2

13. Not terminating, 15 has a factor of 3   15. Terminating, only factors of 2 and 5   17. $\frac{28}{100} = 0.28$ 19. $\frac{3250}{10,000} = 0.325$

21. $0.\overline{4}$ 23. $0.\overline{384615}$ 25. $0.\overline{63}$ 27. $0.\overline{962}$ 29. $0.4\overline{6}$ 31. $1.175 = 1\frac{7}{40}$ 33. $0.275 = \frac{11}{40}$

35. $1.883838 = 1\frac{175}{198}$ 37. $114.58333 = 114\frac{7}{12}$ 39. $21.958333 = 21\frac{23}{24}$

41. $2\frac{3}{8} + 1\frac{1}{2} + 3\frac{5}{16} + \frac{1}{4} = W$; $7\frac{7}{16}$ pounds   43. $64\frac{3}{10} - 48\frac{3}{4} = D$; $15\frac{11}{20}$ miles   45. $22\frac{1}{5} - 11\frac{7}{10} = G$; $10\frac{4}{5}$ gallons

47. Yes, $36\frac{1}{2} - \left(10\frac{5}{32} + \frac{1}{16}\right) = 26.28125 > 26.25 = 26\frac{1}{4}$ 49. $30\frac{1}{2} + 62\frac{1}{8} + 70\frac{5}{8} + 37\frac{1}{16} + 60\frac{13}{16} = T$; $261\frac{1}{8}$ inches

51. Answers will vary for (a) and (b); (c): infinite, infinite.   53. No

55. Answers will vary. One example is: $\frac{2}{3} + \frac{7}{5} = \left(\frac{2 \times 5}{3} + 7\right) \div 5 = 2.0\overline{6} = 2\frac{1}{15}$

Margin Exercises 4.5

1. 1.894167   2. $\frac{784}{713}$ 3. $\frac{17,352}{1539}$ 4. $\frac{2231}{26,937}$ 5. $\frac{37,723}{212,888}$ 6. $\frac{2,327,416}{1,397,412}$ 7. $12\frac{1}{2}$ 8. $16\frac{1}{2}$ 9. $3\frac{3}{5}$

10. $37\frac{23}{153}$ 11. $737\frac{19}{78}$ 12. $\frac{7}{24}$ 13. $\frac{2}{7}$ 14. $\frac{56}{9}$ 15. $\frac{102}{5}$ 16. $\frac{9}{10}$ 17. $2\frac{2}{5}$ 18. 0.9   19. $\frac{4}{9}$

20. $\frac{26}{43}$ 21. $\frac{19}{27}$ 22. $\frac{21.6}{11} = \frac{216}{110} = \frac{108}{55}$

Exercises 4.5

1. $\frac{35}{44}$ 3. $4\frac{1}{2}$ 5. $1\frac{16}{33}$ 7. $\frac{371}{528}$ 9. $8\frac{1}{4}$ 11. $42\frac{1}{2}$ 13. 20   15. $\frac{5}{294}$ 17. $\frac{864}{23,625}$ 19. $1\frac{3}{4}$ 21. 3

23. 3   25. $3\frac{3}{7}$ 27. $\frac{7}{270}$ 29. $42\frac{1}{2}$ 31. $\frac{2}{3} \times \frac{3}{4} = F$; $\frac{1}{2}$ pound   33. $\frac{1}{2} \times \frac{7}{8} = F$; $\frac{7}{16}$ yard

35. A: $2\frac{1}{2} \times 4.40 = A$; \$11.00. B: $4\frac{1}{2} \times 4.40 = A$; \$19.80. C: $\left[8 - \left(4\frac{1}{2} + 2\frac{1}{2}\right)\right] \times 4.40 = A$; \$4.40

37. $48 \div 3\frac{3}{8} = N = 14\frac{2}{9}$; 14 pieces; $\frac{2}{9} \times 3\frac{3}{8} = A$; $\frac{3}{4}$ inches left over   39. $\left(12\frac{1}{2} \times 12.00\right) \div 16\frac{3}{4} = A$, \$8.96 (8.9552238)

41. $11\frac{3}{4} \div 12\frac{1}{2} = R$; $\frac{47}{50} = 0.94$ ohms   43. $8 \div \frac{2}{3} = N$; 12 cars   45. $\left(20,000 \div 7\frac{13}{16}\right) \times \frac{3}{4} = P$; 1920 pounds

47. $\left(7\frac{1}{2} \times 12\right) \div \left(9\frac{3}{8} + \frac{1}{16}\right) = N$; 9 pieces $\left(9\frac{81}{151}\right)$ 49. $\left(13\frac{1}{2} \div 4\right) = D$; $3\frac{3}{8}$ inches   51. 1   53. 4   55. 10 cookies

Review Exercises for Chapter 4

1. $\frac{3}{8}$ 3. $\frac{3}{6}$ 5. A: $\frac{1}{5}$, B: $\frac{3}{5}$, C: $\frac{7}{5}$, D: $\frac{9}{5}$ 7. Proper   9. Improper   11. Proper

13. Answers will vary. One example is: $\frac{6}{16}, \frac{9}{24}, \frac{12}{32}, \frac{15}{40}$ 15. Answers will vary. One example is: $\frac{10}{12}, \frac{15}{18}, \frac{20}{24}, \frac{25}{30}$

17. Answers will vary. One example is: $\frac{14}{50}, \frac{21}{75}, \frac{28}{100}, \frac{35}{125}$ 19. Answers will vary. One example is: $\frac{42}{34}, \frac{63}{51}, \frac{84}{68}, \frac{105}{85}$

21. 8   23. 35   25. 5   27. $\frac{3}{4}$ 29. $\frac{13}{2}$ 31. $\frac{31}{13}$ 33. $\frac{15}{17}$ 35. $\frac{15}{23}$ 37. $\frac{75}{4}$ 39. $\frac{179}{8}$ 41. $\frac{191}{9}$

43. $\frac{774}{17}$ 45. $17\frac{3}{8}$ 47. $16\frac{3}{4}$ 49. $14\frac{2}{9}$ 51. $24\frac{11}{12}$ 53. $1\frac{5}{24}$ 55. $\frac{33}{40}$ 57. $6\frac{11}{12}$ 59. $17\frac{12}{35}$ 61. $\frac{1}{12}$

63. $\frac{1}{12}$ 65. $\frac{1}{2}$ 67. $23\frac{7}{24}$ 69. $13\frac{831}{1105}$ 71. $\frac{5}{16}$ 73. $\frac{14}{45}$ 75. $31\frac{1}{2}$ 77. $13\frac{1}{15}$ 79. $1545\frac{5}{21}$

81. $\frac{10}{21}$ 83. $\frac{4}{5}$ 85. $\frac{2}{3}$ 87. $\frac{1}{200}$ 89. $\frac{81}{254}$ 91. $\frac{2}{3}, \frac{5}{12}, \frac{7}{11}$ 93. $\frac{3}{8}, \frac{9}{16}, \frac{132}{125}$ 95. None

97. $(3 \times 50 \times 2) + \left(6 \times \frac{1}{4}\right) = L$; $301\frac{1}{2}$ feet 99. $\left(2\frac{1}{4} + 1\frac{3}{4} + 2\frac{1}{2} + 3\right) \times 2.50 = A$; \$23.75

101. $\left[5\frac{3}{4} - \left(\frac{7}{16} + \frac{3}{8} + \frac{3}{8} + \frac{9}{16}\right)\right] \div 3 = D$; $1\frac{1}{3}$ inches

103. Food: $\frac{1}{4} \times 8400 = F$; \$2100. Clothing: $\frac{1}{8} \times 8400 = C$; \$1050. Savings: $\frac{1}{5} \times 8400 = S$; \$1680.

Shelter: $\frac{3}{10} \times 8400 = S$; \$2520. Misc.: $8400 - (2100 + 1050 + 1680 + 2520) = M$; \$1050.

105. $1\frac{1}{2} \times \left(2\frac{1}{2} \times 12\right) = C$; 45 cookies

107. (a) $108\frac{1}{2} - 98\frac{3}{4} = W$; $9\frac{3}{4}$ pounds (b) $9\frac{3}{4} \div 6 = A$; $1\frac{5}{8}$ pounds per month 109. $\left(5\frac{3}{4} - 3\frac{7}{8}\right) \div 2 = T$; $\frac{15}{16}$ inches

111. $12\frac{1}{2} \times 1\frac{1}{5} = G$; 15 gallons 113. $\left(5 \times 2\frac{15}{24}\right) + \left(4 \times \frac{1}{16} \times \frac{1}{12}\right) = L$; $13\frac{7}{48}$ feet = 13 feet $1\frac{3}{4}$ inches

115. Length: $6\frac{3}{4} + \frac{1}{4} = L$; 7 inches. Width: $3\frac{1}{4} + \frac{1}{4} = W$; $3\frac{1}{2}$ inches. Height: $1\frac{3}{16} + \frac{1}{4} = H$; $1\frac{7}{16}$ inches.

117. $24\frac{3}{8} - 11\frac{3}{4} + \frac{1}{2} = E$; $13\frac{1}{8}$ inches

119. $16\frac{1}{12} + \left(16\frac{1}{12} + 32\frac{1}{6}\right) + \left(16\frac{1}{12} + 32\frac{1}{6} + 32\frac{1}{6}\right) = D$; $144\frac{3}{4}$ feet = 144 feet 9 inches

121. $11\frac{1}{2} \div \left(1\frac{1}{3} + \frac{1}{2} + \frac{1}{12} + \frac{1}{4}\right) = N = 5\frac{4}{13}$; 5 trips

123. $1 - \left(\frac{2}{7} + \frac{7}{16}\right) = \frac{31}{112}$, $\left(12\frac{2}{5} \div 31\right) \times 112 = L$; $44\frac{4}{5}$ feet = 44 feet $9\frac{3}{5}$ inches

125. (a) $\frac{1}{2}$ (b) $\frac{1}{4}$ (c) $\frac{1}{8}$ (d) $\frac{1}{16}$ (e) $\frac{1}{1024}$ (f) No. There will always be a small distance left to take one-half of.

Just for Fun—Chapter 4

1. 15, 12, 19, 16, 23 3. 100, 150, 125, 175, 150 5. 1815; 18,315; 183,315; 1,833,315; 18,333,315; 183,333,315
7. $(203{,}632) \div (9) = 22{,}625$ Remainder 7
9. Answers will vary. Three examples are: $22 + 23 + 27 + 28 = 100$; $23 + 24 + 26 + 27 = 100$; $21 + 22 + 28 + 29 = 100$; Thirties: no; Tens: no
11. Answers will vary. One example is: $1 \times (2 + 3 + 4 + 5 + 6 + 7) + (8 \times 9) = 99$
13. Answers will vary. One example is: 15. Boise (35108)

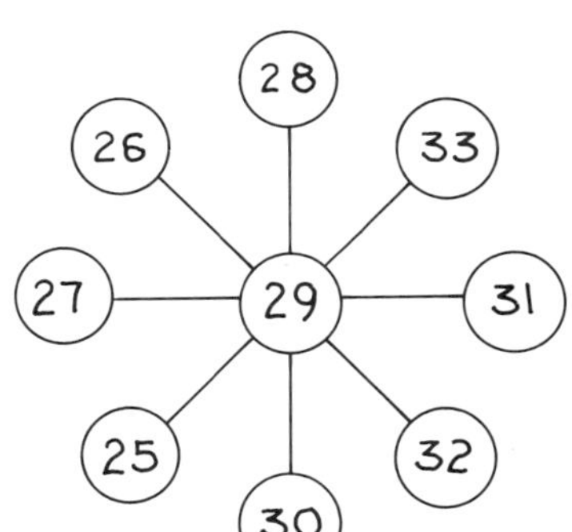

Mathematics and Energy

1. \$20.61 **3.** 136.9 pounds (136.92883) **5.** 20,173 light bulbs (20,173.007) **7.** 40.5¢/gal (0.405295)
9. **(a)** 106,165 kwh (106,164.6) **(b)** 859,869 cubic feet of natural gas (859,868.88)
(c) 6168 gallons #2 heating oil (6168.3312)
11. Answers will vary. **13.** Answers will vary. **15.** **(a)** Oil: \$130.38 **(b)** Natural gas: \$125.92 **(c)** Electricity: \$246.48

CHAPTER 5

Margin Exercises 5.1

1. 12:5 **2.** 20:17 **3.** 25:16 **4.** 15:32 **5.** 9:45 **6.** 7:100 **7.** 7:4, $\frac{7}{4}$ **8.** 4:3, $\frac{4}{3}$ **9.** 3:2, $\frac{3}{2}$

10. 2:25, $\frac{2}{25}$ **11.** 2:3, $\frac{2}{3}$ **12.** 27:28, $\frac{27}{28}$ **13.** $\frac{3}{2}$ **14.** $\frac{1}{3}$ **15.** $\frac{20}{3}$ **16.** $\frac{4}{1}$ **17.** $\frac{12}{1}$ **18.** $\frac{4}{21}$

19. $\frac{45}{1}$ **20.** 45 mph **21.** $\frac{1}{300}$ **22.** 1 gallon per 300 sq ft **23.** $\frac{5}{4}$ **24.** $\frac{2}{5}$ **25.** $\frac{14}{5}$ **26.** $\frac{4}{1}$ **27.** $\frac{9}{102}$

28. $2\frac{75}{49}$

Exercises 5.1

1. $\frac{2}{1}$ **3.** $\frac{1}{7}$ **5.** $\frac{4}{7}$ **7.** $\frac{5}{7}$ **9.** $\frac{28}{25}$ **11.** $\frac{4}{1}$ **13.** $\frac{10}{9}$ **15.** $\frac{26}{41}$ **17.** $\frac{11}{8}$ **19.** $\frac{7}{100}$ **21.** $\frac{5}{12}$ **23.** $\frac{2}{3}$

25. $\frac{25}{264}$ **27.** $\frac{18}{5}$ **29.** $\frac{8}{1}$ **31.** $\frac{1}{4}$ **33.** $\frac{1}{6}$ **35.** $\frac{4}{3}$ **37.** $\frac{264}{1}$ **39.** $\frac{88}{5}$ **41.** **(a)** $\frac{7}{4}$ **(b)** $\frac{7}{11}$ **(c)** $\frac{4}{11}$

43. $\frac{54}{1}, \frac{49.5}{1}$ **45.** **(a)** $\frac{1}{4}$ **(b)** $\frac{3}{4}$ **47.** Answers will vary. **49.** **(a)** $\frac{3}{1}$ **(b)** $\frac{9}{1}$ **51.** Answers will vary.

Margin Exercises 5.2

1. $\frac{2}{3} = \frac{10}{15}$ **2.** $\frac{7}{3} = \frac{21}{9}$ **3.** 16:9 = 48:27 **4.** 18:21 = 24:28 **5.** = **6.** = **7.** ≠ **8.** = **9.** 3 **10.** 2

11. 36 **12.** 675 **13.** 240 **14.** 144 **15.** $\frac{35}{6}, 5\frac{5}{6}$, 5.8333333 **16.** $\frac{645}{76}, 8\frac{37}{76}$, 8.4868421

17. $\frac{1786}{13}, 137\frac{5}{13}$, 137.38461 **18.** $\frac{22{,}932}{11}, 2084\frac{8}{11}$, 2084.7272 **19.** \$2.95 **20.** \$690

21. 9.8 gallons (9.7826086)

Exercises 5.2

1. Equal **3.** Equal **5.** Equal **7.** 0.5 **9.** 1.875 **11.** Equal **13.** Equal **15.** Equal **17.** 35 **19.** 5

21. 4 **23.** $3\frac{3}{5}$ **25.** 36 **27.** 45 **29.** $227\frac{529}{946}$ **31.** $7779\frac{58}{349}$ **33.** $\frac{2}{3} = \frac{N}{24}$, 16 cups of sugar

35. $\frac{9}{13} = \frac{20}{N}$, 28.9 cm ($28.\overline{8}$) **37.** $\frac{35}{140} = \frac{60}{N}$, \$240 **39.** $\frac{45}{56} = \frac{N}{10}$, 8 (8.0357142)

41. $\frac{87.5}{125} = \frac{1000}{N}$, \$1428.57 (1428.5714) **43.** $\frac{3}{0.85} = \frac{N}{5}$, 17 cans (17.647058)

45. $\frac{1}{5280} = \frac{30}{N}$, $N = 158{,}400$; $158{,}400 \div 60 \div 60 = 44$; 44 ft/sec **47.** $\frac{N}{75} = \frac{6}{4}$, 112.5 feet **49.** $\frac{20}{22} = \frac{70}{N}$, 77 feet

51. $\frac{3\frac{1}{2}}{1} = \frac{16}{N}$, 4.6 qt (4.5714285) **53.** $\frac{2\frac{1}{4}}{36} = \frac{15}{N}$, 240 days or about 8 months

55. $\frac{357}{N} = \frac{98}{112}$ (inverse proportion); $N = 408$, $408 - 357 = 51$, 51 reinforcements

Margin Exercises 5.3

1. $\frac{5}{100}$, 0.05 2. $\frac{40}{100}$, 0.40 3. $\frac{24}{100}$, 0.24 4. $\frac{154}{100}$, 1.54 5. $\frac{300}{100}$, 3 6. $\frac{2.5}{100}$, 0.025 7. $\frac{0.5}{100}$, 0.005 8. $\frac{0.6}{100}$, 0.006 9. $\frac{0.003}{100}$, 0.00003 10. $\frac{1}{100}$, 0.01 11. 24% 12. 75% 13. 7.34% 14. 50.34% 15. 0.3% 16. 0.005% 17. 200% 18. 538% 19. 8% 20. 80% 21. 2% 22. 50% 23. 75% 24. 70% 25. 87.5% 26. 31.25% 27. $66\frac{2}{3}\%$ 28. 3.8% 29. 11.5% 30. $241\frac{2}{3}\%$

Exercises 5.3

1. $\frac{25}{100}$, 0.25, 25% 3. $\frac{1}{5}$, $\frac{20}{100}$, 20% 5. $\frac{3}{5}$, 0.60, 60% 7. $\frac{1}{3}$, $\frac{33\frac{1}{3}}{100}$, $33\frac{1}{3}\%$ 9. $\frac{2}{3}$, $\frac{66\frac{2}{3}}{100}$, $66\frac{2}{3}\%$ 11. $\frac{1}{100}$, $\frac{1}{100}$, 1% 13. $\frac{12.5}{100}$, 0.125, 12.5% 15. $\frac{3}{8}$, $\frac{37.5}{100}$, 37.5% 17. $\frac{25}{4}$, 6.25, 625% 19. $\frac{87.5}{100}$, 0.875, 87.5% 21. $\frac{1}{6}$, $0.16\frac{2}{3}$, $16\frac{2}{3}\%$ 23. $\frac{11\frac{1}{9}}{100}$, $0.11\frac{1}{9}$, $11\frac{1}{9}\%$ 25. $\frac{3}{200}$, $\frac{1.5}{100}$, 1.5% 27. $\frac{\frac{1}{3}}{100}$, $0.003\frac{1}{3}$, $\frac{1}{3}\%$ 29. $\frac{1}{2000}$, $\frac{0.05}{100}$, 0.05% 31. 25 33. 1 35. 150 37. 200 39. $\frac{1}{4}$ 41. 0.71 43. 0.93 45. 1.00 47. 0.0001 49. 9.75 51. $\frac{8}{25}$ 53. $\frac{6}{5}$ 55. $\frac{1}{15}$ 57. $\frac{1}{120}$ 59. $\frac{6}{625}$ 61. 52% 63. 30% 65. 31% 67. 6% 69. 7.5% 71. 0.02% 73. 217.8% 75. 1.4% 77. 516% 79. 1000%

Margin Exercises 5.4

1. $\frac{15}{100} = \frac{P}{240}$, 36 2. $\frac{9}{100} = \frac{25.2}{B}$, 280 3. $\frac{37}{100} = \frac{233.1}{B}$, 630 4. $\frac{R}{100} = \frac{114}{950}$, 12% 5. $\frac{R}{100} = \frac{3.87}{129}$, 3% 6. $\frac{0.75}{100} = \frac{P}{850}$, 6.375 7. $\frac{112}{100} = \frac{P}{46}$, 51.52

Exercises 5.4

1. 3.94 3. 12 5. 35 7. 70 9. 256 11. 404.1775 13. 111.66892 15. 491.4831 17. 25% 19. 15.98712% 21. 0.6839% 23. 4.16666% 25. 32 27. 1012.5 29. 436.32075 31. 0.077546 33. 169.93954 35. 700 37. 20% 39. 400 41. 2.6543 43. 1.13527% 45. 2914.1509

Margin Exercises 5.5

1. $P = 0.06 \times 42.89$, 2.5734 2. $P = 0.038 \times 96$, 3.648 3. $11.64 = 0.24 \times B$, 48.5 4. $416.5 = 0.17 \times B$, 2450 5. $7 = r \times 105$, $6\frac{2}{3}\%$ $(0.0\overline{6})$ 6. $39 = r \times 78$, 50% 7. $P = \frac{\frac{1}{3}}{100} \times 396$, 1.32 8. $P = 3.25 \times 420$, 1365

Exercises 5.5

1. 19.7 3. 4 5. 88 7. 160 9. 300 11. 1.4577238 13. 0.383552 15. 303.4949 17. 150% 19. 9.30027% 21. 0.06717% 23. 64 25. $166.\overline{6}$ 27. 196.25425 29. 103,346.15 31. 15.746951 33. 1853.2 35. 3.31659 37. 10.63065% 39. 139.64285 41. 8.83792% 43. 1683.2535% 45. 3.21316%

Margin Exercises 5.6

1. \$412.50 2. \$885 3. 18.75% 4. 5% 5. 400 sheep 6. \$184,000 7. 7650 students 8. 200 pounds

Exercises 5.6

1. 15% **3.** 83.3% ($83.\overline{3}$) **5.** \$4900 **7.** **(a)** \$5.25 **(b)** \$130.25 **9.** 30 **11.** \$80 **13.** 18 games (18.04)
15. 22.2% ($22.\overline{2}$) **17.** \$22,500 **19.** 8.4% **21.** 20 **23.** 1.5 kg **25.** \$203 **27.** 25% **29.** \$2.00
31. 3.6 pounds (3.575) **33.** \$6000 **13.** **(a)** 60.5% **(b)** 153.16% (1.5316455) **37.** \$376.80 **39.** \$21
41. 12.5% **43.** 40 pounds **45.** 7.6% (0.075675) **47.** \$37,835 **49.** 19.4% ($0.19\overline{4}$) **51.** \$1652.36
53. No, Andy \$99; Bert \$100

Review Exercises for Chapter 5

1. $\frac{7}{8}$ **3.** $\frac{5}{7}$ **5.** $\frac{9}{2}$ **7.** $\frac{50}{67}$ **9.** $\frac{8}{1}$ **11.** $\frac{3}{20}$ **13.** $\frac{1}{3}$ **15.** $\frac{7}{10}$ **17.** 24 **19.** 36 **21.** 5000 **23.** 8.2

25. 80 **27.** 70 **29.** 3 **31.** 72 **33.** 685 votes **35.** **(a)** $4\frac{3}{8}$ inches **(b)** $24\frac{1}{2}$ feet

37. **(a)** \$2.00 **(b)** 16 oranges (16.2) **39.** 15 quarts **41.** **(a)** 1,764,596.2 pounds **(b)** 0.068 seconds
43. **(a)** 4 gallons (15.625 quarts) **(b)** 3200 square feet **45.** **(a)** 200 pictures **(b)** 3487.5 horses
47. **(a)** 5.6 g ($5.\overline{5}$) **(b)** 66.7 g (66.6) **49.** **(a)** 10–30 mg **(b)** 4–12 mg ($4.\overline{09}$, $12.\overline{27}$) **(c)** 59–177 g ($590.\overline{90}$, $1772.\overline{72}$)

51. 36 **53.** 6.375 **55.** 4875 **57.** $0.\overline{8}$ **59.** $33\frac{1}{3}$% **61.** 8.54838% **63.** 1152 **65.** 36.8421% **67.** 1

69. 60 **71.** 300% **73.** $86\frac{2}{3}$% **75.** $49\frac{1}{2}$ yards **77.** 1300 **79.** 44.7% (44.69273) **81.** \$32 **83.** 3 dolls

85. 1667 kg ($1666.\overline{6}$) **87.** **(a)** 12.5% **(b)** 800% **89.** **(a)** 11.1% **(b)** Increase 0.9% **91.** 29.7% (29.69691), less by 1.5%
93. 1.14% decrease (1.13825) **95.** Yes; yes

Just for Fun—Chapter 5

1. $\frac{1}{99}, \frac{1}{143}, \frac{1}{195}, \frac{1}{255}, \frac{1}{323}$ **3.** 720; 5040; 40,320; 362,880; 3,628,800 **5.** 49; 4489; 444,889; 44,448,889; 4,444,488,889

7. Answers will vary. Two examples are given: **9.** \$10,519,200 if a leap year occurs

```
  432       231
+ 432     + 231
  864       462
```

11. Answers will vary. One example is:

2	×	=	×	=	×	3	×	3	×	3	×	333	=
	↑		↑		↑		↑		↑		↑		
	1		2		3		4		5		6		

13. Answers will vary. One example is:

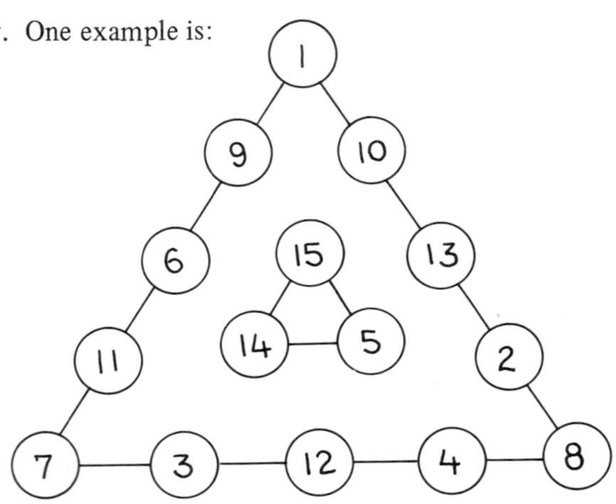

15. Hobo (0.804)

Mathematics and the Automobile

1.

Item / Make	Chevette	Colt	Subaru	Gremlin	Rabbit	Corolla	Mustang
Cost of accessories	$297	$69	$69	$294	$395	$385	$477
HP per pound	0.031	0.040	0.034	0.030	0.040	0.033	0.032
Pounds per HP	32.6	25.0	26.1	33.3	24.9	30.2	31.0
Cruising range:							
195 mi trip mileage	396.5	442.2	396.0	397.5	316.1	343.2	331.5
Open road mileage	429.0	495.0	455.4	442.5	337.9	356.4	344.5
City mileage	273.0	250.8	244.2	262.5	218.0	231.0	195.0
% reduc. or incr. of EPA est.							
Open road	–23.3%	–16.7%	–15.9%	–10.6%	– 5.7%	–12.9%	– 8.6%
City	–32.3%	–34.5%	–33.9%	–16.7%	–16.7%	–30.0%	–28.6%

3.

Make: Chevette	1st year	2nd year	3rd year	4th year	5th year	6th year	7th year	8th year	9th year	10th year
Miles driven	12,000	24,000	36,000	48,000	60,000	72,000	84,000	96,000	108,000	120,000
(a) Value (this year) ($)	3522.00	2465.40	1972.32	1676.47	1425.00	1211.25	1090.12	981.11	883.00	794.70
(b) Depreciation (this year) ($)	1056.60	493.08	295.85	251.47	213.75	121.13	109.01	98.11	88.30	79.47
(c) Cost of gas (this year) ($)	336.23	336.23	336.23	336.23	336.23	336.23	336.23	336.23	336.23	336.23
(d) Cost of repairs (this year) ($)	58.10	120.50	165.10	190.00	274.63	223.96	281.33	218.81	173.03	82.85
(e) Total cost (b + c + d) ($)	1480.93	979.81	827.18	807.70	854.61	681.32	726.57	653.15	597.56	498.55
(f) Total cost (previous years) ($)	–	1480.93	2460.74	3287.92	4095.62	4950.23	5631.55	6358.12	7011.27	7608.83
(g) Total cost ($)	1480.93	2460.74	3287.92	4095.62	4950.23	5631.55	6358.12	7011.27	7608.83	8107.38
(h) Cost per mile	12.3¢	10.3¢	9.1¢	8.5¢	8.3¢	7.8¢	7.6¢	7.3¢	7.0¢	6.8¢

Make: Colt	1st year	2nd year	3rd year	4th year	5th year	6th year	7th year	8th year	9th year	10th year
Miles driven	12,000	24,000	36,000	48,000	60,000	72,000	84,000	96,000	108,000	120,000
(a) Value (this year) ($)	3410.00	2387.00	1909.60	1623.16	1379.69	1172.73	1055.46	949.91	854.92	769.43
(b) Depreciation (this year) ($)	1023.00	477.40	286.44	243.47	206.96	117.27	105.55	94.99	85.49	76.94
(c) Cost of gas (this year) ($)	380.21	380.21	380.21	380.21	380.21	380.21	380.21	380.21	380.21	380.21
(d) Cost of repairs (this year) ($)	58.10	120.50	165.10	190.00	274.63	223.96	281.33	218.81	173.03	82.85
(e) Total cost (b + c + d) ($)	1461.31	978.11	831.75	813.68	861.80	721.44	767.09	694.01	638.73	540.00
(f) Total cost (previous years) ($)	–	1461.31	2439.42	3271.17	4084.85	4946.65	5668.09	6435.18	7129.19	7767.82
(g) Total cost ($)	1461.31	2439.42	3271.17	4084.85	4946.65	5668.09	6435.18	7129.19	7767.92	8307.82
(h) Cost per mile	12.2¢	10.2¢	9.1¢	8.5¢	8.2¢	7.9¢	7.7¢	7.4¢	7.2¢	6.9¢

3. (Cont)

Make: Subaru	1st year	2nd year	3rd year	4th year	5th year	6th year	7th year	8th year	9th year	10th year
Miles driven	12,000	24,000	36,000	48,000	60,000	72,000	84,000	96,000	108,000	120,000
(a) Value (this year) ($)	3368.00	2357.60	1886.08	1603.17	1362.69	1158.29	1042.46	938.21	844.39	759.95
(b) Depreciation (this year) ($)	1010.40	471.52	282.91	240.48	204.40	115.83	104.25	93.82	84.44	76.00
(c) Cost of gas (this year) ($)	396.24	396.24	396.24	396.24	396.24	396.24	396.24	396.24	396.24	396.24
(d) Cost of repairs (this year) ($)	58.10	120.50	165.10	190.00	274.63	223.96	281.33	218.81	173.03	82.85
(e) Total cost (b + c + d) ($)	1464.74	988.26	844.25	826.72	875.27	736.03	781.82	708.87	653.71	555.09
(f) Total cost (previous years) ($)	–	1464.74	2453.00	3297.25	4123.97	4999.24	5735.27	6517.09	7225.96	7879.67
(g) Total cost ($)	1464.74	2453.00	3297.25	4123.97	4999.24	5735.27	6517.09	7225.96	7879.67	8434.76
(h) Cost per mile	12.2¢	10.2¢	9.2¢	8.6¢	8.3¢	8.0¢	7.8¢	7.5¢	7.3¢	7.0¢

Make: Gremlin	1st year	2nd year	3rd year	4th year	5th year	6th year	7th year	8th year	9th year	10th year
Miles driven	12,000	24,000	36,000	48,000	60,000	72,000	84,000	96,000	108,000	120,000
(a) Value (this year) ($)	3542.00	2479.40	1983.52	1685.99	1433.09	1218.13	1096.32	986.69	888.02	799.22
(b) Depreciation (this year) ($)	1062.60	495.88	297.53	252.90	214.96	121.81	109.63	98.67	88.80	79.92
(c) Cost of gas (this year) ($)	430.61	430.61	430.61	430.61	430.61	430.61	430.61	430.61	430.61	430.61
(d) Cost of repairs (this year) ($)	58.10	120.50	165.10	190.00	274.63	223.96	281.33	218.81	173.03	82.85
(e) Total cost (b + c + d) ($)	1551.31	1046.99	893.24	695.80	920.20	776.38	821.57	748.09	692.44	593.38
(f) Total cost (previous years) ($)	–	1551.31	2298.30	3191.54	3887.34	4807.54	5583.92	6405.49	7153.58	7846.02
(g) Total cost ($)	1551.31	2298.30	3191.54	3887.34	4807.54	5583.92	6405.49	7153.58	7846.02	8439.40
(h) Cost per mile	12.9¢	9.6¢	8.9¢	8.1¢	8.0¢	7.8¢	7.6¢	7.5¢	7.3¢	7.0¢

Make: Rabbit	1st year	2nd year	3rd year	4th year	5th year	6th year	7th year	8th year	9th year	10th year
Miles driven	12,000	24,000	36,000	48,000	60,000	72,000	84,000	96,000	108,000	120,000
(a) Value (this year) ($)	4474.00	3131.80	2505.44	2129.62	1810.18	1538.65	1384.78	1246.30	1121.67	1009.50
(b) Depreciation (this year) ($)	1342.20	626.36	375.82	319.44	271.53	153.87	138.48	124.63	112.17	100.95
(c) Cost of gas (this year) ($)	386.13	386.13	386.13	386.13	386.13	386.13	386.13	386.13	386.13	386.13
(d) Cost of repairs (this year) ($)	58.10	120.50	165.10	190.00	274.63	223.96	281.33	218.81	173.03	82.85
(e) Total cost (b + c + d) ($)	1786.43	1132.99	927.05	895.57	490.86	763.96	805.94	729.57	671.33	569.93
(f) Total cost (previous years) ($)	–	1786.43	2919.42	3957.07	4907.94	5454.10	6273.36	7134.60	7919.47	8646.10
(g) Total cost ($)	1786.43	2919.42	3957.07	4907.94	5454.10	6273.36	7134.60	7919.47	8646.10	9271.33
(h) Cost per mile	14.9¢	12.2¢	11.0¢	10.2¢	9.2¢	8.7¢	8.5¢	8.2¢	8.0¢	7.7¢

Make: Corolla	1st year	2nd year	3rd year	4th year	5th year	6th year	7th year	8th year	9th year	10th year
Miles driven	12,000	24,000	36,000	48,000	60,000	72,000	84,000	96,000	108,000	120,000
(a) Value (this year) ($)	4163.00	2914.10	2331.28	1981.59	1684.35	1431.70	1288.53	1159.68	1043.71	939.34
(b) Depreciation (this year) ($)	1248.90	582.82	349.69	297.24	252.65	143.17	128.85	115.97	104.37	93.93
(c) Cost of gas (this year) ($)	441.90	441.90	441.90	441.90	441.90	441.90	441.90	441.90	441.90	441.90
(d) Cost of repairs (this year) ($)	58.10	120.50	165.10	190.00	274.63	223.96	281.33	218.81	173.03	82.85
(e) Total cost (b + c + d) ($)	1748.90	1145.22	956.69	929.14	969.18	809.03	852.08	776.68	719.30	618.68
(f) Total cost (previous years) ($)	–	1748.90	2894.12	3850.81	4779.95	5749.13	6558.16	7410.24	8186.92	8906.22
(g) Total cost ($)	1748.90	2894.12	3850.81	4779.95	5749.13	6558.16	7410.24	8186.92	8906.22	9524.90
(h) Cost per mile	14.6¢	12.1¢	10.7¢	10.0¢	9.6¢	9.1¢	8.8¢	8.5¢	8.2¢	7.9¢

3. **(Cont)**

Make: Mustang	1st year	2nd year	3rd year	4th year	5th year	6th year	7th year	8th year	9th year	10th year
Miles driven	12,000	24,000	36,000	48,000	60,000	72,000	84,000	96,000	108,000	120,000
(a) Value (this year) ($)	4354.00	3047.80	2438.24	2072.50	1761.62	1497.38	1347.64	1212.88	1091.59	982.43
(b) Depreciation (this year) ($)	1306.20	609.56	365.74	310.88	264.24	149.74	134.76	121.29	109.16	98.24
(c) Cost of gas (this year) ($)	495.85	495.85	495.85	495.85	495.85	495.85	495.85	495.85	495.85	495.85
(d) Cost of repairs (this year) ($)	58.10	120.50	165.10	190.00	274.63	223.96	281.33	218.81	173.03	82.85
(e) Total cost (b + c + d) ($)	1860.15	1229.91	1026.69	996.73	1034.72	869.55	911.94	835.95	778.04	676.94
(f) Total cost (previous years) ($)	–	1860.15	3086.06	4112.75	5109.48	6144.20	7013.75	7925.69	8671.64	9449.68
(g) Total cost ($)	1860.15	3086.06	4112.75	5109.48	6144.20	7013.75	7925.69	8671.64	9449.68	10,126.62
(h) Cost per mile	15.5¢	12.9¢	11.4¢	10.6¢	10.2¢	9.7¢	9.4¢	9.1¢	8.7¢	8.4¢

CHAPTER 6

Margin Exercises 6.1

1. $48.05 **2.** 113° F **3.** 6.24 cm^2 **4.** 96.2 feet **5.** 0.2 **6.** 1300 **7.** 25° C **8.** 30 **9.** 625 feet/sec **10.** Thousandth, 5.2004, 5.1995 **11.** 6.24, 1.15 **12.** 7.35, 13.4, 7.44, 13.4, 99.696, 7.35, 12.5, 91.875 **13.** 17.48, 17.5 **14.** 125.4, 125 **15.** 3050, 3047 **16.** 0.0035, 0.00349 **17.** 10.1, 10

Exercises 6.1

1. 0.28 (0.28125) **3.** 7200 **5.** 0.67 (0.6687301) **7.** 1.27 **9.** 54 **11.** 10.4 hours (10.391224) **13.** 201 cm (201.1) **15.** 1043 m (1042.69) **17.** 86.2° F (86.18) **19.** 38° C ($37.\overline{7}$) **21.** 10.1° C ($10.0\overline{5}$) **23.** 54,000 cm (54,022.5) **25.** 43 ohms ($42.\overline{6}$) **27.** 9.8 amps (9.7826) **29.** 508 sq inches (508.024) **31.** 82,000 (82,288.029) **33.** 0.008636 (0.00863578) **35.** 93.6 (93.592063) **37.** 19.1 (19.101661) **39.** 47.6 (47.591595) **41.** 86.5 (86.451429) **43.** 7.4 (7.3786408) **45.** 68.4 (68.392661) **47.** 12 **49.** 10,000 **51.** 20.2 (20.20202) **53.** 1210 pounds **55.** 5 foot candles

Margin Exercises 6.2

1. 4 cm, 38 mm **2.** 21.46 **3.** 62 **4.** 82.7 **5.** 2.98 **6.** 0.023 **7.** 0.0034 **8.** 194.013 **9.** 19.4013 **10.** 1.94013 **11.** 0.194013 **12.** 0.0194013 **13.** 0.00194013 **14.** 8,203,000 **15.** 820,300 **16.** 82,030 **17.** 8203 **18.** 820.3 **19.** 82.03 **20.** 12.1 **21.** 3.122 **22.** 6.743 **23.** 524 **24.** 968 **25.** 780 **26.** 341 **27.** 8.53 **28.** 0.20095 **29.** 523,000 **30.** 24 **31.** 449.6 m **32.** 39.2 cm **33.** 55.08 cm **34.** 125.4 cm **35.** 497.62 m **36.** 19.0 cm

Exercises 6.2

In exercises 1–20 all estimates can vary. **1.** 61 mm **3.** 94 mm **5.** 56 mm **7.** **(b)** 183 mm **(c)** 18.3 cm **(d)** 0.183 mm **9.** **(b)** 235 mm **(c)** 23.5 cm **(d)** 0.235 m **11.** Answers will vary. **13.** Answers will vary. **15.** Answers will vary. **17.** Answers will vary. **19.** **(b)** 27.9 cm **(c)** 279 mm **(d)** 0.279 m **21.** 0.005 **23.** 5000 **25.** 240 **27.** 0.0005 **29.** 50 **31.** 0.1 **33.** 500 **35.** 500,000 **37.** 34.2 cm **39.** 18.49 m **41.** 12.1 km **43.** 142.4 cm **45.** 1660 cm **47.** **(a)** 98.765 mm **(b)** 98.715 mm **49.** 3.52, 3.48 **51.** 142.2 cm **53.** 3.15 cm **55.** **(a)** 118,000 m **(b)** \$9145 **57.** \$10.25 (10.2528) **59.** 835.5 m

61. **(a)**

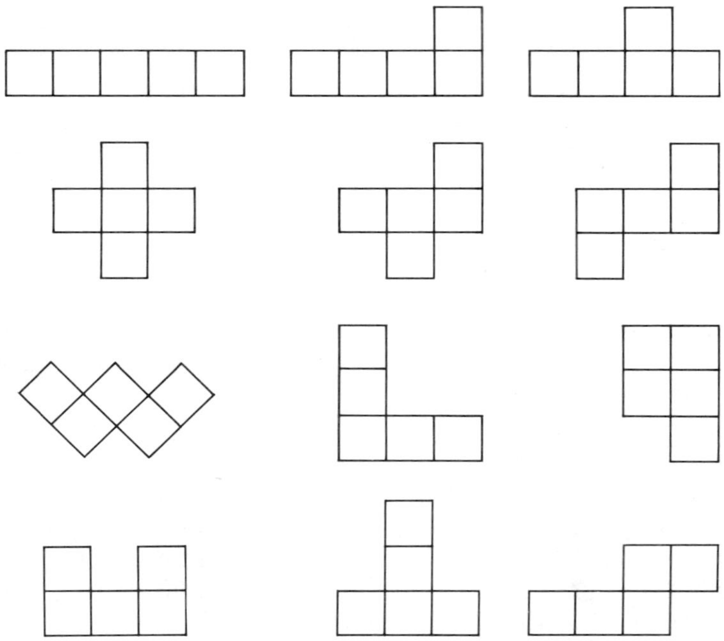

(b)

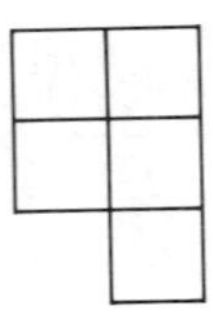

$P = 20$ cm

(c) All others have 24 cm perimeter.

Margin Exercises 6.3

1. 12,100 **2.** 300 **3.** 0.0123 **4.** 0.4527 **5.** 85 **6.** 524 **7.** 0.01 **8.** 0.0075 **9.** 1,000,000 **10.** 0.5 **11.** 0.00000135 **12.** 25,000 **13.** 6 cm^2 **14.** 2 cm^2 **15.** 86 m^2 (85.68) **16.** 60.0 cm^2 (60.04) **17.** 1500 m^2 (1526.61) **18.** 5260 cm^2 (5260.204) **19.** 68.5 cm^2 (68.523563) **20.** 174.7 cm (174.6952) **21.** 110 mm (109.9)

Exercises 6.3

1. 100,000 **3.** 0.00001 **5.** 100 **7.** 0.000001 **9.** 5×10^9 **11.** 0.5 **13.** 1×10^{18} **15.** 1×10^{12} **17.** 1×10^{14} **19.** 1×10^8 **21.** 560 cm^2 (564) **23.** 880 m^2 (882) **25.** 13 m^2 (12.88) **27.** 13.0 km^2 (12.95055) **29.** 85 cm^2 (84.5) **31.** 1160 m^2 (1164.3073) **33.** 171 cm^2 **35.** 17.4 cm^2 **37.** 1896.5 mm^2 **39.** 842.5 cm^2 **41.** 1266.1 cm^2 **43.** \$48.60 **45.** 188 plants (188.275) **47.** **(a)** 6 kg (5.93) **(b)** \$38.73 **49.** 752.4 m^2 **51.** Yes, 0.05 gal too much (0.0452173) **53.** 400 m **55.** 120 m **57.** **(a)** 163.28 m^2 **(b)** \$857.22 **59.** $6\frac{1}{2}$ rolls $(5.28 + \frac{1}{2})$ **61.** \$666.00 **63.** 5.82 m **65.** 69.8 mph (69.8022) **67.** 3.14 m **69.** **(a)** \$1,568,160 **(b)** \$1,567,660

Margin Exercises 6.4

1. 17.7 m^3 (17.671368) **2.** 17,700 l **3.** 1710 cm^3, 1710 ml (1706.13) **4.** 5.0 m (5.0193048) **5.** 724 cm^3 (724.1625) **6.** 2.48 cu inches (2.4830346) **7.** 26 cu inches $(26\frac{5}{7})$

Exercises 6.4

1. 0.5 **3.** 60 **5.** 0.00006 **7.** 50,000 **9.** 0.007 **11.** 105 cm^3
13. **(a)** 34.0 cm^3 (33.99623) **(b)** 66.2 cm^2 (66.1534) **15.** **(a)** 1040 cm^3 (1039.5091) **(b)** 615.7 cm^2 (615.7014)
17. 3.8 m^3 **19.** 1780 m^3 (1780.02) **21.** 76.27 l (76.268107) **23.** 486,000 l (485,976.42)
25. 59,630,000 l (59,631.26 m^3) **27.** 0.966 l (966.12138 cm^3) **29.** 0.86 metric tons (0.857565)
31. 726 m^3 (725.805) **33.** 10,800 l (10,841.785) **35.** 14.8 cm (14.825796) **37.** 29 g (28.736208)
39. 195,000 kg (195,395.53) **41.** 4 trips (3.2977384) **43.** 12 hours 30 minutes **45.** 2.8 l (2.7885878)
47. 116 l (116.32236 **49.** Cut 0.2 cm off height. **51.** 6 units on a side

Margin Exercises 6.5

1. 10.8 g (10.808) **2.** 5 pounds

Exercises 6.5

1. 1 **3.** 3,000,000 **5.** 5,000,000 **7.** 0.15 **9.** 0.12 **11.** 160,000 **13.** 7000 **15.** 18
17. 34.6 kg (34.562) **19.** 295 g (294.78) **21.** 180 kg (181.86288) **23.** 189 g **25.** $275.71 (275.70947)
27. Yes. It weighs too much. Gold weighs 14.475 kg. **29.** 5 g/cm^3 **31.** 68.3 kg (68.299538)
33. Lead, 11.323997 ~ 11.34 **35.** 44 or 45 (44.865942) **37.** 1.18 l **39.** 186 kg (186.2784)
41. 145 cm (145.24163)

Review Exercises for Chapter 6

1. 5.9 (5.8695652) **3.** 4.957 (4.9565461) **5.** 0.206 (0.2058823) **7.** 23.4 (23.4498) **9.** 16.6 (16.614785)
11. 27.3 (27.25531) **13.** 3650 (3649.7084) **15.** 0.170 (0.170476) **17.** 10.5 (10.5043) **19.** 12 **21.** 75
23. **(a)** 50 cm **(b)** 130 cm^2 (126) **25.** **(a)** 53.48 km **(b)** 178.8 km^2 (178.7569) **27.** **(a)** 56.5 m **(b)** 133 m^2 (132.6)
29. **(a)** 61 cm **(b)** 180 cm^2 **31.** **(a)** 26 ft **(b)** 18 sq ft **33.** **(a)** 77.7 m (77.68) **(b)** 420 m^2 (418.08)
35. **(a)** 69.6 m **(b)** 200 m^2 (203.95) **37.** 340 cm^2 (343.17) **39.** 190 m^2 (194.93)
41. **(a)** 6840 sq ft **(b)** 334 ft **43.** 3.6 kg **45.** $15.36 **47.** 15 meters **49.** $3.20 (3.1968)
51. 223 cm (222.94) **53.** 6.50 cm **55.** 282.6 m **57.** Circular skillet **59.** 860 cm^3 **61.** 1800 cm^3 (1846.32)
63. 800 m^3 (803.84) **65.** 96 gallons (95.912727) **67.** 7 m^3 (7.010364) **69.** $6347.53 (6347.5257)
71. 4:1 **73.** 2 units

Just for Fun—Chapter 6

1. 9, $\frac{6}{4}$, 10, $\frac{7}{4}$, 11 **3.** 32, 64, 128, 256, 512 **5.** 8; 188; 2888; 38,888; 488,888; 5,888,888; 68,888,888
7. 9567 + 1085 = 10,652 **2.** $2\frac{1}{2}$ inches **11.** 16 + 17 + 18 + 19 + 20 = 21 + 22 + 23 + 24; yes. Answers will vary.
13. Answers will vary. One solution is: **15.** Lei (137)

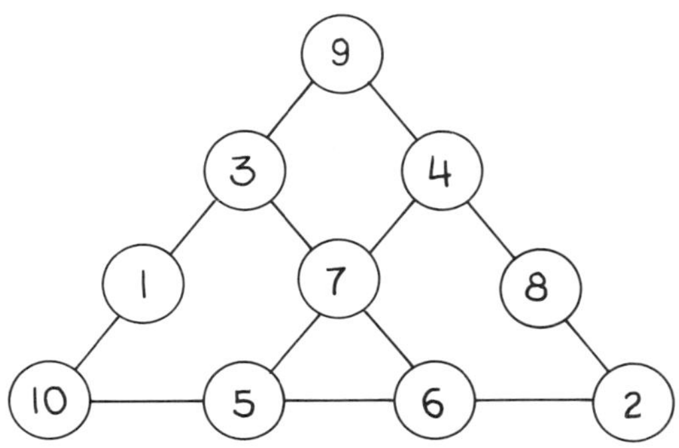

Mathematics and Athletics

1. Buffalo; Buffalo: 88.5% (0.8846153); Detroit: 80.0%
3. Seattle: Burkson, Norwood, Wilkerson, Green, Oleynich—100%; Portland: Gross, Lucas—100%
5. Portland: F.G.: 48.5% (0.4851485); F.T.: 63.2% (63.15789); Seattle: F.G.: 44.4% ($0.\overline{4}$); F.T.: 83.3% ($0.8\overline{3}$)
7. Minnesota: 58.1% (0.5813953); Cincinnati: 40.9% ($0.\overline{409}$); New Orleans: 71.4% (0.7142857); Miami: 55.6% ($0.\overline{5}$); San Francisco: 33.3% ($0.\overline{3}$); Oakland: 56.7% ($0.5\overline{6}$)
9. Parrish (Montreal) 2 hits for 2 times at bat. 1.000.
11.

American		*National*	
Carew	0.375	Parker	0.348
Bastock	0.337	Stennett	0.336
Rivers	0.331	Simmons	0.330
Singleton	0.329	Griffey	0.320
Le Flore	0.320	Templeton	0.320

13. $\text{PCT} = \dfrac{\text{games won}}{\text{games played}}$, $\text{GB} = \dfrac{\text{difference in W} + \text{difference in L}}{2}$

American League (East)

	PCT	GB
New York	0.612	
Boston	0.579	$4\frac{1}{2}$
Baltimore	0.576	5
Detroit	0.489	$16\frac{1}{2}$
Cleveland	0.467	$24\frac{1}{2}$
Milwaukee	0.411	$27\frac{1}{2}$
Toronto	0.344	$35\frac{1}{2}$

National League (East)

	PCT	GB
Philadelphia	0.686	
Pittsburg	0.558	11
New York	0.516	$14\frac{1}{2}$
St. Louis	0.455	20
Chicago	0.404	$24\frac{1}{2}$
Montreal	0.313	$31\frac{1}{2}$

National League (West)

	PCT	GB
Cincinnati	0.626	
Los Angeles	0.549	7
Houston	0.511	$10\frac{1}{2}$
San Diego	0.489	$12\frac{1}{2}$
Atlanta	0.467	$14\frac{1}{2}$
San Francisco	0.419	19

CHAPTER 7

Margin Exercises 7.1

1. 5.9558×10^2 2. 3.7084×10 3. 2.5815×10^3 4. 4.365 5. 5.0627×10^2 6. 3.24783×10^2 7. 1.4523×10 8. 9.89 9. 6.9916×10^4 10. 7.7215×10^2 11. 7×10^{-1} 12. 7×10^{-2} 13. 7×10^{-3} 14. 4.21×10^{-1} 15. 4.21×10^{-2} 16. 4.21×10^{-3} 17. 4.21×10^{-4} 18. 3.756×10^{-1} 19. 1.6877×10^{-4} 20. 5.88×10^{-8} 21. 10^5 22. 10^5 23. 10^7 24. 10^8 25. 10^{-5} 26. 10^{-8} 27. 10^{-2} 28. 10^{-16} 29. 10^{-11} 30. 10^{-11} 31. 10^1 32. 10^4 33. 10^3 34. 10^4 35. 10^{-3} 36. 10^{-1} 37. 10^{-4} 38. 10^2 39. 10^1 40. 10^2 41. 10^4 42. 10^2 43. 10^{-14} 44. 10^{-2} 45. 10^2 46. 10^0

Exercises 7.1

1. 6.89×10^2 3. 2.59×10^5 5. 3.24×10^6 7. 1.94×10^7 9. 4.03×10^{-2} 11. 6.7×10^{-4} 13. 1.5×10^{-1} 15. 1.5×10^{-2} 17. 1.05×10^{-3} 19. 65,000 21. 0.00065 23. 7010 25. 254,000 27. 0.02636 29. 30,470 31. 0.002104 33. 999,990,000 35. 0.066 37. 10^8 39. 10^7 41. 10^1 43. 10^{-1} 45. 10^1 47. 10^9 49. 10^{-15} 51. 10^2 53. 10^{-12} 55. 864,000 57. 2.4×10^{-5} 59. 6.33×10^{16} 61. 341,000,000,000,000,000

Margin Exercises 7.2

1. 5, 4, 66.428287, 66.428287, 5, 4, 66.4428287, 9, 66,428,287,000 2. $21.212297 \times 10^9 = 21{,}212{,}297{,}000$

Exercises 7.2

1. $6.4461953 \times 10^9 = 6{,}446{,}195{,}300$ 3. $30.81656 \times 10^9 = 30{,}816{,}560{,}000$ 5. $26.24695 \times 10^{10} = 262{,}469{,}500{,}000$
7. $37.498651 \times 10^{10} = 374{,}986{,}510{,}000$ 9. $88.771546 \times 10^{11} = 8{,}877{,}154{,}600{,}000$
11. $13.147728 \times 10^7 = 131{,}477{,}280$ feet 13. $4.9582211 \times 10^{10} = \$49{,}582{,}211{,}000$
15. $234.62784 \times 10^{12} = 234{,}627{,}840{,}000{,}000$ miles

Margin Exercises 7.3

1. Overflow symbol 2. 0.0276, 0.000276, 0.00000276, 0.000000276 3. $39.2825 \times 10^{-11} = 0.000000000392825$
4. $42.585312 \times 10^{-9} = 0.000000042585312$ 5. Overflow symbol 6. $12.169312 \times 10^{-9} = 0.000000012169312$
7. 446,616,540 8. $1.7173913 \times 10^{-7} = 0.00000017173913$
9. (a) 3719.0506 (b) 3719.0506 or 3719.0505 depending on your HHC

Exercises 7.3

1. $2.207352 \times 10^{-7} = 0.0000002207352$ 3. $9.546 \times 10^{-5} = 0.00009546$ 5. $7.0103092 \times 10^0 = 7.0103092$
7. $1.104 \times 10^{-4} = 0.0001104$ 9. $4.1401142 \times 10^{-2} = 0.041401142$
11. $4.8404476 \times 10^3 = 4840.4476$ 13. $334.40589 \times 10^{-10} = 0.000000033440589$
15. $8.5019011 \times 10^{-5} = 0.000085019011$ 17. $6 \times 10^{17} = 600{,}000{,}000{,}000{,}000{,}000$ 19. $1.35 \times 10^8 = 135{,}000{,}000$
21. (a) 67.084323 (b) 67.084322 23. 12,755,100 tons 25. 0.0176 feet

Margin Exercises 7.4

1. 1764, 1.5129 2. No 3. Yes 4. No 5. Yes 6. 2.89, 2.9929, 2.999824, 3.0001704, 2.9999972, 2.9999999
7. 1.8, 1.8, 2 8. 1.7888888 9. 3.162276 10. 5.6 feet (5.6035702)

Exercises 7.4

1. 3.3166 3. 6.1644 5. 1.2649 7. 5.1284 9. 1.6031 11. 8.90000 13. 2.3281 15. 133.5665
17. 9.0692 19. 3.9284 21. No. $\sqrt{9} + \sqrt{16} \neq \sqrt{25} = 5 \neq 3 + 4$ 23. Yes. $\sqrt{9} \times \sqrt{16} = \sqrt{9 \times 16} = \sqrt{144} = 12 = 3 \times 4$
25. 13.00 27. 59.82 (59.816385) 29. 663.32 (663.31968) 31. 90.03 (90.034715)
33. (a) 67.28 feet (b) 63.74 feet 35. 8.94 feet (8.9442719) 37. 17.32 inches (17.320508)
39. 26.25 feet (26.248809) 41. 115.26 miles (115.25623) 43. 258.07 feet (258.0698)
45. 1577.62 miles (1577.6247)

Margin Exercises 7.5

1. \$6.36 2. \$6.74 (6.7416) 3. \$22.50 4. \$24.23 5. \$1.73 6. Answers will vary. 7. 1.191016; answers will vary.
8. \$239.66 9. \$1190.89 10. 417.03 (417.025) 11. \$328.75 (328.75335) 12. 6.17% (6.16772)

Exercises 7.5

1. (a) \$976.63 (b) \$156.63 (c) 6% 3. (a) \$729.99 (b) \$129.99 (c) 4% 5. (a) \$4665.60 (b) \$665.60 (c) 8%
7. (a) \$4686.64 (b) \$686.64 (c) 8.24% 9. (a) \$525.53 (b) \$25.53 (c) 5.11%
11. (a) \$3041.63 (b) \$411.63 (c) 4% 13. (a) \$8741.78 (b) \$2241.78 (c) 5.06%
15. (a) \$1960.64 (b) \$60.64 (c) 6.49% 17. (a) \$3612.22 (b) \$1612.22 (c) 6.09%
19. (a) \$1618.62 (b) \$418.62 (c) 6.17% 21. \$29,386.56 23. \$29,718.93 25. \$114.06 27. 38¢
29. 349,448,840 31. \$134,147.28 33. \$1850.84 35. \$10,074.80 37. 19 years 39. \$47,418.16

Review Exercises for Chapter 7

1. 3.765×10^7 **3.** 8.572×10^{11} **5.** 7.396×10^{-9} **7.** 3.1×10^{-4} **9.** 4.887642×10^4 **11.** 1.1×10^{-5} **13.** 3.0×10^{10} **15.** 9.5×10^{-7} **17.** 4,670,000 **19.** 9400 **21.** 0.000000085 **23.** 0.00435 **25.** 0.00000000000000000000000167 **27.** 0.0000015 **29.** 25,000 **31.** $10.16889 \times 10^5 = 1{,}016{,}889$ **33.** $42.424857 \times 10^6 = 42{,}424{,}857$ **35.** $3.4224598 \times 10^{-6} = 0.0000034224598$ **37.** $2.5091911 \times 10^{-2} = 0.025091911$ **39.** $0.256\overline{9} \times 10^{-1} = 0.0256\overline{9}$ **41.** 5900 angstroms **43.** 6350 gallons **45. (a)** 2.536783×10^{98} **(b)** $2.536783 \times 10^{98} > 5 \times 10^9$ **47.** 9.7980 **49.** 0.3521 **51.** 0.7195 **53.** 71.9514 **55.** 8.4629 **57.** 2203.27 feet (2203.2702) **59.** 17.67 feet **61.** \$1124.86 **63.** \$5307.99 **65.** \$954.22 **67.** \$1411.80 **69.** \$2290.72 **71.** \$1218.99 **73.** $4\frac{1}{2}$%: \$124.62; 4%: \$121.90 **75.** 12 years, \$835.65; 18 years, \$1476.95 **77.** 8.24% ($8.\dot{2}4321$) **79.** \$3364.86

Just for Fun—Chapter 7

1. 3.1, 3.7, 4.3, 4.9, 5.5 **3.** $\frac{13}{6}, \frac{15}{6}, \frac{18}{6}, \frac{20}{6}, \frac{23}{6}$ **5.** 11; 111; 1111; 11,111; 111,111; 1,111,111; 11,111,111

7. 5473 − 2194 = 3279 **9.** Both sons cross; one son comes back; father crosses; one son comes back; both sons cross.
11. (a) The answer gets closer and closer to 2. **(b)** Answers will vary. **(c)** Answers will vary.

13.

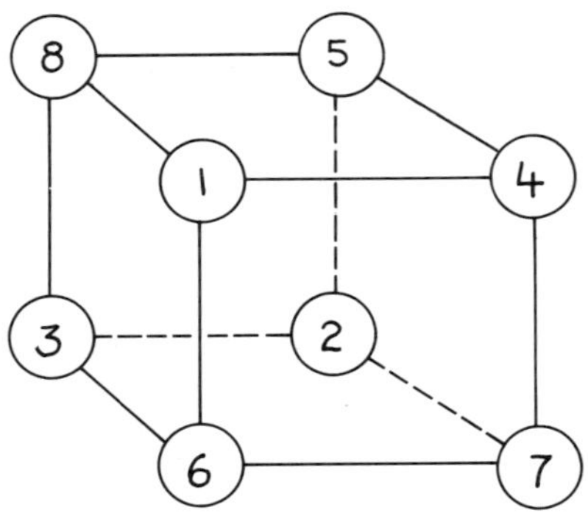

15. Shill (77145)

Mathematics and Your Dog and Cat

1.

Brand	A	B	C	D	E	F
Amount in ounces	105	60	43.5	800	800	800
Amount in pounds	6.56	3.75	2.72	50.00	50.00	50.00
Total cost	\$1.00	\$1.00	\$1.00	\$7.98	\$6.99	\$10.89
Total calories	3281.3	1875.0	1359.4	75,000	75,000	75,000
Cost per oz (cents)	0.95¢	1.67¢	2.3¢	1.0¢	0.87¢	1.36¢
Cost per lb (cents)	15.23¢	26.67¢	36.78¢	15.96¢	43.98¢	21.78¢
Cost per 100 calories	3.05¢	5.33¢	7.36¢	1.06¢	0.93¢	1.45¢
Calories per dollar	3281.3	1875.0	1359.4	9398.5	10,729.6	6887.1

3. 1.10 oz **5. (a)** \$2.69 **(b)** \$0.53 **7. (a)** \$1.88 **(b)** 37.5¢ **9. (a)** 2.88 pounds **(b)** 0.96 pounds
11. Package greater by 255 calories

Index